Advances in Protein Purification

Advances in Protein Purification

Edited by **Caroline Gardner**

New York

Published by Callisto Reference,
106 Park Avenue, Suite 200,
New York, NY 10016, USA
www.callistoreference.com

Advances in Protein Purification
Edited by Caroline Gardner

International Standard Book Number: 978-1-63239-054-7 (Hardback)

Printed in the United States of America.

Contents

Preface

This book provides an in-depth knowledge of protein purification. The book is designed and written to expedite speedy access to important information regarding protein purification and its various methodologies. It intends to present a summary on the latest methods for the purification, examination and quantification of proteins in composite samples utilizing various enrichment techniques.

This book has been the outcome of endless efforts put in by authors and researchers on various issues and topics within the field. The book is a comprehensive collection of significant researches that are addressed in a variety of chapters. It will surely enhance the knowledge of the field among readers across the globe.

It is indeed an immense pleasure to thank our researchers and authors for their efforts to submit their piece of writing before the deadlines. Finally in the end, I would like to thank my family and colleagues who have been a great source of inspiration and support.

Editor

The Isolation of Invertase from Baker's Yeast – An Introduction to Protein Purification Strategies

Anthony P. Timerman
University of Wisconsin-Stevens Point,
USA

1. Introduction

The vast number of roles that proteins serve in cell structure and function is unrivaled by any other type of compound found in nature and, in this sense, it can be argued that proteins are the most functionally and structurally diverse class of all known substances. Despite this vast diversity, each one of the hundreds to thousands of different proteins expressed in a cell possesses the same general composition: they are all un-branched polymers constructed from a common set of twenty amino acid building blocks. Because of this remarkable similarity in composition, the task of selectively isolating a functionally active protein from its biological source in the amounts required to study its function and shape might appear at first consideration to be a hopeless, if not an impossible, separation to achieve at the lab bench. However, by carefully taking advantage of variations in many physical properties between different types of proteins (including their mass, length, solubility, charge, and ability to selectively bind to specific ligands) the technology and art of protein purification has evolved into both a routine and rewarding exercise.

The purpose of this chapter is to provide a broad overview of some common concerns and corresponding strategies used to purify a protein from a natural source by way of a case study entitled, "The Isolation of Invertase from Baker's Yeast" which is a series of three-hour laboratory exercises designed to introduce methods of protein purification to a large group of undergraduate students. The reader should note that the specific, procedural details of each part of this project can be found in a methods paper recently published in the Journal of Chemical Education (Timerman et al., 2009) and, for the most part, will not be repeated here. Instead, this chapter focuses on the rationale and development of the sequence of steps used in this project as well as a brief description of each method used to measure, extract, isolate and characterize the enzyme. Other chapters of this book will provide more complete information and details regarding many of these and other selected topics in protein purification.

2. Function and properties of invertase

Invertase is the common name of the enzyme that catalyzes the hydrolysis of table sugar (i.e. sucrose) into a much sweeter, equimolar mixture of glucose and fructose called "invert" sugar (equation 1 and figure 1). Because invert sugar is a key ingredient in a number of sweets and confectionary products, the bakery industry provides one of the most important commercial applications of this enzyme reaction. For this reason, the enzyme has been extensively characterized and commercial sources of pure invertase are readily available.

$$\text{Sucrose (aq)} + H_2O \xrightarrow{\textit{Invertase}} \text{Fructose (aq)} + \text{glucose (aq) ("invert" sugar)} \qquad (1)$$

While aqueous solutions of either pure sucrose or glucose display weakly dextrorotatory behavior, meaning they cause a slight right-handed rotation of plane polarized light, solutions of pure fructose are strongly levorotatory and cause a much greater left-handed rotation of the light. The enzyme reaction, therefore, catalyzes the inversion of the right-handed rotation of polarized light observed through sucrose solutions to the left-handed rotation observed for solutions of "invert" sugar, hence the enzyme's common name of "invertase". For similar reasons, the common monosaccharides glucose and fructose are also known as dextrose and levulose, respectively.

Because enzymes are systematically named and classified by the substrate and subclass of reaction that they catalyze, the systematic name of invertase is "sucrose glycosidase" implying that it is a member of the subclass of enzymes that hydrolyze glycosidic (or acetal) linkages with a substrate specificity for sucrose. The yeast form of the enzyme has been assigned the unique four digit enzyme classification code (EC) number of 3.2.1.26 and it is also commonly called β-fructofuranosidase or sucrase. The intestinal enzyme lactose glycosidase (or lactase, EC 3.2.1.108), which hydrolyzes milk sugar into an equimolar mixture of galactose and glucose, is a related member of this enzyme subclass that may be more familiar because a deficiency of this enzyme is associated with symptoms related to lactose intolerance.

In yeast cells, invertase is classified as an extra-cellular, glycoprotein which is localized to the thin volume of space that exists between the yeast's plasma membrane and its outer cell wall (this peripheral volume is often called the periplasmic space). The enzyme serves the important biological function of cleaving sucrose on the outside of the cell into monosaccharides that can be transported (and subsequently metabolized) in the cytoplasm. That is, in the absence of invertase, yeast would have a difficult time utilizing table sugar as an energy source. Kinetic studies indicate that this extracellular form of invertase has a pH and temperature optima of about 4.8 and 40° C, respectively, and the Km for its substrate is about 5 mM sucrose. The enzyme's native mass of about 270 kiloDaltons is constructed from two identical and heavily glycosylated subunits with a molecular weight of about 135 kiloDaltons (Neumann & Lampen, 1967). Because extracellular proteins are typically conjugated with oligosaccharide chains (i.e. glycosides) by post-translational modification before they are exported from eukaryotic cells, it is not surprising that the periplasmic form of yeast invertase is indeed a glycoprotein. However, invertase is unusual in that the numerous oligosaccharide chains attached to the two identical subunits account for nearly 50% of enzyme's native mass (Lampen, 1971).

These cellular and structural features of yeast invertase offer several advantages in this purification project: (i) first, the enzyme can be gently and selectively extracted from yeast cells by using conditions that disrupt the cell wall while leaving the plasma membrane intact; (ii) the high oligosaccharide content increases the stability of the extracted enzyme (either by preventing protein aggregation or reducing its susceptibility to attack by proteases and other undesirable reactions) (Schulke & Schmid, 1988); and (iii) variations in the sugar content of each subunit causes them to migrate as a smeared band that is easy to detect during SDS-PAGE analysis (Moreno et al., 1980). On the other hand, this unusually high sugar content also reduces the ability of the protein to bind to Coomassie brilliant blue, the key component of the Bradford dye-binding protein assay. For this reason, solutions of pure commercial invertase prepared by dissolving a weighed mass of the solid enzyme to a final concentration of 1 mg per mL are observed to have a relative or equivalent concentration of only 0.10 mg per mL when compared to bovine serum albumin as the standard, reference protein in the Bradford assay.

3. Measurement of invertase activity

In order to selectively purify a specific protein from a mixture containing many other different proteins, it is essential to be able to selectively identify and measure the amount of the target protein without interference from others in the sample. For this reason, the first and perhaps most crucial step of any protein purification project is the **detection assay** used to monitor and compare the amount of a specific target protein contained in different mixtures.

Because enzymes catalyze very specific cellular reactions, the relative amount of an enzyme in a sample is traditionally defined by the amount it increases the reaction rate under a strictly defined set of conditions. By convention, relative amounts of enzyme are expressed in terms of **"International Enzyme Units"** where **"1-enzyme unit"** is defined as the amount of enzyme required to either: (i) consume 1.00 μmole of reactant per minute; or (ii) produce 1.00 μmole of product per minute. Furthermore, because the units used to describe enzyme activity rates (μmole • min^{-1}) are different than the traditional units used to describe chemical reaction rates (μmoles • mL^{-1} • min^{-1}), enzyme units are conveniently calculated by simply multiplying the concentration change observed in a solution by its volume (mL).

In the case of invertase, 1-unit of activity is therefore defined as the mass of enzyme required to either: (i) hydrolyze 1-μmole of sucrose per minute, or (ii) produce 1-μmole of invert sugar per minute under a precisely defined set of reaction conditions. According to this convention, a sample with 100 total units of invertase activity contains twice the mass (and thus twice the number of moles) of invertase as a sample with only 50 units of activity. Because enzyme reaction rates can change significantly with very small changes in temperature, pH, or substrate concentration (that is, enzyme activity units are a conditional property of matter) the precise value of each of these three conditions must be unambiguously described for the detection assay. For invertase, the enzyme detection assay is typically performed near its optimal pH (4.8) and temperature (40° C) at initial sucrose concentrations in excess of enzyme's substrate Km of 5 mM sucrose.

3.1 Spectroscopic measurement of enzyme activity units

Recall that Beer's law states that the absorbance (**ABS**) of a solution is proportional to the product of (**a•b•c**) where (**a**) represents the millimolar absorptivity constant (or extinction coefficient) of the solute (expressed in units of mM^{-1}•cm^{-1}) at a specified wavelength of light; while (**b**) defines the distance of the light-path through the cuvette (expressed in cm units);

and **(c)** is the solute concentration (expressed in milli-molar, mM units). Therefore, if the **a** and **b** terms of the equation are held at a known and constant value, then rearrangement of Beer's law indicates that any rate of change observed in the absorbance of the solution **(ΔABS/min)** must correspond to a proportional rate of change in the millimolar concentration **(ΔC/min)** of the colored solute, as outlined in equation (2):

$$\Delta C/min = \frac{\Delta ABS/min}{a \cdot b} = \frac{1/min}{(mM \cdot cm)^{-1}(cm)} = \frac{mM}{min} = \frac{mmole}{L \cdot min} = \frac{\mu moles\ solute}{mL \cdot min} \qquad (2)$$

And, the corresponding number of enzyme units (μmoles of solute per minute) contained in any reaction that produces (or consumes) of a colored solute is simply calculated from the rate of the absorbance change **(Δ Abs/min)** and the fixed volume **(VmL)** of the colored solution in the cuvette with a fixed light path, as summarized in equation 3.

$$Enzyme\ units = \frac{\Delta ABS/min}{a \cdot b} \bullet VmL = \frac{\mu moles\ solute}{mL \cdot min} \bullet mL = \frac{\mu moles\ solute}{min} \qquad (3)$$

3.2 The standard 5-minute "stop-assay" of invertase activity

Because both the reactants and products of the invertase reaction are colorless, the reaction is followed spectrophotometrically by using the monosaccharides produced from the reaction to subsequently reduce brightly yellow-colored solutions of 3, 5-dinitrosalicylate (DNS) to dark orange-colored solutions of 3-amino-5-nitrosalicylate which can be detected with an inexpensive spectrophotometer at a wavelength of 540 nm (Melius, 1971 and Sumner & Sisler, 1944). In this manner, the standard 5-minute stop assay of invertase activity used in this project can be summarized in four-steps.

Step 1: Invertase catalyzed hydrolysis of sucrose: At 30 second intervals, small aliquots (0.10 mL or less) of each enzyme-containing fraction (or diluted samples of each fraction) are quantitatively transferred and gently mixed into 1.0 mL of substrate solution (composed of 20 mM sucrose in 40 mM sodium acetate buffer at pH 4.80 and an ambient temperature of 20-24° C) contained in separate 13 x 100 mm disposable test-tubes where the invertase reaction (see Figure 1) is allowed to proceed for a precise time of 5.0 minutes.

$$Sucrose + H_2O \xrightarrow{Invertase} D\text{-}Glucose + D\text{-}Fructose$$

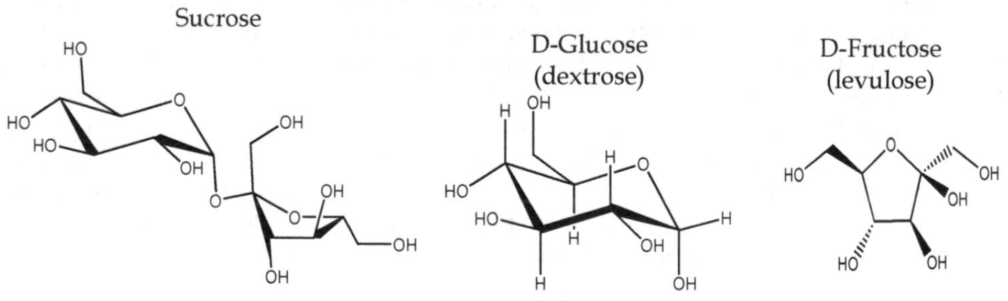

Fig. 1. Structures of sucrose and the cyclic (anomeric) conformations of each monosaccharide produced in step 1 of the invertase detection assay. The reaction proceeds for precisely 5.0 minutes in 20 mM sucrose at pH 4.8 and ambient temperatures of 20-24° C before it is abruptly stopped by denaturation of the enzyme with the rapid addition of an alkaline solution of 3,5 dinitrosalicylate (DNS).

Step 2: Denaturation of invertase and the reduction of DNS: At the end of each 5.0 minute reaction period, the invertase activity in each tube is instantaneously stopped at 30 second intervals by denaturation with the rapid addition of 2.0 mL of alkaline DNS solution (composed of 0.20 M NaOH, 23 mM 3, 5-dinitrosalicylate, and 0.53 M sodium potassium tartrate). After each reaction is "stopped", the entire set of assay tubes is transferred to a boiling water bath for 7-8 minutes. Under these conditions, the D-Fructose produced in step 1 rapidly isomerizes to D-glucose (see figure 2) which, along with the other mole of glucose produced by the enzyme reaction, reduces the brightly yellow-colored 3,5-dinitrosalicylate to dark orange-colored solutions of 3-amino-5-nitrosalicylate (see figure 3). The entire set of assay tubes is then cooled in a beaker of tap water before each solution is diluted to a final volume of 6.1 mL with the addition of 3.0 mL of 50 mM sodium acetate buffer at pH 4.8.

$$\text{D-Fructose} \xrightarrow{\textit{NaOH \& heat}} \text{D-Glucose}$$

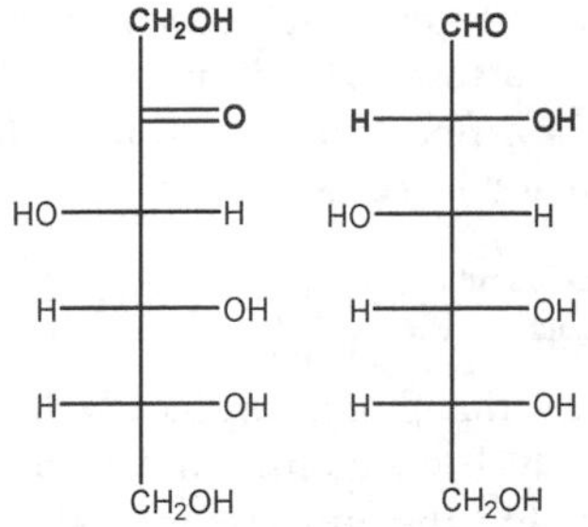

Fig. 2. Fisher projections of the open-chain conformations of the reactants and products of the base-catalyzed isomerization of D-fructose (a non-reducing keto-hexose) to D-glucose (a reducing aldo-hexose) which proceeds in a boiling water bath for about 7-8 minutes during step 2 of the invertase detection assay.

$$3\ \text{Glucose} + \text{3,5-dinitrosalicylate} + 3\ OH^{-1} \xrightarrow{\textit{Heat}} 3\ \text{Gluconate} + \text{3-amino-5-nitrosalicylate} + 2\ H_2O$$

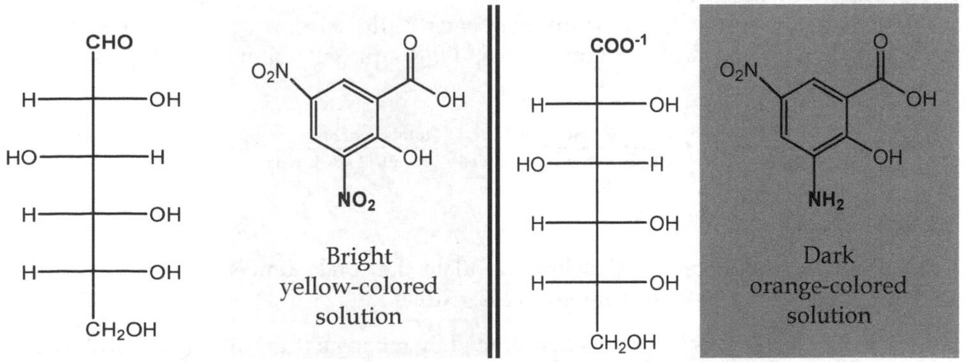

Fig. 3. The structures of the reactants and products for the base catalyzed reduction of brightly yellow-colored solutions of 3,5-dinitrosalicylate into dark orange-colored solutions of 3-amino-5-nitrosalicylate by glucose which is oxidized to gluconate. The reaction proceeds for about 7-8 minutes in a boiling water bath during step 2 of the invertase detection assay.

Step 3: Calculation of invertase activity in each assay tube: A spectrophotometer is used to read the absorbance of the solution in each disposable test tube at 540 nm against a reagent blank prepared exactly as one of the assay tubes in steps 1 and 2 except for the volume of enzyme added to the substrate solution in step 1 is replaced with water (or simply just omitted). Equation 4 is then used to calculate the number of units of invertase activity contained in each tube based upon the absorbance change over the reagent blank (ΔABS) that resulted from the 5.0 minute reaction period and measured in a tube with a 1.0 cm light path and a final volume of 6.1 mL:

$$\text{Units of invertase activity in each assay tube} = \frac{\Delta ABS/5.0 \text{ min}}{a \cdot 1.0 \text{ cm}} \bullet 6.1 \text{ mL} \qquad (4)$$

The value of the milli-molar absorptivity constant **(a)** in equation (4) is determined from the absorbance change (ΔABS) of a calibration tube prepared exactly as one of the assay tubes in steps 1 and 2, except that the volume of enzyme added to the substrate solution in step 1 is replaced with 0.10 mL of a 20 mM stock solution of invert sugar (i.e. a mixture that contains both 20 mM glucose and 20 mM fructose). The dilution of 0.10 mL of 20 mM invert sugar to a final volume of 6.1 mL yields a final concentration of 0.33 mM invert sugar and the absorptivity constant (a) is calculated by equation (5):

$$a = \frac{\Delta ABS \text{ of calibration tube}}{0.33 \text{ mM invert sugar} \cdot 1 \text{ cm}} = mM^{-1} \bullet cm^{-1} = \frac{mL}{\mu mole \text{ invert sugar} \bullet cm} \qquad (5)$$

The calibration tubes prepared in this manner typically have absorbance values of about 0.66 (above the reagent blank) which yields millimolar absorptivity constant values **(a)** of about 2.0 $mM^{-1} \bullet cm^{-1}$ which, for the purpose of calculating enzyme units, is more conveniently expressed as 2.0 $mL \bullet \mu mole^{-1} \bullet cm^{-1}$.

Step 4: Calculation of total invertase activity in each fraction: The final step of the detection assay is to calculate the total number of enzyme units contained in each fraction collected during the purification procedure. This is essentially a straight-forward application of the factor-label method from general chemistry where the number of enzyme units per assay tube (calculated in equation 4) must be converted to the total number of units in an entire fraction by accounting for: (i) the volume of diluted enzyme solution added to the substrate solution in step 1; (ii) the dilution factor of the enzyme solution; and finally, (iii) the total volume of the stock fraction. To keep life simple, all volumes are expressed in mL units:

$$\text{Total units} = \frac{\text{\# of units}}{6.1 \text{ ml assay}} \bullet \frac{6.1 \text{ mL assay}}{\text{mL of dilute enzyme added}} \bullet \frac{\text{total mL of dilute enzyme}}{\text{mL of stock added}} \bullet \frac{\text{mL of stock fraction}}{1} \qquad (6)$$

3.3 Sample calculation

Using the information provided below, calculate the total number of units of invertase activity contained in a 25.0 mL sample of yeast extract.

Data: The 25.0 mL of yeast extract is first diluted by mixing a 0.250 mL sample with 4.75 mL of buffer (for a total dilution volume of 5.00 mL). The dilute extract is then analyzed for invertase activity using the standard 5-minute stop assay described above. Briefly, 0.100 mL of the dilution is mixed into 1.0 mL of substrate solution at 22° C. The reaction is stopped at precisely 5.0 minutes with the addition of 2.0 mL of DNS solution, heated in a boiling water bath for about 7 minutes and diluted with 3.0 mL of acetate buffer to a final volume of 6.1 mL. The solution in the assay tube has an absorbance value at 540 nm of 0.744 (above a reagent blank) compared to an absorbance value of 0.620 for a calibration tube containing a final concentration of 0.330 mM invert sugar.

Solution:

Step 1, use equation 5 to solve for the value of the millimolar absorptivity constant (a = 1.86 mL•µmole^{-1}•cm^{-1}).

Step 2, use equation 4 to calculate the number of units of invertase activity in the assay tube (0.488 units).

Step 3, finally, use equation 6 to calculate the total number of units in 25.0 mL of extract (2440 units).

4. Extraction of Invertase from yeast

After the method for measuring the amount of a specific protein in different samples is established, the second order of business is to optimize a procedure for extracting the functionally active protein from the initial raw material into a defined aqueous solution (the extraction medium). The objectives of this process are straight-forward: First, the tissue must be homogenized or disrupted adequately enough for the protein to be released as a soluble form into the extraction medium. Second, the homogenate must be filtered and centrifuged sufficiently enough to remove any solid, cellular debris from the soluble extract.

The apparent simplicity of this procedure may suggest that tissue extractions are the least complicated task of the entire project; however, in reality, the success of this step, and therefore the outcome of the entire purification, often depends upon the precise control of a surprisingly large number of variables. Some of the more common concerns include: (1) proteins can be denatured by the shearing forces (or chemicals) used to disrupt the initial raw material; (2) proteins can be degraded by digestive enzymes co-extracted from the tissue; (3) proteins can become insoluble or inactivated due to differences between the composition of the cellular fluid and extraction medium (including pH, ionic strength, and the concentration of reducing agents or other specific solutes). In many cases, it is not an exaggeration to state that the success of an entire purification procedure requires a very specific set of extraction conditions including the amount of force, length of time or temperature in which the tissue is homogenized or the precise pH, ionic strength, or concentration of specific supplements (such as protease inhibitors) included in the extraction medium.

While specific details of each extraction must be optimized empirically, there are several rules of thumb that apply in most cases (Scopes, 1982). First, the initial raw material should be the simplest biological system that contains the greatest concentration of the target protein in order to both (i) maximize the final yield and (ii) reduce the amount of other proteins extracted during the homogenization step, especially digestive enzymes or related, but unwanted, isoforms of the target protein. This first point is simply stating the obvious fact that it is always easier to work with as pure and as large of an amount of a protein as possible. For example, in order to isolate an enzyme compartmentalized to bovine heart mitochondria, it would be advantageous to use isolated beef heart mitochondria as the initial raw material rather than an entire beef heart (or an entire cow for that matter).

Second, the initial raw material should be disrupted as gently as possible in order to reduce the risk of protein denaturation. For this reason, protein purification labs are typically equipped with a host of instruments designed to provide a wide range of shearing forces required to physically disrupt different types of tissues (including glass-teflon or glass-glass hand-held homogenizers, electric blenders, freeze clamps, sonicators, and French presses). In addition to these physical methods of disruption, a wide variety of the three general classes of detergents (non-ionic, ionic, or zwitterionic) are available to solubilize (i.e. chemically extract) specific proteins from different tissues and isolated organelles. Third, all of the materials, solutions and equipment should be kept as cold as possible (ideally between 0-4° C) in order to decrease the rate of undesirable proteolytic and other unwanted side reactions. For this reason, preparative (i.e. large scale) operations are often performed in a dedicated cold room while scaled-down procedures can be carried out on buckets of ice. Regardless of the scale of the operation, refrigerated centrifuges (ranging from low to ultra speed devices) are almost always used for separating the soluble extract from the insoluble fraction of the homogenate. Finally, the pH, ionic strength, and other components of the extraction medium are often adjusted to match that of the original cellular conditions in order to maximize both the solubility and stability of the extracted protein.

Since jars of dried baker's yeast can be stored on grocery store shelves for several months at room temperature, it is not surprising that many proteins extracted from dried yeast are also quite stable at ambient temperatures for prolonged periods of time. The isolation of invertase in this project provides the additional benefit that extracellular proteins, in general, tend to be much more stable than intracellular proteins. This observation has been attributed, in part, to the role of protein glycosylation and, since oligosaccharides account for nearly 50% of the composition of invertase, may explain the observation that very little, if any, loss of invertase activity is detected in yeast extracts stored for up to five weeks at 0-4° C. Furthermore, because the yeast cell wall is selectively lysed in dilute solutions of sodium bicarbonate, the extracellular form of invertase is gently extracted from the periplasmic space by simply mixing the contents of a 113 gram (4 oz.) jar of dried yeast from the grocery store into 400 mL of 0.10 M $NaHCO_3$ and incubating the suspension in a tightly sealed reagent bottle for about 15 hours at 35° C. The insoluble debris (which accounts for roughly one-half of the volume of the lysate) is then separated from the soluble fraction by centrifugation for 30 minutes at 15,000 x g and 4° C. The invertase enriched extract is then simply poured off of the solid pellet, diluted with extraction medium to a final volume of 250 mL and stored at 0-4° C.

Two common types of yeast currently stocked in many grocery stores include: (i) traditional "active-dry" strains which have been used in baking for many generations; and (ii) "bread-machine" strains that have been recently selected or engineered for the purpose of reducing the length of time required for dough to rise in electronic bread makers. Because the dough rising reaction is fueled primarily by table sugar (sucrose), it seems plausible that the new "bread-machine" strains might be characterized by higher concentrations of invertase in order to increase the rate of sucrose digestion and, therefore, allow the dough to rise more rapidly. For this reason, the invertase content of extracts prepared from different commercial brands of "active-dry" and "bread-machine" yeast strains was analyzed to determine the best choice of raw material to use in this project. Not surprisingly, this survey (summarized in table 1) demonstrated that extracts prepared from the bread-machine strains from two different companies contained significantly higher concentrations of invertase activity (90 to 300 units per mL) compared to the extracts prepared from their corresponding active-dry strains (16 to 82 units per mL). Furthermore, the Red-Star brand of bread-machine yeast was clearly the top choice for the raw material in this project because it consistently produced extracts that contained 2-3 times higher concentrations of invertase activity (190 to 300 units per mL) compared to extracts prepared from Fleischmann's bread-machine yeast (90 to 140 units per mL).

In summary, the isolation of yeast invertase is an especially attractive target for a protein purification project designed for a large group of undergraduate students working in a laboratory at ambient temperatures not only because the starting material is readily available and relatively inexpensive but also because the exceptionally uncomplicated extraction procedure produces a large volume of a solution that is enriched with a remarkably stable enzyme!

Commercial Brand of Yeast	Concentration of Invertase activity in extracts prepared from Active-Dry Yeast (units per mL)	Concentration of invertase activity in extracts prepared from Bread-Machine Yeast (units per mL)
Fleischmann's	44-82 (n = 4)	90-140 (n = 4)
Red-Star	16-20 (n = 3)	190-300 (n = 3)

Table 1. Comparison of the concentration of invertase activity contained in extracts prepared from different commercial sources of dried baker's yeast.

Each extract was prepared by incubating the contents of a 113 g jar of yeast in 400 mL of a 0.10 M aqueous solution of sodium bicarbonate for about 15 hours at 35° C. Each lysate was cooled in an ice-bath and centrifuged for 30 minutes at 15,000 x g and 4° C before the soluble extract was poured off of the pellet, diluted with extraction medium to a final volume of 250 mL and stored at 4° C. One-unit of invertase activity is defined as the hydrolysis of 1-µmole of sucrose per minute at ambient temperatures of 20-22° C in a 40 mM solution of sodium acetate at pH 4.8 containing an initial substrate concentration of 20 mM sucrose. The results in the table represent the range of values measured from extracts prepared from (n) number of different lots of each yeast strain.

5. Purification of invertase from yeast extract

The next goal of the project is to selectively isolate a single, specific protein from the tissue extract while removing as many of the other polypeptides as possible, a task that is traditionally accomplished with a series of sequential **isolation steps** that take advantage of differences in two or more physical properties between individual proteins, such as variations in their size, charge, and solubility. A minimum of two "back-to-back" isolation steps is usually required because, while it may be common for many proteins to have a similar size or a similar charge, it is far less likely for two different proteins to possess both the same size and charge (or some other combination of physical properties). In summary, the overall objective of this stage of the purification is to obtain as pure or homogenous of a sample as possible in the fewest number of isolation steps. In this project, invertase is purified from a 25 mL sample of fresh yeast extract by (i) differential precipitation with ethanol; (ii) gel filtration; and (iii) ion-exchange chromatography.

5.1 Precipitation of invertase with ethanol

The selective precipitation of a protein from an aqueous solution is one of the oldest, most effective, and technically simple isolation steps used in protein purification. Furthermore, the solid protein precipitated from a large volume of extract can also be concentrated by dissolving it back into a much smaller volume of solvent that is more convenient to apply to a variety of chromatography columns used in subsequent isolation steps. For this reason, it is not uncommon for protein precipitation to be used as the initial isolation step of many purification procedures.

In order for a large protein molecule to become "solvated" or dissolved in an aqueous solution, the majority of its surface must be able to form complexes with an enormous number of water molecules producing a large "hydration shell" that is energetically stabilized by ion-dipole, hydrogen bonds, and dipole-dipole attractive forces between the water molecules and side chains of polar amino acids exposed on the protein's surface. Because these interactions with the water molecules (i.e. the hydration shell energy) required to keep a protein solvated are sensitive to the pH, ionic strength, and polarity of the solvent; AND, because each protein has a unique hydration shell network, it is possible to selectively entice the surfaces of specific proteins in a mixture to become less and less hydrated, such that their newly exposed surfaces are forced to stick or clump together into an insoluble aggregate or precipitate by simply adding a high enough concentration of an acid (to lower the pH of the solvent), or a salt (to increase the ionic strength of the solvent), or an organic liquid (to decrease the polarity of the solvent). It should be emphasized that protein precipitation is very different from protein denaturation in the vital sense that a protein contained in the solid aggregate retains its native, three-dimensional shape so that it is possible to fully restore its biological function by simply dissolving it back into solution.

In summary, it is possible to partially purify a protein from a mixture by adding a precipitating agent which selectively perturbs the complex set of interactions between its surfaces with the large excess of water molecules from the solvent. The two most common types of precipitating agents used for this purpose are sulfate salts (especially ammonium sulfate) and miscible organic liquids (such as acetone or ethanol). Because salts and organic liquids have a different affect on protein solubility, salts are better precipitating agents for some proteins while organic solvents are better for others (Scopes, 1982). In theory, salts are expected to be more efficient for precipitating proteins which contain larger areas of hydrophobic patches on their surfaces while organic solvents are better for those with surfaces that are almost exclusively dominated by polar amino acid side chains and other hydrophilic groups (such as the carbohydrate chains of a glycoprotein like invertase). Despite these differences in their physical behavior, the same straight-forward and simple set of steps is used by each type of precipitating agent in the procedure: (i) A precipitating agent is added to the extract above the "threshold-concentration" required for the desired protein to become insoluble; (ii) The mixture is incubated for a short period of time to allow the precipitation reaction to go to completion; (iii) The mixture is centrifuged into both a soluble fraction (the decantate) that contains other proteins but is devoid of the target protein and an insoluble fraction (the solid pellet) which is enriched in the desired protein; and, finally, (iv) The decantate is carefully poured off the pellet which is then dissolved into a much smaller volume of a desired solvent.

In this project, invertase is enriched and concentrated from the initial extract using a 2-step (or differential) method of precipitation with ethanol. In the first precipitation reaction, 10.0 mL of ethanol is added to 25.0 mL of fresh yeast extract, for a final ethanol concentration of 29% (by volume) which is below the threshold concentration required to precipitate invertase activity. After centrifugation, the first decantate (which contains nearly all of the initial invertase activity) is poured off and saved while the first pellet (of contaminating proteins, lipid complexes and other cellular debris) is discarded. In the second precipitation reaction, another 7.0 mL of ethanol is added to the first decantate, for a final volume of about 42 mL and ethanol concentration of about 40% (by volume) which now exceeds the required threshold concentration. Following a short incubation period, the mixture is centrifuged so that the second decantate can be poured off of the invertase enriched pellet which is dissolved in 2.0 mL of gel filtration column buffer composed of 5 mM each NaH_2PO_4, Na_2HPO_4, and NaN_3 (included as an anti-bacterial agent) at pH 7.0. In summary, this first isolation step both partially purifies the invertase activity contained in a 25.0 mL sample of yeast extract (fraction 1) and concentrates it about 10-fold in gel filtration column buffer to a final volume of 2-2.5 mL (fraction 2).

5.2 Separation of Invertase by gel filtration

Gel filtration (or size-exclusion) chromatography is a powerful method commonly used to separate proteins based upon their differences in size (McLoughlin, 1992 and Melius, 1971)). More specifically, gel filtration separates particles according to the length of their **"Stokes radius"** which defines the "rotational volume" occupied by the particle as it spins freely in solution. In this sense, separations by gel filtration are affected by particle shape because an elongated, rod-shaped protein will have a much longer Stokes radius (and corresponding rotational volume) than a spherically shaped protein with the same molecular weight. However, because a large percentage of soluble proteins extracted from tissues are characterized as 'spherically-shaped' globular proteins, a reasonable correlation often exists between their Stoke's radius and molecular weight.

In the first step of this procedure, a small volume of a concentrated protein mixture is carefully loaded on to the top of a long column that is packed with a size-exclusion resin composed of tiny, porous beads with a defined mesh-size or cut-off limit. After the sample is loaded, additional buffer is pumped through the resin which drives each particle through the length of the column which acts as a filter that forces larger particles to migrate much more rapidly through the column than smaller proteins. In this manner, it is possible to collect proteins of different sizes into separate fractionation tubes as they elute from the end of the column in the order of larger proteins that come off first in a lower elution volume followed by smaller proteins that are collected in higher elution volumes. Figures 4A and 4B provide a display of the gel filtration apparatus used by students in this project as well as the separation observed shortly after a 1.0 mL solution of blue dextran (a large polysaccharide with a molecular weight of 2,000 kiloDaltons) and hemoglobin (a red colored protein with a molecular weight of 65-70 kilodaltons) is loaded on to the top of the column.

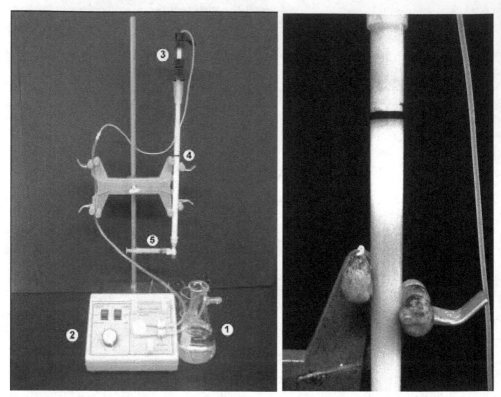

Fig. 4A. (left). The apparatus for purification of yeast invertase by gel filtration chromatography consists of: (1) a reservoir of column buffer contained in the Erlenmeyer flask; (2) a peristaltic pump; (3) An adjustable column flow adapter; (4) a 1 x 42 cm packed column of Sephacryl HR-300 (a shorter column is presented in the photo); and (5) a syringe connected to a three way Luer lock positioned at the bottom of the column.

Fig. 4B. (right). Separation of blue dextran from hemoglobin by gel filtration chromatography shortly after a 1.0 mL solution of 0.75 mg/mL of blue dextran and 1.0 mg/mL hemoglobin was loaded onto a 1 x 42 cm column of Sephacryl HR-300 and separated at a flow rate of about 1.2 mL per minute.

The mesh-size or cut-off limit of the size-exclusion resin is an important parameter of a gel filtration column because it defines the size of the largest, spherically shaped particle that can penetrate the surface of the porous beads and equilibrate into the "internal volume"(V_i) of the column (Scopes, 1982). Since this equilibration is responsible for the slower migration rate of particles that are small enough to be retained by the filtering action of the size-exclusion resin, all particles that are larger than this cut-off limit cannot penetrate into the beads and rapidly migrate through the entire length of the column in the interstitial spaces between the beads called the excluded or "void volume" (V_o). Meanwhile, any protein on the other extreme side of this size-spectrum that is small enough to freely equilibrate between the internal and void volumes must migrate through the column's total volume

(Vt) which can be calculated from its bed-height (H) and inner diameter (d) $[Vt = H \pi(\frac{1}{2} d)^2]$ and is also equivalent to the sum of the void and internal column volumes ($Vt = Vo + Vi$). Due to the restrictions summarized above, the effective range of sizes that can be actually separated on a gel filtration column is limited to moderately sized particles that elute in the internal volume of the column ($Vi = Vt - Vo$)) because they are too small to elute in the void volume (Vo) and yet too large to elute in the total volume (Vt). For globular proteins within this limited size range, a linear relationship is observed between the log of their molecular weight (log MW) and the peak of their elution volume (Ve). This relationship has a practical application in that it allows for the native mass of unknown proteins to be estimated by comparing their elution volumes from a size-column that has been calibrated against the elution volumes of standard proteins with known molecular weights (as the one demonstrated in figure 5).

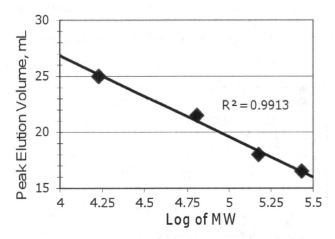

Fig. 5. Sephacryl HR-300 calibration curve. Correlation of the log of the molecular weight and peak elution volumes (Ve) observed for four different proteins collected from a 1 x 42 cm column of Sephacryl HR-300 at a flow rate of about 1.2 mL per minute. The four proteins analyzed (from left to right) are (i) myoglobin, MW 17 kiloDaltons; (ii) hemoglobin, MW 64 kiloDaltons; (iii) alcohol dehydrogenase, MW 150 kiloDaltons; and (iv) yeast invertase, MW 270 kilodaltons.

In order to prepare a gel filtration column that provides an adequate level of resolution for a specific separation, a large number of size-exclusion resins with a wide range of mesh sizes and mechanical flow properties are commercially available. These resins are typically composed of very tiny beads constructed from specific polymers (such as dextrans used to synthesize Sephadex resins and polyacrylamide contained in Sephacryl resins) that are chemically cross-linked to a defined mesh-size. In order to pack the column, a small sample of the dry beads is first hydrated and equilibrated in column buffer which causes the volume of the beads to swell by a factor of ten or more. The mixture is shaken to allow the slurry of resin to be poured into a column where the swollen beads eventually settle by gravity to a desired packed height or column length. Since the actual separation of particles by gel filtration (as mentioned above) is confined to the columns internal volume (Vi), the resolution or degree of separation achieved between two proteins can be improved by increasing its internal volume which, for a column with a fixed diameter, is proportional to the length of the column. In short, resolution tends to increase as the length of the gel filtration column increases (Scopes, 1982). On the other hand, increasing the length of the column also generates a number of procedural complications (including higher column pressures, lower flow rates, and sample dilution due to zone-broadening) that must be considered in determining the most practical column length to use for the separation.

In this project, the yeast invertase from fraction 2 is further purified by gel filtration chromatography on a (1 x 42 cm) packed column of Sephacryl HR-300, a resin composed of poly-acrylamide beads with a mesh-size designed to exclude globular proteins with molar masses in excess of 300 kilodaltons and engineered to withstand the column pressures encountered at the relatively high rates, HR, of flow of 1-mL per minute. Because invertase is a colorless solute with a native mass of about 250 kiloDaltons, a 1.0 mL sample from fraction 2 (approximately one-half of the 2.0 mL fraction obtained by ethanol precipitation) is first spiked with three visual column markers before it is loaded onto the size-column: (i) blue dextran, the blue colored polymer with a molecular weight of over 2,000 kiloDaltons

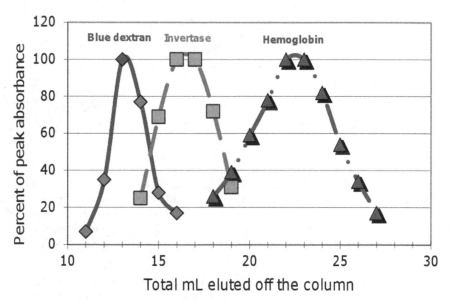

Fig. 6. Elution profiles for the separation of blue dextran, invertase activity, and hemoglobin by gel filtration chromatography. A 1.0 mL sample of F2 (29-40% ethanol cut) was spiked with 0.050 mL each of 15 mg/mL blue dextran and 20 mg/mL hemoglobin. The mixture was then separated on a 1 x 42 cm column of Sephacryl HR-300 at a flow rate of 1.2 mL per minute. The blue dextran in fractions 11-16 was detected on a Novaspec spectrophotometer at 620 nm with a peak absorbance value of 0.295 in fraction 13. The red tinted hemoglobin fractions (18-27) were detected at 420 nm with a peak absorbance value of 0.800 in fractions 22 and 23. Invertase activity in 10.0 μL samples of fractions 14-19 was monitored by the reduction of DNS by glucose to produce the orange-colored product (3-nitro-5-amino salicylate) which is detected at 540 nm.

used to mark the void volume (Vo) and is expected to elute ahead of the smaller, colorless molecules of invertase; (ii) hemoglobin, a red colored protein with a native mass that is about one-fourth the size of invertase that is expected to elute from the column after invertase. In this manner, the colorless molecules of invertase are expected to elute from the column in between the peak elution volumes observed for blue dextran and hemoglobin; and finally, (iii) dinitrophenol aspartic acid (DNP-Asp) a yellow-colored derivative of a small amino acid which marks the total volume of the column (Vt). After loading, the sample is separated on the column at a flow rate of 1.2-1.3 mL per minute while hand collecting 1-mL fractions from the end of the column in small disposable test-tubes. During the run, the three visual markers are observed to separate (as in figure 4B) and then eventually elute from the column in peak volumes of 13 mL for blue dextran, followed by 22-23 mL for hemoglobin, and finally over 30 mL for DNP-Asp. Small samples (10 µL) from each of the 1-mL fractions collected between the peak tubes of blue dextran and hemoglobin are analyzed for invertase activity (using the standard 5-minute stop assay) so that the four fractionation tubes containing the most invertase activity (typically tubes 15-18) can be pooled together to obtain a new 4-mL fraction of purified invertase (fraction 3). Figure 6 shows the results of a typical elution profile for blue dextran, invertase and hemoglobin from the gel filtration column.

5.3 Isolation of Invertase by ion exchange (Adsorption) chromatography

While ion exchange chromatography is commonly used to separate proteins on the basis of their charge differences, it is a specific example of a more general separation process called adsorption chromatography that also includes both hydrophobic-interaction chromatography and affinity (or ligand-binding) chromatography. In each case, the protein of interest selectively binds or adsorbs to a solid resin packed in a column while other proteins simply flow through or are washed off before the target protein is selectively dissociated from the column (Scopes, 1982). In addition to being a powerful purification tool, adsorption chromatography also provides a practical method of concentrating dilute protein solutions. For example, a protein contained in 100 mL of solution can be concentrated 50-fold by adsorbing it to a column with a 1.0 mL bed volume and eluting it off in a final volume of 2.0 mL. The general procedure for each type of adsorption chromatography method can be divided into four common steps:

Packing: the resin is first mixed with a small volume of equilibration solution (which allows the protein of interest to bind the column) to form a slurry that can be poured into a column and allowed to settle by gravity to a desired bed height or column volume.

Loading: the sample (ideally contained in the same solution used to pack the column) is then pumped through the resin at a flow rate that is slow enough to permit the specific adsorption reaction to come to completion. To insure maximal binding, the flow through is often collected and re-applied to the column (or continuously pumped through the column for a defined period of time). Batch loading refers to an alternative method of packing and loading a column in which the packing slurry is prepared by mixing the resin directly into the sample solution prior to packing the column.

Washing: after loading the sample, the column is washed by pumping a minimum of two to three bed volumes of solution to remove both (i) non-adsorbed proteins trapped in the interstitial spaces of the resin, and (ii) other proteins adsorbed to the column less tightly than the target protein.

Dissociating: the protein of interest is finally eluted from the column using 2-3 bed volumes of solution and a slow enough flow rate to maximize its yield off the column.

Separations by ion exchange chromatography are sub-classified into anion vs. cation exchange chromatography based upon the charge of the particles that bind to the resin. In the case of anion exchange chromatography used in this project, negatively charged proteins in the sample are first absorbed via salt bridges (i.e. electrostatic attractive forces) to a positively charged anion–exchange resin commercially produced by conjugating an insoluble matrix (such as cellulose) with a quaternary amine or other positively charged functional group. In contrast, cation-exchange resins are prepared by the incorporation of a negatively charged functional group (such as a carboxylate or sulfate) to the matrix. The loaded anion-exchange column is then washed with increasing concentrations of salt, typically NaCl, until the chloride ion reaches a high enough concentration to displace each protein from the column by replacing the salt bridge formed between the resin and protein (hence the term anion exchange chromatography). In this manner, proteins are sequentially eluted from the ion exchange column in order of the weakest to strongest binding affinities to the resin which, in turn, is a function of the charge density contained on localized regions of the protein's surface.

Therefore, the purpose of the third and final isolation step of this procedure is to use an anion-exchange column to further purify and concentrate the dilute solution of invertase contained in fraction 3. For this purpose, 3.5 mL of the blue–tinted, 4.0 mL sample pooled from the gel filtration column is slowly loaded onto a 0.5 mL packed bed volume of DEAE cellulose (i.e. an anion exchange resin composed of cellulose derivitized with diethyl-amino-ethane). After loading, the non-adsorbed particles are washed from the one-half milliliter column with 2.0 mL (i.e. four bed volumes) of gel filtration column buffer before the final fraction of invertase activity (fraction 4) is obtained by dissociating it from the column with 1.0 mL (or 2 bed volumes) of column buffer supplemented with 50 mM NaCl. For demonstrative purposes, proteins that bind to the DEAE-cellulose column more tightly than invertase are then dissociated with 1.0 mL of 250 mM NaCl and also saved for analysis by SDS-PAGE).

6. Assessment of invertase purity

In the last stage of this project, the relative purity of the initial yeast extract (fraction 1) and each fraction obtained from the three sequential isolation steps (fractions 2-4) is compared to that of a commercial source of invertase (Sigma Product Information Bulletin I4504) using both a visual, qualitative method (SDS-PAGE) in addition to the more precise quantitative evaluation of specific activity measurements.

6.1 SDS-PAGE analysis

SDS-PAGE is an acronym for sodium dodecyl sulfate polyacrylamide gel electrophoresis which is a common technique used to separate, visualize, and therefore compare the relative amount of individual polypeptide chains contained in different fractions (Jisnusun &

Bhinyo,1977 and Roberts et al., 1977). In practice, the procedure takes place in three general steps:

Denaturation: a small sample from each fraction is first mixed with an excess of SDS (sodium dodecyl sulfate, an ionic detergent), βME (β-mercaptoethanol, a thiol reducing agent), and bromophenol blue (an intensely colored dye with a negative charge and low molecular weight that is used to track the progress of each sample through the gel during electrophoresis) (Laemmli, 1970). Each mixture is then placed in a boiling water bath for several minutes where the βME reduces all disulfide bridges to sulfhydryl groups (disrupting tertiary protein structure) and the dodecyl sulfate anions bind to the backbone of each polypeptide at a ratio of about 1 anion for every 2 residues (disturbing secondary, tertiary and quaternary levels of protein structure). In the end, each polypeptide chain is unfolded into a negatively-charged, rod-shaped complex with a relatively constant charge to mass ratio (that is, the native charge of the polypeptide in the complex is essentially masked by the huge excess of detergent anions).

Electrophoresis: a small aliquot from each denatured sample is transferred to separate wells of a solid, rectangular gel of polyacrylamide that is crosslinked to a desired mesh or size-exclusion limit. An electric field is applied across the loaded gel with the positive pole positioned on the opposite side of the samples in order to drive the negatively-charged complexes through the mesh of polyacrylamide which effectively filters them according to the length of their Stokes radius as they wiggle through the porous matrix to the opposite side of the gel. Since small particles move through the gel more rapidly than larger ones, the electrophoresis is visually followed by the movement of the small bromophenol blue tracking dye to insure that the power supply is shut off before any protein in the sample reaches the bottom of the gel.

Staining and destaining: following electrophoresis, the entire gel is soaked in a stain solution containing a dye that tightly binds to the backbone of each denatured polypeptide chain (usually Coomassie brilliant blue). Excess stain is then washed from the gel by soaking it in a destain solution long enough for the blue-stained bands of each polypeptide to be visualized against the clear, colorless background of the gel.

Comparison of separations by gel filtration vs. SDS-PAGE: Because the separations achieved by SDS-PAGE and gel filtration chromatography are based upon the same physical property, i.e. differences in the Stokes radius of each particle, it is not surprising that a linear relationship is observed between the log of the protein's molecular weight and its relative mobility through a size column or polyacrylamide gel. Despite this similarity, there are two crucially important differences between the separations obtained by these two methods. First, larger proteins migrate down gel filtration columns more rapidly than smaller ones because proteins with a large Stokes radius are less likely to enter the internal volume of the resin and are restricted to the interstitial spaces between the beads. In SDS-PAGE, however, this relationship is exactly reversed as smaller proteins move through a gel more rapidly than larger ones because they are being force to migrate through a single, continuous barrier instead of a column of packed beads. Since there is no exit route around the barrier in SDS-PAGE (i.e. there is no equivalent of the void volume), the larger proteins are simply forced to stack up on the top of the gel while exceptionally small proteins migrate towards the bottom of the gel very closely to the small tracking dye.

The second important difference is that while the separation of native proteins by gel filtration is affected by the shape of the particle (elongated proteins have a larger Stokes radius and move through the column more rapidly and appear to be larger than spherically shaped proteins of the same mass), shape is not a factor in SDS-PAGE because the proteins have been completely denatured into complexes with similar shapes and charge densities prior to electrophoresis. For this reason, protein separations by SDS-PAGE are based solely upon the length (i.e. the number of amino acids) of each polypeptide chain, which is, in turn, essentially proportional to its molecular weight. In fact, under ideal conditions, SDS-PAGE can effectively separate two polypeptide chains that differ in length by just a few amino acids, a degree of separation that is simply impossible to achieve by gel filtration chromatography.

For these reasons, gel filtration and SDS-PAGE provide different, but related, information about the mass of purified multimeric proteins that are composed of two or more polypeptide chains (Scopes, 1982). That is, gel filtration data is used to estimate the mass of the entire set of subunits that comprise the quaternary structure of the functional molecule while SDS-PAGE yields the mass of each individual subunit derived from the denatured complex. In the end, the mass of the native structure estimated by gel filtration must be equal to the sum of its individual polypeptide chains determined by SDS-PAGE. Hemoglobin, a hetero-tetrameric oxygen transport protein composed of two alpha and two beta subunits (Perutz, et al, 1960), provides a classic example of the mass information obtained by these two methods. Gel filtration experiments indicate the functional hemoglobin molecule has a native mass of about 65 kiloDaltons while SDS-PAGE analysis reveals the pure protein is composed of two different polypeptide chains with masses of about 16-17 kiloDaltons. Because the color intensity of the two bands observed in the gel is very similar, the results further suggest that the native molecule contains a similar mass of each subunit such that the quaternary structure must contain two copies of each polypeptide in order to account for a native mass of 65-70 kiloDaltons.

In summary, SDS-PAGE analysis has developed into a vital assessment tool in protein purification because it provides a rapid and visual comparison of both the relative purity and the amount of a specific protein contained in different fractions collected during the isolation procedure. The degree of purity between different fractions, and therefore the effectiveness of each isolation step, is evaluated by simply comparing the total number of blue stained bands observed in each fraction. That is, with each isolation step in the procedure, one expects each new fraction to yield fewer and fewer bands on the gel until a homogeneous fraction (composed of just one protein) is obtained. For the simplest case of proteins composed of just one single type of polypeptide chain, the point of homogeneity in the purification is defined by the observance of just a single band in the gel. Furthermore, since the color intensity or darkness of each band is proportional to protein mass, the relative amount of a specific protein loaded onto the gel from each different fraction is estimated by simply comparing the darkness of each band. Finally, the molecular weight of each polypeptide in the gel can be estimated by simply comparing its distance of migration or relative mobility (Rf) with that of a set of proteins of known molecular weight (i.e. molecular weight markers) loaded onto a separate well of the gel.

The results of SDS-PAGE analysis on a typical set of fractions collected by students during the isolation of invertase from baker's yeast is presented below in figure 7. The first lane on the left (labeled M) contains a set of molecular weight markers used to estimate the mass of proteins contained in samples prepared from the following fractions (from left to right): F1 (the initial 25 mL of fresh yeast extract); F2 (the 2.0 mL fraction obtained by precipitation in 29-40% ethanol); F3 (the 4 mL peak of invertase activity collected off the gel filtration column); FT (the flow through of proteins that did not absorb to the DEAE resin); F4 (the 1 mL invertase enriched fraction eluted from the DEAE resin with 50 mM NaCl); HSW (the 1 mL high salt wash of proteins eluted from the DEAE column with 250 mM NaCl); and, finally, Σ (a solution of commercial invertase purchased from Sigma Chemical Company prepared by dissolving a weighed mass of the solid protein in column buffer to a final concentration of 1 mg per mL).

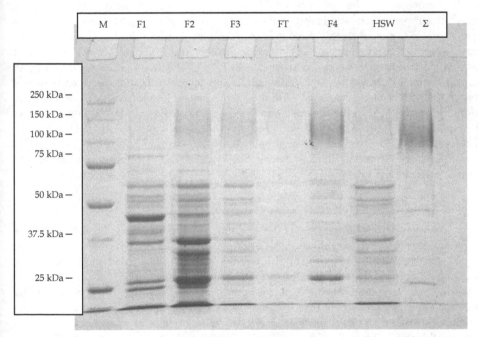

Fig. 7. SDS-PAGE analysis of fractions obtained in the isolation of invertase from baker's yeast. All samples were denatured (Laemmli, 1970) and reduced by mixing with an excess of SDS, β-mercaptoethanol and heated in a boiling water bath prior to separation by electrophoresis at 2 W of constant power on a discontinuous gel (composed of an 8% acrylamide resolving gel and a 4.5% stacking gel each with an acrylamide to bis-acrylamide ratio of 1:29).
M = Biorad Precision Plus Molecular Weight Markers with masses listed for the seven visible bands (as shown from the top to bottom) equal to 250 kDa, 150 kDa, 100 kDa, 75 kDa, 50 kDa, 37.5 kDa and 25 kDa.
Other lane labels are described in the text.

The lane containing the commercial sample of invertase (Σ) on the far right is clearly dominated by the presence of a large band that is both intensely stained and smeared as it migrates (compared to all other bands) over an unusually wide range of molecular weights extending from 100 to 150 kDa. This observation is consistent with previous work characterizing yeast invertase as a large protein with a native mass of 270 kDa composed of two identical and heavily glycoslyated subunits with masses of about 135 kDa. For an undergraduate exercise, this unusually smudged appearance of the invertase band (which is due to the extensive but variable amount of sugars attached to the subunits) is a benefit that allows an inexperienced observed to readily identify the position of invertase in each sample and visually assess the relative degree of purity in each fraction by comparing the darkness of the invertase band vs. that of the other bands in each lane.

The SDS-PAGE analysis unambiguously demonstrates that an impressive and significant degree of purification is achieved between the initial yeast extract (F1) and final fraction (F4) which is dominated by the presence of the sane large, smeared band observed in the commercial sample which was barely visible in the initial extract (F1). Furthermore, this analysis also indicates that this final fraction (F4) is significantly less pure than the commercial source of the enzyme (Σ) as evidenced by the difference in the number of contaminating bands with masses between 25-75 kDa in the two fractions. On the other hand, it is difficult to assess the sequential increase in purity anticipated with each step of the procedure (from F1-F4) by SDS-PAGE analysis because, in addition to being purified, the proteins precipitated in the first isolation step were also concentrated 10-fold in F2 and then subsequently diluted 4-fold in the second isolation step by gel filtration chromatography in F3. In spite of this complication with large increases and decreases in concentrations, the expected trend in which the invertase band represents a larger and larger percentage of the total protein in each fraction seems to be apparent on the gel.

6.2 Specific activity measurements

Specific activity is defined as the ratio of the total number of units of enzyme activity vs. the total mass of protein contained in a sample and is most commonly expressed as "enzyme units per mg of protein" (or more simply as units/mg). Because the overall goal of the purification is to obviously maximize the yield of the total number of enzyme units from the initial tissue extract while removing as much other protein as possible, the specific activity of each fraction is expected to increase with each isolation step until it plateaus to a maximal value that is a characteristic, intrinsic property (independent of the concentration or amount) of the purified enzyme. That is, it is impossible for the specific activity to increase beyond the homogeneity point of the purification procedure because any additional isolation step will result in a proportional loss of both total protein (which is now 100% enzyme) and enzyme activity to produce a constant specific activity ratio.

Therefore, the final task of this project is to quantitatively assess the effectiveness of the isolation procedure by carefully measuring the specific activity of each isolated fraction and comparing each one against the specific activity of commercial invertase which, according to the previous analysis by SDS-PAGE analysis, represents a nearly homogenous sample of the enzyme. In the end, a reasonable agreement is expected between the analysis by SDS-PAGE and specific activity measurements. That is, with each isolation step in the procedure, the specific activity should increase as the number of contaminating bands in the gel decreases until the homogeneity point is detected by a plateau in the specific activity measurement and the appearance of just a single protein in the gel. In short, specific activity is an important index by which the relative purity of an enzyme in different samples can be compared. That is, a fraction with a specific activity of 50 units per mg is described to be five-times more pure than a fraction with a specific activity of 10 units per mg. For this reason, the effectiveness of each isolation step in the procedure can be assessed by simply determining the extent by which it increases the specific activity of the enzyme.

In practice, the result of each isolation step used in a procedure is traditionally presented as a **purification table**, which is simply a summary of the total amount of protein, enzyme activity and specific activity contained in each isolated fraction. However, before the specific activity values of different samples are compared, it is important to recognize that specific activity represents a conditional property of an enzyme (an issue that can be very frustrating for undergraduate students accustomed to measuring non-conditional properties of matter such as length or mass). That is, because the methods used to measure both activity units and protein mass are dependent upon many variables that must be precisely controlled in each assay (including temperature, pH, and reagent concentrations which are, in turn, affected by the total assay volume), it is very plausible for two different groups to honestly report widely different specific activity values for the exact same set of samples by simply changing the assay conditions! For this reason, purification tables are required to include a detailed list of the precise set of conditions used to measure both enzyme activity and protein content (or a reference to this list). Perhaps most important in terms of laboratory procedures, this conditional nature is an important issue in order to understand and appreciate the vital role that sampling consistency plays in reducing many experimental errors that affect specific activity measurements. For this reason, whenever possible, the assays for each fraction reported in a purification table should be performed at the same time, using the same reagents, instruments, equipment, and, perhaps, even the same person.

Table 2 presents a traditional purification table summarizing the average results of specific activity measurements obtained by eight different groups of students for the same initial yeast extract (F1), the three fractions obtained from each isolation step (F2-F4), and a stock solution of commercial invertase prepared by dissolving solid protein with column buffer to a final concentration of 1.0 mg per mL (Σ). In terms of percent yields, the table clearly shows a sequential decrease in the total mass of protein in each fraction (which was measured indirectly by a modification of the Bradford dye binding technique using bovine serum albumin as the standard reference protein) starting with an average of 39 milligrams in the initial 25 mL of yeast extract to only 0.16 milligrams in the final 1-mL fraction (F4) for a combined percent yield of just 0.41% of the initial protein. Likewise, the total number of enzyme units (as measured spectrophotometrically using the 5-minute stop assay described previously) also dropped with each step from an initial value of 1070 units to 72 units in F4 which is an overall percent yield of just 6.7 % of the initial enzyme activity (but still 16 times higher than the percent yield of total protein). However, in contrast to the loses of both total protein and activity units, the table indicates a sequential and significant increase in the specific activity of each fraction starting from 28 units per mg in the initial extract to 490 units per mg in the final fraction (F4) and an overall n-fold purification of 17.5 (that is, the invertase contained in the final fraction, F4, is, on average, 17.5 times more pure than the invertase in the initial yeast extract). These results simply confirm (as previously observed by SDS-PAGE analysis) that each step of the procedure succeeded in selectively retaining a higher percentage of the invertase compared to other proteins in each sample.

Fraction	Volume (mL)	Total Protein (mg)	Total Activity (units)	Specific activity (units / mg)	n-fold purification (vs. F1)
F1	25.0	39 ($\pm$ 6)	1070 ($\pm$ 95)	28 ($\pm$ 5)	1.0
F2	2.0	8.2 ($\pm$ 2.6)	717 ($\pm$ 100)	94 ($\pm$ 24)	3.4
F3	4.0	0.72 ($\pm$ .33)	145 ($\pm$ 60)	210 ($\pm$ 80)	7.5
F4	1.0	0.16 ($\pm$ 0.08)	72 ($\pm$ 27)	490 ($\pm$ 180)	17.5
Σ	1.0	0.11 ($\pm$ 6)	115 ($\pm$ 15)	1100 ($\pm$ 300)	39

Table 2. Purification table for the isolation of invertase from Baker's Yeast

All data reflect the average values ($\pm$ standard deviation) reported by 8 different groups of students using the same yeast extract (F1) prepared from Red-Star bread machine yeast. Protein values were measured by a modification of the Bradford dye binding method using bovine serum albumin as the reference protein (Zor & Selinger, 1996). Invertase activity was monitored at pH 4.8, ambient temperature (20-24° C), and 20 mM sucrose using the standard 5-minute stop assay to reduce 3,5 dinitrosalicylate to 3-nitro-5-amino salicylate which is detected at 540 nm . F1= Yeast extract; F2 = 29-40% ethanol cut; F3 = gel filtration peak; F4 = DEAE enriched invertase; Σ = Commercial invertase (Sigma chemical company) dissolved in gel filtration column buffer to a final concentration of 1 mg per mL.

In comparing the effectiveness of the three isolation steps used in the procedure, the precipitation with ethanol appears to have been the most efficient step by providing a 3.5-fold increase in the average specific activity of invertase compared to the additional increases of 2.2- and 2.3-fold obtained by gel filtration and anion exchange, respectively. Within this context, it should be pointed out that because the specific activity of the nearly-homogenous sample of commercial invertase (as judged by SDS-PAGE) is equal to 1100 units per mg, that the maximum n-fold purification possible from this extract (at 28 units per mg) is a 39-fold increase. Therefore, the combined 17.5-fold enrichment obtained at the end of the three steps in this project accounts for about 45% of the upper limit. Likewise, by comparing the specific activity of the commercial enzyme (1100 units per mg) with the total number of units in the initial extract (1070 units), one can estimate that invertase accounts for about 1 out of the 39 mg of total protein (or 2.5%) contained in the initial 25 mL of extract prepared from the Red-Star bread machine yeast.

Finally, the results of this project demonstrate excellent agreement between the SDS-PAGE analysis and specific activity measurements in comparing the purity of the final fraction (F4) with that of the commercial invertase (Σ). That is, the commercial enzyme appears much more pure on the gel because it contains far fewer contaminating bands than that observed in DEAE enriched fraction of invertase (F4). Likewise, the specific activity of the commercial enzyme (1100 units per mg) is about 2.2-times higher than the final fraction (490 units per mg). While it may seem disappointing that, after all of this work, the students do not obtain as pure of a sample as the commercial enzyme, it must be remembered that these steps were not optimized to maximize either the yield or purity of invertase. Instead the steps were designed to introduce general methods of protein purification to large groups of undergraduate students over the course of several three-hour laboratory periods. If a more pure fraction of invertase is desired, each of the following modifications to this procedure would be worth considering: (i) first and foremost, start with a much larger volume of the initial extract (such as 250 – 1000 mL) because it is always better to work with a larger vs. smaller amount of the desired protein; (ii) use a much larger and longer gel filtration column in order to be able to both load larger samples and maximize the separation between invertase and other proteins on the column; (iii) raise the concentration of salt in the gel filtration column buffer in order to reduce the chance that other proteins might adsorb-to and co-elute with invertase off the size-column; (iv) monitor each one of the invertase enriched fractions eluted from the size-column by SDS-PAGE in order to discard the most contaminated fractions from the ones that are pooled together and applied to the anion-exchange column; (v) improve the efficiency of the ion-exchange column by using a salt gradient to gradually increase the sodium chloride concentration and dissociate the enzyme from the column more selectively; (vi) incorporate an additional or different isolation step into the procedure, such as lectin affinity chromatography (lectins, such as conconavalin A and wheat germ agglutinin, are sugar binding proteins that, when immobilized to a solid column support are useful in the purification of glycoproteins).

7. Conclusion

The exercises described in this chapter provide a practical, hands-on introduction to many general considerations and corresponding strategies encountered during the course of isolating a specific protein from its initial biological source. Furthermore, the project is especially well suited for incorporation into an undergraduate laboratory curriculum for several reasons: First, the starting material, dried baker's yeast, is relatively inexpensive and can be obtained from most local grocery stores. In addition, the procedure required to extract invertase from the periplasmic space of yeast cells is both exceptionally uncomplicated and yields a large volume of an extract that is enriched with an uncommonly stable glycoprotein that is easy to detect by SDS-PAGE analysis. Second, the materials and equipment required for the detection assay, isolation steps, and final analysis are also reasonably inexpensive and currently stocked in most undergraduate laboratories. Finally, the isolation steps are designed to infuse a large number of visual cues into the isolation of a colorless enzyme from a natural product (and therefore it represents a more typical real life example compared to isolating colored proteins, such as green fluorescent protein or myoglobin (Miller et al. 2010), that are over expressed in genetically modified sources). These visual cues include the following: (i) the 2-step precipitation in ethanol requires students to consider where their enzyme is located at all times (the decantate or pellet); (ii) the addition of visual markers to the sample prior to gel filtration lets them know where their enzyme is located on and off the column; (iii) contaminating blue dextran in the gel filtration enriched fraction of invertase is removed in the last isolation step by adsorption to the DEAE cellulose column, in this way they know the column is working; (iv) The fractions are first visually analyzed by SDS-PAGE, prior to performing the more technically challenging and less intuitive estimate of purity by specific activity measurements; and, finally (v) commercial sources of the pure enzyme to serve as a gold standard of comparison against their own results is readily available and inexpensive.

8. References

Jisnuson, S. & Bhinyo, P. (1977). SDS-Polyacrylamide Gel Electrophoreses: A Simple Explaination Why it Works, *Journal of Chemical Education*, Volume 54, No. 9 (September 1977), pp 560-562, ISSN 0003-2697

Laemmli, U.K. (1970). Cleavage of the Structural Proteins During the Assembly of the Head of Bacteriophage T4, *Nature*, Vol. 227, (August 15 1970), pp 680-685, ISSN 0028-0836

Lampen, J.O. (1971)Yeast and Neurospora Invertases, In *The Enzymes*, Boyer, P.D., Ed; pp 291-305, Academic Press, ISBN 0121227022 New York

McLoughlin, D.J. (1992). Size Exclusion Chromatography: Separating Large Molecules From Small, *Journal of Chemical Education*, Vol. 69, No. 12 (December, 1992), pp 993-995, ISSN 0003-2697

Melius, P. (1971). The Isolation of Yeast Invertase by Sephadex Gel Chromatography. *Journal of Chemical Education*, Vol. 48, No. 11 (November, 1971), ISSN 0003-2697

Miller, S., Indivero, V., & Burkhard, C. (2010). Expression and Purification of Sperm Whale Myoglobin, *Jounal of Chemical Education*, Vol. 87, No. 3 (March 2010) pp 303-305, ISSN 0003-2697

Moreno, S., Sanchez, Y., & Rodriquez, L. (1980). Purification and Characterization of Invertase from Schizosaccharomyces pombe, *Biochemical Journal*, Vol. 267, pp 697-702, ISSN 0264-6021

N.P. Neumann & J.O. Lampen, (1967). Purification and Properties of Yeast Invertase, *Biochemistry*, Vol. 6, No. 2 (February 1967), pp 468-475, ISSN 0006-2960

Perutz MF, Rossmann MG, Cullis MG, Muirhead H, Will G., North ACT (1960). Structure of haemoglobin. A three-dimensional Fourier synthesis at 5.5Å resolution, obtained by X-ray analysis, *Nature, Vol* 185 pp 416–422 ISSN 0028-0836

Roberts, C.A., Spencer E.J., Bowman G.C., & Blackman D. (1976). Polyacrylamide Gel Electrophoresis of Yeast Invertase: A Biochemistry Laboratory Experiment *Journal of Chemical Education*, Vol. 53, No. 1 (January 1976), pp 62-63 ISSN 0003-267.

Schulke, N. & Schmid F.X. (1988). The Stability of Yeast Invertase Is Not Significantly Influenced By Glycosylation, *The Journal of Biological Chemistry*, Vol. 263, No. 8 (June 25 1988), pp 8827-8831, ISSN 0021-9266

Scopes, R., (1982) *Protein Purification: Principles and Practice*, Springer-Verlag, ISBN 0-387-90726-2 New York.

Sigma Product Information Bulletin Number I4504, Sigma Chemical Company, St. Louis, MO, 2002.

Sumner, J.B. & Sisler, E.B. (1944). A Simple Method For Blood Sugar, *Archives of Biochemistry*. Vol. 4 ,pp 333-336, ISSN 0096-9621

Timerman, A.P., Fenrick, A.M., & Zamis, T.M. (2009). The Isolation of Invertase from Baker's Yeast, A Four Part Exercise in Protein Purification and Characterization, *Journal of Chemical Education*, Vol. 86, No. 3 (March 2009), pp 379-381, ISSN 0003-2697

Zor, T. & Selinger, Z. (1996). Linearization of the Bradford Protein Assay Increases Its Sensitivity: Theoretical and Experimental Studies, *Anal. Biochem.* Vol. 236, No. 2, (May 1996) pp 302- 308, ISSN 0003-2697

2

The Art of Protein Purification

William Ward

Rutgers University, New Brunswick NJ,
USA

1. Introduction

Describing, in words, the details of protein purification to a relative novice in the field is not unlike explaining on paper the steps required to turn a set of colored oils into a beautiful pastoral scene on sheet of stretched canvas. Playing the oboe in a sophisticated metropolitan orchestra or performing a solo aria in a Gilbert & Sullivan operetta are accepted artistic endeavors that command great mastery of technique. Each of these art forms requires years of experience and endless experimentation and refinement of technique. Protein purification is no different. It is an art form. Like all other art forms, perfecting the art of protein purification requires a long apprenticeship. But, like all other art forms, protein purification is aesthetically rewarding to the practitioner. Every day brings new challenges, new insights, new hurdles, and new successes. Art is a process, not a destination. Protein purification fits the same definition.

Perfecting the skills of protein purification can take many years of hands-on experience as well as periodic upgrading of those skills. Perhaps the most important part of protein purification is the set of pre-column steps that precede column chromatography. Pre-column steps are not covered as much in the protein purification literature as column chromatography, HPLC, and electrophoresis. So, I have chosen to focus much of my attention on the earlier stages of protein purification. More than column chromatography, pre-column steps are highly diverse and highly creative. Here the artistic aspects of protein purification are most apparent. But, still, there are basic guiding principles that can be communicated fairly effectively in written form. The purpose of this chapter is to outline many of these principles and techniques such that a relatively inexperienced biochemist can get started. Getting started is never easy. Inertia always seems to get in the way. When I think of the problem of overcoming inertia, I am reminded of the words of my first graduate school mentor. He chose to explain overcoming inertia with a metaphor based upon physical chemistry, "The function of education is to help others overcome their own energy barriers." In part, overcoming energy barriers is what I hope to accomplish in this chapter.

2. Protein purification in the analytical field

The words in my introductory paragraphs are more relevant to preparative techniques of protein purification than they are to analytical methods. Most of my research career has been focused upon preparative methods—the approach I liken to other art forms. Analytical methods of protein purification are less likely to encompass the artistic range I ascribe to preparative methods.

The focus of analytical methods is usually to make a large number of precise measurements in a short period of time. One version of analytical methodology used extensively in the biopharmaceutical industry is called high throughput screening (HTS). Most commonly, HTS is used in drug screening. But HTS and other high throughput methods are applicable to analytical protein purification as well. But, as HTS is, by its very definition, a very rapid process, extensive protein purification is not possible by this method. Complex, multistep processes are almost always precluded. To meet time demands, just one simple and rapid purification step may be all that is permitted. Often this means that fast "sample cleanup" is the major goal of analytical processes. This "cleanup" may require nothing more than the removal of a particular interfering substance—an endogenous enzyme inhibitor, for example. External effector molecules may give falsely high assay values, or, more commonly, may inhibit enzyme activity, lowering an assay value, significantly. If, for example, one has a large number of relatively impure samples for which accurate values of the glucose oxidase activity is needed, it may be necessary to separate all other oxidoreductases from glucose oxidase. Alternatively, it may be sufficient to remove all endogenous sources of glucose. These types of separations are done routinely in clinical, medical, and pharmaceutical diagnostics laboratories. Sometimes, microliter samples are robotically introduced into small HPLC (high performance liquid chromatography) columns followed by on-line analysis of the protein of interest. On other occasions, machine-processed samples are introduced robotically into multi-well microtitre plates. Then, built-in robotic components introduce enzyme substrates and cofactors as the plates are stacked up by thousands to be measured after a precise incubation period.

In such analytical operations, the art is in the design of robust sample handling methods including electronic, mechanical, and robotic components. Optimization of protein separation may be an integral part of system design, but once the entire system is on-line, only routine validation tests along with periodic trouble-shooting of the overall system are required. Once the creative aspects of system design have been completed, everything now devolves into system maintenance.

3. Preparative protein purification

General Strategy

The greatest differences between analytical-scale and preparative-scale protein purification processes are that preparative methods (1) usually involve much larger volumes of starting material, (2) generally take much longer to carry out (days, weeks, or months), (3) usually require a variety of different purification methods or techniques (sometimes repeated), and (4) almost always have, as the primary goal, achieving very high purity (rather than high throughput). Sometimes, the amount of desired protein is so small, and the amount of macromolecular contaminant is so high, that one needs to employ nearly every "trick of the trade" to achieve high purity. Imagine wanting to isolate milligrams of a precious protein from thousands of liters of crude jellyfish extract. Our research group has done this for almost 3 decades (Roth, 1985, Johnson and Shimomura, 1972, Blinks, et. al., 1976, and others). Sometimes, purifying a protein to homogeneity, from such large volumes of highly viscous starting material, may involve separating one milligram of the protein-of-interest (POI) from

100 mg of initial total protein. This is called a 100-fold purification. In other cases the required purification factor may be on the order of 1000-fold or 10,000-fold. My most difficult purification project was to isolate microgram amounts of green-fluorescent protein (GFP) from the homogenates of whole sea pens. In this instance, not only was the GFP present at about 1 part in 100,000 of total protein, but the proteoglycan-derived viscosity in the crude extract was so great that a magnetic stir bar failed to rotate (Ward and Cormier, 1978). So the issues facing a scientist working on a difficult protein purification project are many. Among these issues are those shown in Table 1.

1	Choosing or developing a sensitive, reproducible, and selective assay for the protein-of- interest (POI).
2	Establishing conditions under which the POI is stable and biologically active.
3	Finding conditions under which the POI can be stored safely between steps.
4	Choosing the best biological starting material (natural source or recombinant).
5	Developing or choosing appropriate methods for gross extraction.
6	Decreasing viscosity of crude extracts and removing particulates from those extracts.
7	Reducing volume.
8	Finding the substrate(s), inhibitors, activators, allosteric effectors, etc., if the protein-of- interest is an enzyme.

Table 1. Early steps in designing protein purification strategies

Some very useful information can be acquired, unambiguously, if a small sample of pure protein can be obtained. A former professor of mine said to our group of graduate students, "Don't waste clean thinking on a dirty enzyme." It is so easy to make major errors if you try to over-analyze a crude sample. Acquiring a pure sample of the protein-of-interest may be difficult (if the specific purification methods have not been optimized). But, obtaining a small amount of pure protein can be very useful for future optimization of purification. Table 2 lists a few of the characteristics of a pure POI that can be used to design a more effective purification strategy. Unless the protein-of-interest is pure, data on its characteristics can be very misleading (Karkhanis and Cormier, 1971).

a	Solubility in water, salt solutions, organic solvents, etc.
b	Presence of isoforms or isoenzymes.
c	Molecular weight.
d	Degree of oligomerization (monomer, dimer, tetramer, aggregation, etc)
e	Isoelectric point.
f	Partial amino acid sequence (needed if the mRNA is to be found).
g	Post translational alterations (phosphorylation, glycosylation, blocked N-terminus, etc).
h	Amino acid analysis.
i	Relative hydrophobicity (as determined by HIC trials or ammonium sulfate precipitation).
j	Antibodies to the protein-of-interest
k	Essential cofactors, prosthetic groups, stabilizing agents, etc.

Table 2. Physical and chemical properties of a pure sample that may be needed to effectively design a purification strategy.

Where to Begin

It is difficult to suggest a logical order of steps leading to a successful protein purification project. Proteins are very different from each other (and so are the mixtures of other components in which the protein-of-interest is found). So there is no common approach. Perhaps the best way to introduce protein purification is by example. I will do this by showing some of the intimate details of how one protein, *Aequorea victoria* GFP, has been purified in our academic lab at Rutgers University (Roth, 1985, Ward and Swiatek, 2009). In parallel, I will discuss the similarities and differences that accompany purification of another protein, soybean hull peroxidase. The latter has been purified in our Rutgers spin-off, start-up company, Brighter Ideas, Inc. (Holman, C., manuscript in preparation, Ward, 2012). I will not discuss, in detail, purification methods employed with recombinant proteins. These methods are much simpler and much more straight-forward (requiring considerably less "art" once the molecular biology has been completed).

The Assay

Before a protein purification process can begin, there must be a way to identify the protein-of-interest (POI). The means for identification is called an assay. For enzymes, the assay is usually a measure of enzyme activity. For proteins with distinctive chromophores, spectroscopic measurements of the chromophore help to distinguish the POI from other proteins. Sometimes all that one knows about the protein-of-interest is its molecular weight. In such cases the POI can be followed by SDS gel electrophoresis. Sometimes a protein is assayed by its immune response. Sometimes immune response is all that the scientist knows in the beginning. The protein, calmodulin, was discovered in brain tissue solely on the basis of its ability to bind radioactive calcium (Cheung, 1971). Binding calcium was all that was known about calmodulin in the earliest stages of its purification. But, the more one knows about alternate ways to detect the protein of interest, the easier the chore is likely to be.

GFP is not an enzyme, so there is no enzymatic assay. But, it has a spectroscopically measurable, covalently-bound chromophore (Fig. 1) that absorbs light maximally at 397 nm (Ward, 2005). GFP fluoresces brilliantly (emission peak at 509 nm) when excited in the UV. A hand-held, 365 nm, mercury vapor lamp ("black light") becomes a convenient, portable detector. Molar extinction coefficient at 397 is 27,300, but that value varies 5-10% depending upon the degree of dimerization of the protein (Ward, 2005, Ward, *et. al.*, 1982). Fluorescence quantum yield is 80%. With all proteins, measurements by absorbance or fluorescence requires samples with VERY low turbidity (light scatter). Even partially clarified crude extracts have far too much scatter to measure any protein accurately by UV/Vis spectrophotometery (Fig. 2). Sometimes it takes a few purification steps before the level of GFP, for example, can be measured with any reliability.

Soybean peroxidase (SBP), like GFP, has a chromophore—a heme group that absorbs maximally at 403 nm. Absorbance at this wavelength can be used to quantitate the enzyme. But many other substances in crude soybean hull extracts absorb strongly at the same wavelength. So, the enzyme needs to be highly purified before this measurement is useful. Another assay is needed. Peroxidases, in general, bind to hydrogen peroxide, creating an active oxygen species that can then attack another molecule. In our case, the other molecule is ABTS (2,2'-azino-bis(3-ethylbenzthiazolene-6-sulfonic acid) available from the Sigma Chemical Co. ABTS, dissolved in a pH 5 buffer with added hydrogen peroxide, has only a very slight visible absorbance. But in the presence of peroxidase, the active oxygen attacks the ABTS producing a teal colored solution. As with many other colorimetric assays, attention must be paid to the stability of the assay solution and the kinetics of the reaction.

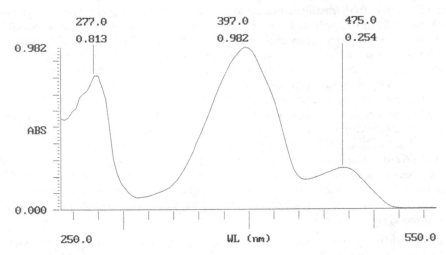

Fig. 1. Absorption spectrum of pure green fluorescent protein. Optical density vs. wavelength in nanometers.

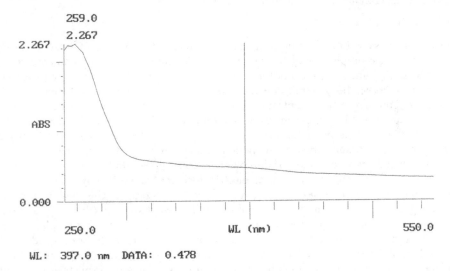

Fig. 2. Spectrum of diluted crude *E.coli* suspension. Apparent optical density (mostly scatter artifact) vs. wavelength in nanometers.

Stability

Probably the second most important characteristic for an effective protein purification scheme is the protein's stability, especially stability to heat and pH. But, just determining conditions of high stability at the outset of purification is seldom sufficient. Some proteins are more stable in the crude form and others more stable when pure. So, at each step along the way, stability needs to be checked.

GFP and SBP are both thermally stable, up to 65 C for GFP (Bokman and Ward, 1981, Ward and Bokman, 1982) and nearly 90 C for SBP (Holman, C, manuscript in preparation). GFP is stable to proteases and aqueous alcohol solutions (Roth, 1985). The C-terminal 8 amino acid tail of native jellyfish GFP is protease labile, so we usually keep the crude extracts cold. We use sodium azide to inhibit microbial growth and phenylmethyl sulfonyl fluoride (PMSF) to inhibit the activity of serine proteases (Ward, 2005). Circular dichroism measurements confirm that the native secondary structure of GFP (predominantly beta pleated sheet and just a small amount of alpha helix) is directly proportional to the protein's fluorescence (Ward, et. al., 1982). GFP retains its fluorescence and its secondary and tertiary structure at elevated pH (up to 12.2) but loses fluorescence at pH 12.3 (and simultaneously loses its CD signature) (Ward, 2005). Under acidic conditions (pH 6 and below) GFP fluorescence also fades as does the CD signal. Under the right conditions, GFP will recover most of its fluorescence after denaturation in acid, base, and guanidine hydrochloride (Bokman and Ward, 1981). The only known detergent to destroy GFP fluorescence, permanently, is sodium dodecyl sulfate (SDS).

Soybean peroxidase is stable over a wider range of pH and a wider range of temperature than GFP. But, its activity is inhibited by sodium azide and other agents that react with heme proteins. Instead of sodium azide, we use 10 % ethanol as a preservative for SBP. However, not all enzymes are stable in the presence of alcohol.

Storage Conditions

It is usually necessary, in multi-step purification protocols, to store the POI between steps. Generally, this is accomplished by freezing the protein solution. Freezing and cold storage work for both GFP and SBP, but not for all proteins. Some multisubunit proteins are cold labile. In such cases, the subunits are held together by hydrophobic interactions. Such hydrophobic bonding can be entropy driven, as structured water (surrounding the monomers in an ordered way) becomes released (and more disordered) when subunits bind to each other. The ΔS term in the equation: $\Delta G = \Delta H - T\Delta S$ increases with increasing temperature. GFP, SBP, and most monomeric proteins, are not cold sensitive. In addition, based upon its long-term retention of fluorescence, GFP appears to be stable for months at room temperature (Roth, 1985). But, isoelectric focusing of GFP may show extensive microheterogeneity after prolonged room temperature storage. The highly protease-sensitive eight amino acid C-terminal segment of native jellyfish GFP, (that extends from a protease-resistant beta barrel) is easily clipped by proteases—often in different places (Roth, 1985). When the recombinant protein is C-terminally tagged with hexa-histidine (for eventual immobilized metal affinity chromatography (IMAC), now both the naturally occurring octapeptide and the added hexapeptide are susceptible to cleavage at many sites by a variety of proteases.

Starting Material

In some cases, one has a choice of starting material. Luciferase, for example, can be isolated from a variety of fireflies and beetles. But, some firefly luciferases are very hard to purify while others are much easier. The sea pansy, *Renilla reniformis* (Wampler, et. al., 1971,

Matthews, *et. al.*, 1977, Prendergast and Mann, 1978, Ward and Cormier, 1979) and the jellyfish, *Acquoria victoria*, (Morise, *et. al.*, 1974, Roth, 1985, Ward, 2005) were chosen as the starting materials for isolating and purifying GFP. In part, the selection of organisms was based upon their geographical locations, the availability of nearby laboratory facilities, and the means for collecting the animals. The shallow waters off the coast of Georgia proved to be a good location for collecting sea pansies and there was a local shrimper only too willing to do the collecting before the shrimp season began. The University of Georgia had a primitive laboratory on Sapelo Island, but early stage processing did not require sophisticated facilities. *Aequorea* jellyfish were abundant for decades at the University of Washington's Friday Harbor Labs (FHL) and the lab facilities were excellent. Excellent facilities were essential, as extensive floating docks were needed to provide close access to the water (so that the jellyfish could be scooped up with pool skimming nets). Processing involved holding the jellyfish (sometimes 10,000 per collection day) in large, circulating sea water aquaria. The FHL facilities include many circulating sea water aquaria, a walk-in coldroom, and a Sorvall centrifuge for further sample processing. The FHL staff was particularly supportive and encouraging.

While peroxidases can be isolated from many sources including horseradish, potatoes, sweet potatoes, and other plants, we chose soybean hulls as our starting material. The choice was based primarily upon easy access and low price. Perdue Farms processes huge quantities of soybeans for chicken feed. The hulls, a byproduct of their processing the more valuable soybean oil and soybean meal, are usually shipped to multi-grain bread manufacturers. To reduce storage and shipping volume, the hulls are crushed, on the Perdue site, into finer particles ranging down to the micrometer range. The bread producers apparently pay very little for an otherwise "throw-away" byproduct of the soybean. We, for example, ordered 2000 lbs of hulls, paying $400 for hulls. The price included seven 55-gal barrels plus shipping. While access, ease of acquisition, and facilities were more than adequate for early, on-site processing of sea pansies, jellyfish, and soybean hulls, later laboratory processing was VERY demanding. This leads us into the next section, "Extraction".

Extraction

In the case of the sea pansy, extraction of GFP was accomplished by first anesthetizing the animals in a bath of the calcium-chelating agent EGTA plus magnesium sulfate. This was to preserve a luciferin binding protein, easily triggered to luminesce with calcium ions. Grinding the sea pansies with protein-saturating levels of ammonium sulfate came next, followed by acetone precipitation and rapid drying of the organic solvent. The powder that resulted, largely ammonium sulfate, was stored in chest freezers until processing time (Matthews, *et. al.*, Ward and Cormier, 1979).

GFP isolation from the jellyfish was entirely different. A single jellyfish has a volume of about 35 ml. On days when we collected 10,000 animals, the volume we needed to process reached 350 liters. However, all of the luminescent tissue is found in a very narrow strip along the margin of the "bell" (Fig. 3). Special dissecting tables were constructed, allowing a small team of workers to dissect up to 10,000 animals in one collecting day. Dissection reduced the volume to about 15 liters. Next, the tissue was shaken vigorously, 500 ml at a time, in 3 liters of sea water (in a 4-liter flask). Seventy-five shakes released most of the photocytes into suspension. After crude filtration, the photocyte suspension was trapped in a large cake of celite (diatomaceous earth) held in a large Buchner funnel. After a wash with

75% saturated ammonium sulfate solution (containing EDTA to chelate calcium), the photocytes were lysed with dilute EDTA solution. A gentle vacuum applied to the suction flask released an amazingly bright stream of fluorescence that was captured in the 4-liter vacuum flask. The extract was precipitated with solid ammonium sulfate — the precipitated protein being trapped on a smaller cake of celite or collected by centrifugation. These procedures were developed by Dr. John Blinks (Blinks, *et. al.*, 1976).

Fig. 3. Underwater photograph of the jellyfish *Aequorea victoria*.
Photograph is courtesy of R. Shimek of the University of Washington's Friday Harbor Laboratories.

Soybean peroxidase extraction just requires that the pulverized hulls be stirred in five volumes of distilled water for one hour.

Viscosity Reduction and Particle Removal

As one might imagine, extracts of whole coelenterates or coelenterate tissues (jellyfish or sea pansies) present a huge problem with viscosity. Aside from water, the animals are almost entirely composed of connective tissue and very high molecular weight proteoglycans. For 17 seasons, we solved the viscosity problem by passing crude extracts of jellyfish photocytes (and surrounding tissues) through an 8-liter gel filtration column of P-100 BioGel (our next step after ammonium sulfate precipitation). The void volume fraction (calibrated to have a molecular weight of 40 million Daltons or greater) contained most of the viscosity and none of the GFP. But, while this 3-day procedure worked quite well as a viscosity reduction method, each gel filtration run could handle, one at a time, only 5% of a season's collection. Larger amounts of extract invariably fouled the column. If one includes the frequent column washes, required to maintain reasonable flow, it takes 5-6 months to pass a season's worth of jellyfish extract through the column. It was not without trying many alternative methods that we settled on this highly unusual first chromatography step (Fig 4). Gel filtration is generally reserved as a late-stage polishing step. Much later in our work, we discovered that simple passage through a column of Celite easily solved the viscosity problem (W. Ward, unpublished). Diatomaceous earth is so inexpensive that the column contents could be discarded after the desired protein easily passed through. The above example illustrates one of the great dilemmas in selecting steps for a protein purification protocol. When do you decide that you have spent enough time searching for a better way to do things? When do you give up trying to search for a better procedure by settling on a brute force method? The expression, "Are you going to fish or cut bait?" seems appropriate here.

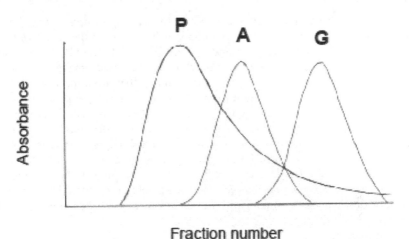

Fig. 4. P-100 Biogel profile of crude jellyfish extract. **P** marks the absorbance profile of total protein at 280nm. **A** marks the activity of Aequorin protein. **G** marks the GFP fluorescence.

Soybean peroxidase crude extracts are fairly low in viscosity, but the hull extracts present a very significant problem with particulates. The crude extracts include large particles (millimeter size) as well as tiny particles in the micrometer range—some as colloidal suspensions. Large fragments of hulls are easily filtered away with fine mesh nylon nets, but this leaves a very cloudy suspension of fine to very fine particles. Centrifugation has been ruled out because of the large volumes of extract produced and the high centrifugal forces needed to pellet the finest particles. Even continuous flow centrifugation trials have failed, repeatedly, because most of the particulates, including colloidal materials, have failed to sediment during the short interval of time it takes for liquid to traverse the centrifugation path. After trying everything we could imagine and after investing money in a variety of expensive filter devices (G. Swiatek and M. Browning, personal communication), we suspended this project for several years. Then we happened upon an ion exchange method normally applied to water purification. We found a company called ResinTech that provides, at very low cost, a high capacity polystyrene-based anion exchanger. The beads are large (1 mm) and dense, so, after stirring, they quickly settle to the bottom of a large container. Binding kinetics, however, are slow, because of the large size of the beads and relatively small pore size (access to the interior is slow and limited to proteins of MW 50 kdal or lower. So, notwithstanding the slow kinetics of binding and elution, these beads are useful for batch ion exchange applications—in our case, to trap the highly anionic soybean peroxidase (C. Holman, manuscript in progress, Ward, 2012). A provisional patent for our unique SBP purification method has been filed with Rutgers University. The fine particles of soybean hull extract (much too fine to settle on their own) are, however, too large to enter the ResinTech pores. So the bound SBP can be separated from these fine particles. But, much to our surprise, we found that the fine particles, as soon as stirring ceases, immediately aggregate into a dense gelatinous mass that settles above the beads. By aspiration, this gelatinous mass is easily separated from the beads that now containing nearly all of the SBP.

Volume Reduction

In a typical academic or start-up corporate laboratory, the starting sample of crude protein might range in volume from a few milliliters to tens or hundreds of liters. In commercial operations, liquid volumes may reach thousands or hundreds of thousands of liters. Here, I focus on moderately large volumes that require much more effort than smaller volumes. The volume of starting sample dictates, in a sense, the methods that are appropriate for early stages of purification. Large aqueous volumes require an early stage trapping step—a step that eliminates large quantities of water while binding (or otherwise retaining) the protein-of-interest. The focus is not on separating a variety of macromolecules from each other. The focus is to reduce aqueous volume to a more reasonable level. Higher resolution methods can come later. Generic trapping can be accomplished by tangential flow ultrafiltration (Scopes, 1994), so long as the feed stock is not so viscous as to plug the membrane pores with large particles, colloidal materials, or slimy DNA or polysaccharides. Such membrane fouling will slow down (even halt, altogether) the trans-membrane penetration of water, salts, and small molecules.

Alternative methods include ion exchange or hydrophobic interaction. If ion exchange is chosen, the adsorbent should have relatively large particle size (several hundred micrometers to 1 millimeter in diameter). Large size ion exchange beads or fibers are preferable when trapping proteins from large volumes of dirty samples. It is advisable to save, for later, the higher resolution ion exchange materials, (such as positively charged DEAE Sepharose Fast Flow or negatively charged CM Sepharose Fast Flow—GE Healthcare). It is only after viscosity and the presence of particulates have been greatly reduced that high resolution ion exchangers can be expected to deliver superior flow with relatively little fouling. Crude starting materials are best processed in batch mode rather than by axial flow chromatography. Radial flow columns offer much greater surface area, but even these columns can clog if the feedstock has high viscosity (from DNA, polysaccharides, or lipid micelles). Turbid samples containing small particles or colloidal suspensions can be as troublesome as samples with high viscosity. Frequent stirring in batch mode overcomes this problem by allowing the POI to bind to the matrix, without the problems of column fouling. However, highly acidic DNA and sulfonated or carboxylated polysaccharides will also bind to anion exchange materials, such as DEAE. While batch adsorption to DEAE can work well, the viscosity problems may return if the POI and the highly acidic biopolymers come off the anion exchanger together. But, DNA and acidic polysaccharides generally bind to DEAE, or other anion exchangers, much more tightly than the POI. When this is the case, the desired protein will elute from the anion exchanger at much lower concentrations of aqueous salt solutions than the highly acidic biopolymers. DNA and anionic polysaccharides will remain bound to the anion exchange material, while the protein-of-interest elutes with greatly reduced viscosity.

Hydrophobic protein-binding materials, like Phenyl Sepharose (GE Healthcare), are excellent trapping agents for most proteins. This method is called hydrophobic interaction

(HIC). Just a few exposed hydrophobic amino acid R-groups are needed for binding to the phenyl group. The amino acids having R-groups that are strongly attracted to an HIC matrix include: phenylalanine, tyrosine, tryptophan, methionine, leucine, isolucine, valine, proline, and lysine. It may be surprising that lysine is included as a very hydrophobic amino acid because lysine carries a positive charge at all pH values below 10. Hydrophobic interaction is not favored when charged residues are present. There is an exception when oppositely charged groups, within hydrophobic patches, are close enough to each other to bond electrostatically. Under these conditions, the electrostatic bond is exceedingly strong. Independent of electrostatic bonds, in which lysine could participate, the R-group of lysine is frequently exposed to the exterior (lysine has the greatest exposure of all amino acids, as its long string of methylene groups extends far into the aqueous medium). Hydrophobic interaction is not with the epsilon amine of lysine at the end of this string, but with the four methylene groups, themselves, to which the amine is attached. HIC and IEX media are available as very soft beads made of cross-linked dextran polymers or polyacylamide, or they come in a more rigid form that is agarose-based. An agarose-based HIC medium, such as Phenyl Sepharose, is more pressure-tolerant and more robust than the older style, softer beads. Additionally, the agarose pores are larger, allowing very large proteins to enter the internal spaces. Despite the fact that some nucleic acids and some anionic polysaccharides could enter agarose beads, this does not happen with HIC media. In the case with ion exchange trapping chemistry, DNA and other acidic biopolymers may compete with, or displace, an anionic protein-of-interest. But, highly charged nucleic acids, as well as acidic and neutral polysaccharides, are not sufficiently hydrophobic to bind tightly to Phenyl Sepharose and related HIC materials. So, they easily separate from a protein-of-interest having a few exposed, hydrophobic amino acid side chains. On the downside, HIC, as a trapping step, can become very expensive if the volume of crude extract is large. HIC gels are expensive. There is an additional economic downside to HIC when large volumes must be processed. Highly purified ammonium sulfate is fairly expensive and the cost of disposal may be even higher. Many kilograms of ammonium sulfate may be required to trap proteins by HIC, especially if the protein of interest is fairly hydrophilic (highly water soluble). Proteins that are quite hydrophilic may require a very large amount ammonium sulfate to induce binding to the HIC resins.

For a protein that is very stable at its isoelectric point (pI), isoelectric precipitation can provide an excellent, inexpensive trapping step (Scopes, 1994). Almost always this method requires a very low salt concentration, as electrostatically-driven protein-protein interaction is the mechanism that promotes precipitation. The flocculated protein may settle to the bottom of the container. If not, it may be pelleted in a centrifuge or collected by simple filtration on beds Celite. Resolubilization is accomplished by raising or lowering the pH or by adding salt. For proteins that remain soluble at their pI values, addition of a water-soluble organic solvent (generally a small aliphatic alcohol) may be used to promote isoelectric precipitation. Addition of a somewhat non-polar solvent lowers the dielectric constant of water, promoting charge-charge interactions among protein molecules. If this does not work, lowering the pH below the protein pI with simple addition of acetic acid, phosphoric acid, or HCl may cause precipitation. Occasionally, one finds that diatomaceous earth, alone, will bind certain proteins quite selectively. Because Celite is so inexpensive (available in 50 lb bags at pool supply stores), it makes sense to try Celite as a trapping agent.

With native *Aequorea* GFP, we never encountered a huge volume reduction problem because the dissection step and the trapping of whole photocytes on Celite greatly reduced the volume. But, soybean peroxidase is a different matter. See the section: "Viscosity Reduction and Particle Removal." We found that one volume of soybean hull powder requires 5 volumes of water for efficient extraction. For 2000 lbs. of hulls, the amount of water required for extraction has been determined to be 16,000 liters (G. Swiatek, personal communication). Even if scaled down to 20 lbs. of hulls per batch, 160 liters of water would be required. Volume reduction is accomplished very effectively by trapping the SBP on ResinTech anion exchange beads. When we compared binding capacity of ResinTech beads with that of DEAE Sepharose Fast Flow, both exchangers bound the same amount of pure GFP (38 mg of protein per milliliter of swollen gel). Binding capacity of ResinTech beads with larger proteins, such as rabbit IgG, is considerably lower, as the ResinTech pores are much smaller than those of DEAE Sepharose.

Chromatographic Methods

On Table 1 and Table 2 are shown the categories of basic information generally needed to facilitate early stages of protein purification. The properties of a POI that should be known are listed here in no particular order of importance. In fact, almost never is the order of information discovery the same for any two proteins. In the course of developing a start-to-finish protocol for any given protein, unexpected information is uncovered along the way. Long after developing a working protocol, one may discover, for example, that the POI is glycosylated. Following this discovery, one might want to experiment with affinity chromatography using an immobilized lectin or may wish to try a boronate column that binds vicinal hydroxyl groups on sugar residues (Scopes, 1994). The message is that no purification protocol is ever final. There are always alternate ways that could improve or streamline an earlier protocol. This is one of many places that the artistry of protein purification comes into play.

Ion Exchange Chromatography (IEX)

Once viscosity has been largely eliminated and once the crude protein sample is particle free, it may be time to use ion exchange chromatography (IEX) — the most frequently employed chromatographic method for proteins. Early, small-scale testing with a relatively salt-free sample is advised. There are simple, syringe-operated ion exchange columns available from Pall Corporation or GE Healthcare — both anion exchange columns and cation exchange columns. These columns can be used to determine (within one-half of a pH unit) the isoelectric point of the protein. This is accomplished by equilibrating the two columns with low ionic strength buffers of varying pH values. The most common cation exchange functional group is carboxymethyl, abbreviated CM. CM is essentially immobilized acetic acid and, like acetic acid, CM takes on a negative charge at pH values of 4 and above. CM is designated a weak cation exchanger as it has little binding capacity below pH 4. Sulfonated or phosphorylated exchangers are called strong cation exchangers because they can be used at pH 2. For the POI to bind to CM, the protein must be positively charged (below its isoelectric point). CM is not satisfactory for GFP purification as GFP is unstable below its pI of 5.3. When GFP takes on a positive charge (below pH 5.3) the protein slowly denatures, losing its fluorescence. So, it is not possible to use CM with GFP in any slow process like column chromatography. But, if GFP exposure time is kept at a minimum, the pI of GFP can be estimated by its binding to CM at pH's below 5.3. Diethylaminoethyl (DEAE) is the most commonly used anion exchanger. The DEAE functional group is a tertiary amine, protonated (and positively charged) at pH values below 10. DEAE is designated a weak anion exchanger as it cannot be used effectively above pH 10. But, a bead-bound quarternary amine extends the range of anion exchange to pH 12. So any medium designated Q (or QAE, for quarternary amino ethyl) is called a strong anion exchanger. All four of these types of these media (weak and strong cation exchangers and weak and strong anion exchangers) are available in small, syringe-operated columns. If one of these DEAE columns is equilibrated at a variety of pH values (10, 9, 8, 7, 6, 5, and 4), GFP will bind from pH 10 to pH 5.5, but not at pH 5, indicating that the pI of GFP is below 5.5.

Once the pI has been determined and the anion exchanger has been chosen, a preparative column can now be poured. Most ion exchangers can bind 30 to 50 mg of protein per 1 ml of swollen gel. One can estimate the total amount of protein in the sample by absorbance at 280 nm, ascribing one absorbance unit to one mg/ml of protein. But, high levels of DNA and moderate turbidity (Fig. 2) will artificially elevate this absorbance number (sometimes greatly). It is good practice to test, experimentally, the capacity of an ion exchange material in a small trial. Using 1 ml of swollen gel, add crude extract in successive 100 microliter volumes until the gel becomes saturated with protein. The saturation limit can be determined by taking POI activity measurements after each incremental addition of extract. For enzymatic measurement, remove just a few microliters of the supernatant after the gel settles (so the aqueous volume remains about the same). When the activity appears in the supernatant, you will have determined the saturation point in terms of mg of extract per ml of gel. Now fill a chromatography column with at least 5-times as much gel as your preliminary testing indicates you will need for total binding. Short, stout columns are usually better than long thin ones. Resolution comes not from column dimensions, but from the rate at which the eluting strength of the salt (usually sodium chloride) is raised in the elution phase. Take note of the fact that an ion exchanger is an excellent buffer, so pH equilibration of the gel requires many column volumes of dilute buffer solution. Alternatively, a very high concentration of buffer may be used to titrate the column, first. But, after titration, at least one column volume of the dilute (low ionic strength) buffer must be passed through the column. It is also necessary to use a buffering salt that has the same charge as the ion exchange gel. When using positively charged DEAE columns, positively charged Tris(hydroxymethyl aminomethane) buffer in the chloride form (generally abbreviated as Tris) is commonly used. For negatively charged CM, negatively charged sodium phosphate buffers are recommended. The protein of interest should be equilibrated in the same dilute buffer. For best resolution, a shallow, continuous gradient (50 column volumes or greater) from 0.0 M NaCl to 0.5 M NaCl is recommended. To achieve near base line resolution of 5 GFP isoforms (differing from each other by one or two amino acids), I have eluted a 100 ml DEAE column with 80 column volumes (8 liters) of sodium chloride solution from 0.05 to 0.25 M (Ward, 2009). In this case (and in all other cases) the salt solutions need to be prepared in the same buffer used to equilibrate the column.

Hydrophobic Interaction Chromatography (HIC)

HIC media are available in several strengths. The hydrophobic ligands are usually attached to the porous hydrophilic gels via a 3-carbon spacer based on epichlorohydrin chemistry (Scopes, 1994). From strongest binding to weakest binding ligands, the order is Phenyl > Octyl > Butyl > Methyl. Strongly hydrophobic ligands are appropriate for weakly hydrophobic proteins and weakly hydrophobic ligands for strongly hydrophobic proteins. Early testing, calculation of gel volume, and choice of column dimensions are carried out in a similar fashion as the protocols used for ion exchange. Hydrophobic binding is favored by very high salt concentration (up to 3 molar ammonium sulfate, in some cases). Elution is accomplished by lowering the salt concentration in increments (step gradient) or by applying a continuous linear gradient of decreasing salt concentration. Be aware that gradients of ammonium sulfate produce gradients of refractive index, easily confused by a spectrophotometer as a higher UV absorbance value or a lower UV absorbance value. If precise 280 nm absorbance measurements are desired following gradient elution of proteins from an HIC column, it is necessary to have a continuously changing blank that closely matches the salt concentrations of the samples. An advantage to having HIC follow IEX is that one need not remove the NaCl in the fractions eluted from the IEX column. NaCl neither favors nor inhibits hydrophobic interaction nor does it interfere with spectroscopic measurements as much as ammonium sulfate. If the two steps are reversed, ammonium sulfate must be removed entirely before going on to IEX.

Affinity Chromatography

Some prefer to use affinity chromatography very early in a protein purification process — as a "one-step purification method" (Scopes, 1994). I use quotation marks because, despite frequent claims, affinity chromatography is seldom a one-step method. Often contaminants remain in affinity-purified proteins. Commonly, those contaminants are large protein aggregates that result from the almost inevitable leaching of " bound" ligand. That released ligand then forms a high molecular weight complex with the protein-of–interest. When we purify 'anti-GFP' antibodies on an immobilized GFP affinity column we almost always detect , by SEC-HPLC, a high molecular weight aggregate that is distinctly fluorescent, suggesting that an antigen(GFP)-antibody complex has formed. Because most affinity columns are quite expensive and could be plugged by crude starting samples, I prefer to use affinity chromatography late in a protocol. The principle is easy. Take for example, that a ligand, recognized by an enzyme, is covalently bound to the matrix (usually agarose). That ligand may be a pseudo-substrate, a cofactor, an inhibitor, or an antibody. Binding is easy, but elution may be difficult. It is preferable to use, as the eluting solvent, a solution containing a competing ligand (the pseudo-substrate, cofactor, inhibitor, or antibody). But, sometimes the competing ligand is very expensive, unavailable, or irreversibly bound to the enzyme. In such cases, other eluting solvents must be used. Dilute solutions of ethylene glycol in buffer are sometimes used. So are buffers of low pH, a variety of salts, metal chelators, etc. Many other forms of affinity chromatography exist. We purify anti-GFP antibodies on a column to which GFP is covalently immobilized. We normally elute with a concentrated pH 3.0 solution of sodium citrate. The pH 3 buffer temporarily denatures both the antibody and the GFP. Both column-bound GFP and the eluted antibody are rapidly renatured with a strong pH 8 buffer. Based upon analytical techniques (including size exclusion (SEC), HPLC, SDS gel electrophoresis, UV absorption spectroscopy and western blotting) purity of GFP-specific antibody can approach 99% (see Fig 5. a, b, c, d, and e).

However, if purity greater than 99% is desired, affinity chromatography requires a follow-up step. Most commonly we use preparative gel filtration to remove protein aggregates that may form when a small quantity of bound ligand leaches from the column.

For recombinant proteins, the favorite affinity column is an immobilized (chelated) metal ion column (abbreviated IMAC for immobilized metal ion affinity chromatography) (Scopes, 1994). In IMAC columns, nickel ions or cobalt ions are bound to the column in a chelation complex. The column-bound chelator is usually nitrilotriacetic acid. The metal ion, chelated to the IMAC column, can be co-chelated, non-specifically, by the R-groups of histidine, cysteine, and tryptophan. Binding may occur if one or more of these amino acids are exposed on the surface of the protein-of-interest (or any protein contaminant in the mixture). Almost universally, recombinant proteins that are subjected to generic affinity chromatography are processed by IMAC. But to achieve specificity (and tight binding), the recombinant proteins are genetically modified by the addition of a string of 6 histidine residues, sometimes on the C-terminus, sometimes on the N-terminus, and sometimes within exposed loop regions. The string of 6 histidines (the HIS-tag) is a strong co-chelator and the tag is sufficiently exposed that the His-tag almost always out-competes any naturally occurring co-chelators found in high abundance on the surface of a protein contaminant. The method is carried out at pH of 8 or higher and it must be performed in the absence of other metal chelators such as EDTA, citrate, oxalate, ammonium ion, etc. Concentrated solutions of imidazole are usually used for elution.

In my experience, all affinity chromatography columns, each time they are used, leach a bit of their covalently bound ligand, often as high molecular weight complexes with the POI. That ligand winds up in the fractions that have eluted from the column. So, in every case in which IMAC is used, it is wise to follow this step with a gel filtration run.

Gel Filtration Chromatography

Low pressure gel filtration is the easiest chromatographic method in principle, but it is the hardest method to administer properly. Because gel filtration seems so straight forward, liberties are sometimes taken in utilizing the method. For best results, attention to detail is essential. Gel filtration (or size exclusion as the method is called in HPLC) separates macromolecules by size. Size exclusion chromatography (SEC) is generally used as an analytical HPLC method while gel filtration is used primarily in preparative protein separations. Size exclusion HPLC utilizes small, rigid, uniform, spherical beads of 5 micrometer or 10 micrometer diameter. Small, porous, silica beads used in SEC provide higher resolution than low pressure gel filtration. But, the price per ml of HPLC column packing material is much higher than that of any soft gel used in low pressure applications. For further discussion of HPLC, refer to the HPLC section later in this chapter.

Low pressure gels are comprised of small (20 to 300 micrometers) porous beads which, unlike Fast Flow adsorption beads, have blind cul-de-sacs that provide differential flow paths through the column. The largest molecules are unable to enter any pores, so they must travel around the beads. This means that large molecules exit first while smaller molecules spend some time inside the beads, so they exit later. The volume in which the very large molecules exit (DNA, proteoglycans, ribosomes, lipid micelles, and protein complexes) is called the void volume. The void volume, usually 25% of the total column volume, is often measured by the elution position of Blue Dextran (GE Healthcare), a covalently-dyed sugar polymer having a molecular weight of 2 million Daltons. So, if the column volume ($\pi\, r^2\, h$) is 200 cubic centimeters (200 ml), the center of the Blue Dextran-calibrated void volume peak will appear close to the 50 ml mark. The next 25% of the column volume (the second 50 ml in this example) is the resolving zone, accessible to moderate size proteins. The final 50% of the column volume (100 ml) is the zone in which peptides, very small proteins, oligonucleotides, other small molecules and salt ions will elute. The total liquid volume in the column (salt volume) is accepted as being either the total volume of the column, as calculated from $\pi\, r^2\, h$, or it is the volume measured by adding a measurable salt to the applied sample. The salt can be sodium chloride, detectable by conductivity, or sodium nitrite, detected by its fairly strong absorbance at 280 nm.

Gel filtration is, intrinsically, a low resolution separation method for proteins, yet it is frequently used in protein purification. Gel filtration is gentle to the sample and it is the best preparative method for fractionating native proteins by size and shape. Passage though the partially accessible pores in the beads will generate broad elution bands, each band lying within just 25 percent of the total column volume, thus the intrinsically low resolution of the method. Generally, the highest resolving columns, containing very small beads of soft gel materials, like Sephadex G-100 Superfine (GE Healthcare) or BioGel P-100 minus 400 mesh (BioRad Laboratories), operate under low gravitational force fields (50 cm pressure head, or smaller). Beads used for relatively large proteins must have low degrees of cross-linking, making the gels soft and highly compressible. For the most compressible beads (G-200 Sephadex, for example), pressure heads may need to be as small as 15 cm. In general, gel filtration columns are able to give baseline resolution for no more than 4 proteins, each differing in molecular weight by a factor of 2. So, under the best of conditions, a mixture of globular proteins of MW 200,000, 100,000, 50,000, and 25,000 Daltons can be baseline resolved.

Listed in Table 3 is a set of "best conditions," — those that give maximum resolution by gel filtration.

1	Sample volume divided by column volume must be in the range of 1-2 %.
2	Sample must be applied very carefully to avoid channeling.
3	Beads must be very small (20-50 micron size range).
4	Flow rate must be very low (<2 ml/cm2 per hour), a rate which requires >47 hours for 121 cm X 1 cm columns.
5	The biological sample applied to the column must be low in viscosity.
6	At least 100 fractions should be collected, preferably in the protein resolving zone.
7	The pressure must be low, so as not to collapse the beads.

Table 3. Best conditions for maximum resolution in gel filtration

Physical Set-ups in Column Chromatography

Those new to column chromatography often ask, "What size column should I use and what are the most appropriate dimensions of length and width"? Clearly there is no one correct answer. But there are some appropriate generalities that can help with column selection in adsorption chromatography. Adsorption chromatography includes ion exchange, hydrophobic interaction, affinity chromatography, and all other forms of chromatography in which the analyte binds to the stationary phase (all methods other than gel filtration and SEC). Protein resolution in adsorption chromatography depends upon the rate of change of the eluting solvent, not upon the length or width of the column. Better resolution results from gradual change in the strength of eluting solvent. The limit of "slow rate of change" is no change at all. In chromatography, we call "no change at all" isocratic elution. Isocratic elution, at the right solvent concentration, generates, the highest possible resolution, but peak spreading will be greater in isocratic elution than in gradient elution. In general, the amount of adsorbent in a column should have the capacity to bind three-to-five-times the amount of protein being loaded. The length and width of the column are not critical. It is not unreasonable to use a short, stout column for adsorption chromatography—a column with length 2- to 3-times the column diameter, for example. Such columns allow very high flow rates, so a large volume of eluent can be used in a fairly short period of time. But, one should not greatly extend column width at the expense of length (eg. dimensions of a cake pan are problematic). A wide diameter column, where the eluent exits through a port at the center of the cylinder, provides early elution of protein that happen to migrate down the center of the column. Protein (of the same type) that migrates near the circumference of the column will exit significantly later. This differential elution (side vs center of the cylinder) produces smearing of a band that might otherwise be sharper (if the column had more "normal" dimensions). An exceedingly long and thin column is not desirable either. Flow rates will be very slow, especially if the gel is soft. If, in an attempt to speed up flow, pressure is increased, the gel may compress and flow will slow down. In extreme cases, flow may stop altogether. Even if the adsorbent is rigid and non-compressible (as in size exclusion HPLC) a column with a small cross-sectional area may over-pressure if particles collect on the surface. It is common to use long, narrow columns for gel filtration, but here as well, columns need not have such extreme dimensions. The problem of particles collecting on the surface of the column may still occur. But even if the sample is particle free, flow rate may be much slower than desired. A 50 cm column, with a diameter of 2.5 cm, can give excellent resolution in gel filtration as well as in adsorption chromatography. But, to achieve maximum resolution in gel filtration (in columns of such dimensions), the conditions listed in Table 3 must still be observed.

HPLC

The term HPLC stands for high performance liquid chromatography. Those with limited budgets prefer to substitute the word, "price," for "performance." Some use the word, "pressure." But, pressure is not what distinguishes HPLC from other forms of column chromatography. The fundamental difference between HPLC and columns containing relatively soft gels (Sephadex, BioGel, agarose, cellulose, etc.), is that the beads in HPLC columns are considerably smaller. HPLC beads are usually 5 micrometers in diameter. Columns with such small beads will not flow by gravitational pressure, nor will they flow with the pressure generated by a peristaltic pump. So, as a consequence of small beads, mechanical pumps capable of pressures as high as 7000 psi are needed. But, in practice, pressures greater than 2000 psi are seldom used. Even pressures of 2000 psi require very strong columns, usually of stainless steel. Tubing down-stream from the pump must also tolerate very high pressures. Very rigid gels are required or the beads will collapse under the high pressures generated in HPLC. The most common of the rigid HPLC beads are made of porous silica.

More than 90% of HPLC columns in use are reverse phase columns (RPC). Reverse phase media is made of porous silica, but is functionally similar to low pressure hydrophobic interaction media made from soft gels. The greatest difference between RPC and HIC (other than tolerance for high pressure) is that reverse phase beads are much more hydrophobic than HIC beads. RPC beads have long aliphatic chains or aromatic groups bonded to the silicon dioxide media. The relative hydrophobicity of RPC columns is related to both carbon chain length and the carbon load. Carbon load (that can reach 20%) reflects the density of hydrocarbon substitution on the silica beads. The reported percentage is the ratio of the weight of bound hydrocarbon to the weight of silica. So, not only are the bonded phase hydrocarbons more hydrophobic than the ligands in HIC, but the density of hydrophobic ligands is also greater in RPC. The name "reverse phase" comes from the fact that the polarity of the mobile phase (the solvent) and that of the stationary phase (the beads) have been reversed. The original silica-based columns used unmodified silica that is polar and highly charged. So, this "normal phase" chromatographic method used a polar stationary phase and a non-polar mobile phase. RPC reverses the phases.

Typically, samples are introduced into an HPLC column through an injection valve that maintains atmospheric pressure on the outside and high pressure down-stream. So, a standard syringe can be used to load a sample while solvent continues flowing at very high pressure. RPC is more appropriate for small polar molecules (amino acids, peptides, oligonucleodes and polar lipids) than for native proteins. Most proteins bind too strongly and may bind irreversibly or become denatured.

Batch Methods for Protein Purification

Occasionally one finds a batch method that works as well in purifying a particular protein as a variety of chromatographic methods. Batch methods are particularly useful in early stages when a sample is highly viscous or full of fine colloidal material. Such batch methods include ammonium sulfate precipitation, precipitation from other salt solutions, from aqueous solutions at low pH, or from organic solvents (usually acetone, ethanol, ethylene glycol, or polyethylene glycol). In some cases, recrystallization from salt solutions is possible. Even if crystals do not form, differential precipitation can be an effective purification method. A particularly effective batch method is isoelectric precipitation in which the pH of a dilute aqueous buffer is adjusted to the isoelectric point (pI) of the POI. The protein-of-interest, or contaminants in the POI mixture, can be adsorbed to Celite, alumina gels, calcium phosphate gels (hydroxyapatite), and other media. If antibodies are available, the protein of interest can be selectively bound to those antibodies. If aggregates form upon such treatments, the aggregate can be collected by centrifugation and then dissociated into free antibody and free POI by a variety of methods including application of low pH buffers.

A-Free IgG

Recently, I have been exploring, with repeated rounds of ammonium sulfate precipitation, the purification of rabbit-derived antibodies, goat anti-rabbit IgG, and chicken IgY. Because this process does not utilize Protein-A, I call the method "A-Free." For rabbit-derived antibodies, the "A-free IgG" procedure works at least as well as chromatography on columns of Protein-A. Goat-derived antibodies, that are not as amenable to purification on Protein-A columns, and chicken-derived IgY, that cannot be purified on Protein-A at all, respond equally well to the "A-Free" method. Although very commonly used in purifying therapeutic monoclonal antibodies, Protein-A is quite expensive. Despite its being covalently bound to the affinity column matrix, Protein-A is able to leach from the column matrix during the elution phase. Traces of Protein-A in therapeutic monoclonals could present a health hazard, as Protein-A may bind to other essential antibodies in the patient. We have not found formal regulations limiting the use of Protein-A in purifying therapeutic monoclonals, but manufacturers might prefer a safer, more cost-effective method. We have a satisfactory replacement for Protein-A in the method we call "A-Free IgG." This method has been submitted, through Rutgers University, as a provisional patent application.

As often occurs in experimental science, the "A-Free IgG" method of antibody purification arose from an accident. I am primarily a bench scientist. But, with all the other things I must do, I get too little bench time to satisfy my urges to discover and create. As a consequence, when I have a bit of research time, I tend to rush through projects, sometime binging well into the night. Often this means cutting corners to save time. Such was the case with developing the "A-Free IgG" method—an accident created by my hasty experimentation.

In the course of purifying IgG from rabbit serum by a traditional single round of ammonium sulfate fractionation, I made a mistake that was picked up by size exclusion HPLC. The SEC profile showed more contaminants than I had seen previously. So, to remove those additional contaminants, I repeated the entire process. To my surprise, the second round of ammonium sulfate precipitation produced a cleaner IgG sample than I had previously seen with just one round of precipitation. But the redissolved pellet was still slightly pink (not all the transferrin was removed), and the HPLC profile still showed a tiny shoulder of albumin. So, I did the same precipitation process a third time. This time, the HPLC profile showed 99% pure IgG—virtually no high molecular weight contaminant and no indication of any albumin (Fig. 5 a, b, c, and d). The SDS gel profile showed strongly stained heavy chain and more weakly stained light chain (normal for IgG) and a very weakly staining contaminant or two (Fig. 5 e). These side-by-side experiments show that the "A-Free IgG" method actually out-performs Protein-A affinity chromatography. The time involvement is similar and the price is much lower. On occasion we perform a 4th ammonium sulfate precipitation, obtaining a sample marginally cleaner than that resulting from three rounds of precipitation.

The method works equally well with goat anti-rabbit IgG, an antibody less amenable to Protein-A purification. We have a large supply of chicken egg yolk containing anti-GFP antibodies (IgY) for which Protein-A is totally ineffective. The A-Free method is suitable with IgY so long as the large amount of lipid has been removed by a freeze-thaw method.

Three-Phase Partitioning (TPP)

The most exciting method we have used for protein purification is three-phase partitioning (TPP). TPP was developed in the 1990's (Dennison and Louvrien, 1997) and rediscovered in

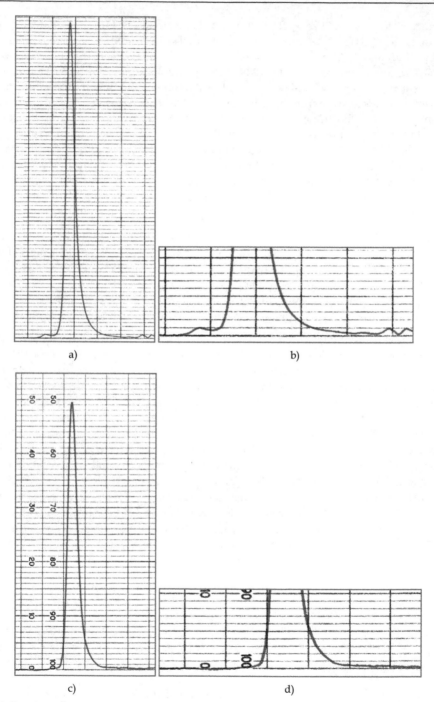

Fig. 5. Continued

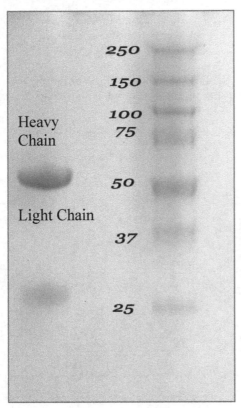

e)

Fig. 5. (a) HPLC profile of rabbit IgG purified by standard Protein-A chromatography. (b) Expanded view of **Fig. 5 (a)** showing small contaminants. (c) HPLC profile of rabbit IgG purified by Brighter Ideas Inc's new proprietary method, "A-Free IgG". (d) Expanded view of **Fig. 5 (c)** showing no visible contaminants. (e) SDS- PAGE gel showing heavy chain and light chain bands of rabbit IgG prepared by "A-Free IgG". Right hand lane shows molecular weight markers in kDa.

our Rutgers University lab in 1998. We happened upon this method by accident in 1998— not by reading the paper, but by experiencing the method ourselves. The process is so elegant that we can purify recombinant or native jellyfish GFP to 80% purity in less than half a day. In the early years of our research, we purified GFP from jellyfish extracts by traditional methods, spending 6 months to reach 80% purity. TPP provides about a 3000-fold savings in time and significant savings in equipment use and materials expenses. What is the magic?

Our adaptation of the TPP method for purifying recombinant GFP begins with whole, unlysed *E. coli* cells transformed with the gene for GFP. Three-phase partitioning works very well with GFP-containing cell extracts, but it works even better if the process begins with unlysed cells. Entire companies are built around releasing recombinant proteins from whole

E. coli cells (Glens Mills, for example). Huge French presses, sonication baths, day-long, repeated freeze-thaw cycles, treatment with lysozyme, or use of a bead mill are some of the standard methods for rupturing *E. coli* cells (Scopes, 1994). Fig. 6 shows an SDS gel electrophoretic profile for a sample prepared by TPP as compared to identical samples extracted by three other standard methods. TPP accomplishes the release of recombinant proteins in seconds, using the simplest of standard equipment. Described below are three stages in the process:

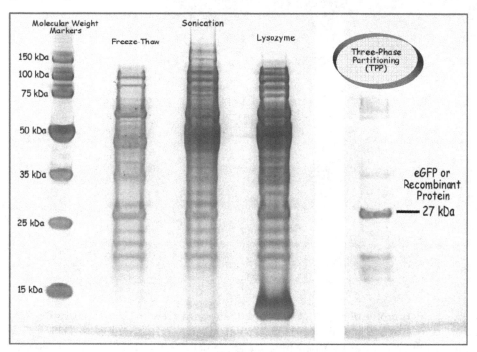

Fig. 6. SDS- PAGE gel showing the released *E.coli* proteins resulting from routine lysis methods compared with the non-lysis Three-Phase Partitioning method as applied to the same amount of starting material.

Stage I. To release GFP and to perform the first stage of TPP purification, we treat whole *E. coli* cells with 1.6 M ammonium sulfate with shaking. Then we add one volume of tertiary butanol. If we do this in a 50 ml Falcon tube, we pour a suspension of the cells (in 25 ml of 1.6 M ammonium sulfate, pH 8.0) into the tube and then we add 25 ml of t-butanol. After about 1 minute of vigorous shaking, the Falcon tube is centrifuged in a moderate speed (3000 rpm) table-top centrifuge for 15 minutes at room temperature. Although t-butanol is completely miscible with water, it is quite insoluble in aqueous solutions having high concentrations of salt, especially ammonium sulfate solutions. Three phases separate during centrifugation (Fig. 7) (or, for large scale operations, by settling in a tank by gravity alone). The upper phase contains t-butanol which expands to 30 ml, having taken up 5 ml of water. Release of 5 ml of water from the lower aqueous phase raises the ammonium sulfate concentration from 1.6 M to 2.0 M.

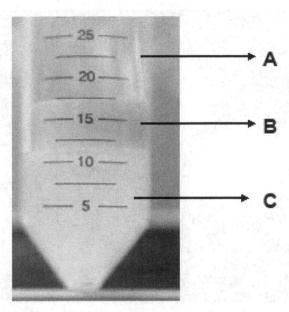

Fig. 7. Tube showing Stage- I of Three-Phase Partitioning. Three phases are formed after centrifugation. Layer A contains t- butanol, B is a thick "pancake" layer of precipitated material and C is aqueous ammonium sulfate solution containing GFP.

Meanwhile, membrane phospholipids, triacylglycerols, pigments, dyes, cholesterol and other steroids, fats, oils, and miscellaneous lipids become dissolved in the upper layer. Exposure of a complex macromolecular mixture to both t-butanol and the now higher salt concentration causes massive precipitation. The precipitate settles below the organic layer as a thick "pancake" of congealed protein, nucleic acids, cell walls, and other unwanted materials (Fig. 8). While the t-butanol, under the influence of high salt, has dissolved the cell membrane, it has not affected the cell wall. Normally, with membrane dissolved, nearly every macromolecule in the cell can escape to the outside through the cell wall. But, "stressed" by the high concentration of ammonium sulfate, t-butanol binds to anything that is even slightly hydrophobic, causing massive precipitation of most of the proteins, and virtually all chromosomal DNA. These aggregates are too large to exit through the cell wall pores, so they remain entombed inside the cell, behind the cell wall barrier. The binding of t-butanol to these macromolecules (whether they have remained within the cell or escaped to the outside) lowers the density of the precipitated macromolecules to such a point that the still intact cells, with their entombed macromolecules, easily float above the ammonium sulfate solution below. The whole cell mass forms a thick rubbery mat that floats above the salt solution (Fig. 8). As centrifugation simply speeds up the formation of three layers, the process can be scaled up to almost any volume by simple gravitational settling in a large tank. There is no limit to the scale-up potential. After loss of 20 % of the water to the overlying alcohol layer, GFP is still soluble in the aqueous ammonium sulfate (now at a concentration of 2.0 M. Because GFP remains soluble, it escapes easily through the pores in the cell wall and enters the aqueous layer.

Fig. 8. Recovered "plugs" of precipitated material from Stage I of Three-Phase Partitioning.

Stage II. The alcohol layer is removed by aspiration and the floating disk of precipitated cells and macromolecules is also removed (almost as easily as flipping a pancake with a spatula). To the 20 ml of aqueous solution remaining in the tube, we add 30 ml of fresh t-butanol, (with vigorous shaking once again). Fresh t-butanol causes further dehydration of the lower liquid phase, creating a saltier aqueous solution. The saltier solution now favors precipitation of the GFP, already coated with many molecules of t-butanol. So, like the aggregated molecules in stage one, the now precipitated GFP (with its bound cage of t-butanol) moves into the organic-aqueous interface as a fine disk, compressed by centrifugal force between the alcohol layer and the aqueous ammonium sulfate layer. Both liquid layers are then carefully removed.

Stage III. The GFP disk, remaining after the liquid phases have been removed, is taken up in a series of very small volumes of 25% saturated ammonium sulfate which, when added to the very salty GFP disk, raises the ammonium sulfate concentration to 1.6 M.. One at a time, these suspensions of GFP are serially transferred into one or more microfuge tubes. Serial transfer allows virtually 100% of the GFP to be transferred in a minimum volume. Volume is kept at a minimum because GFP is incredibly soluble, even in 1.6 M ammonium sulfate. When GFP has just barely gone into solution, the tube(s) is spun in a microcentrifuge. Those remaining contaminants, having lower solubility than GFP, now collect as a pellet at the bottom of the tube(s). There is usually a tiny floating disk of contaminant and a small volume of overlying alcohol. In a sense, Stage I of TPP has been repeated in Stage III. The final GFP product of TPP is pipetted from the microfuge(s) as a bright, crystal-clear green liquid. On average, the GFP has been purified from its original milieu by a factor of 100-fold and concentrated by a factor of 50. The amazing effectiveness of TPP is also shown in the before and after absorption spectra seen in Fig. 9.

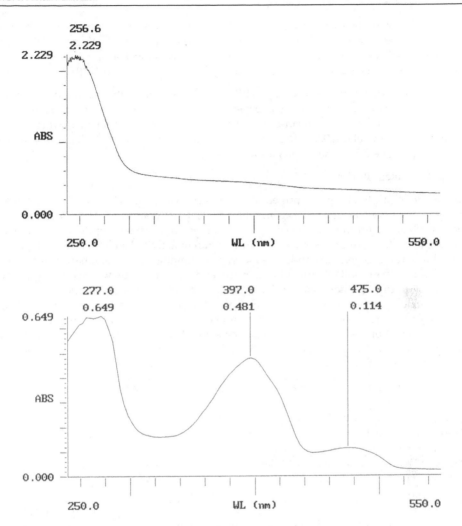

Fig. 9. Spectrum of GFP-containing crude cell extract **(a)** before and **(b)** after Three-Phase Partitioning, now 60% pure. Final purity has reached 60%.

A Rutgers University patent was issued in 2008 for a protein mini-prep kit (based upon our work with TPP). The patent calls for a mixture of two organic solvents (a mixture of t-butanol and isopropanol). Included in the patent description is the use of a microbiological dye, previously added to a very concentrated ammonium sulfate stock solution. The purpose of the dye is to facilitate detection of the solvent interface, as all of the dye leaves the aqueous layer and travels into the alcohol layer. We have explored sixteen water-soluble microbiological dyes, each of which partitions effectively into the organic phase. The boundary between the colored organic layer and the colorless aqueous layer provides a visible means for separation of the two layers. Visualization of this boundary is especially useful when tiny quantities of protein are being prepared. Surprisingly, TPP works almost equally well when small concentrations of protein are processed. Such very dilute protein solutions of protein are almost never amenable to ordinary ammonium sulfate precipitation. What's more, TPP works, not only on crude extracts and whole *E. coli* cells, but it works well as a polishing step. Even 90% pure protein can be taken to near homogeneity by a second round of TPP.

Other proteins may be purified by TPP, but often the initial ammonium sulfate concentration must be adjusted on a protein-by-protein basis so as to maximize both recovery and purity of the protein-of-interest. The salt concentration over which TPP is effective ranges from about 0.6 M to 2.6 M ammonium sulfate. Below 0.6 M, the two solvents are miscible. Above 2.6 M, the salt begins to precipitate.

Criteria for Protein Purity

Demonstrating purity of a given protein is not an easy task. But, without achieving protein homogeneity, serious errors and experimental artifacts may arise. Even a 1% contaminant may contribute to erroneous observations. A minor contaminant (protein or otherwise) could significantly raise or lower an enzyme's apparent activity level. If an impure protein-of-interest is used to generate antibodies, a very immunogenic contaminant could induce more antibody than the POI. Some biochemists and some journals will accept, as the sole criterion of purity, a photograph or a densitometry trace of a Coomassie-stained SDS polyacrylamide gel that shows one stained band. But, I know of a case in which a "single band" on an SDS gel, accepted by a prominent journal as proof of purity, turned out to be a 97% contaminant of the protein-of-interest. The actual POI represented only 1% of the total "pure protein" (Karkanis and Cormier, 1971). Errors of this magnitude can be avoided by using a variety of different criteria for evaluating protein purity.

1	Constant specific activity across a broad portion of the peak in the final preparative chromatography column.
2	Single, symmetric band by size exclusion HPLC.
3	Single band, in the correct MW region, on an SDS gel (or, for hetero-oligomers, the appropriate number of bands in the correct positions.
4	Unambiguous, single amino acid detected in N-terminal amino acid analysis. Inability to detect an N-terminal amino acid may also be taken as evidence of purity – not a very strong criterion as many other proteins have blocked N-terminal amino acids.
5	Single band on a native polyacrylamide gradient gel (or appropriate number of bands of correct MW for heterodimers, heterotetramers, etc).
6	Single, sharp band by isoelectric focusing in an acrylamide gel or in a capillary isoelectric focusing system (or the appropriate number of bands for hetero-oligomeric proteins).
7	Unambiguous N-terminal peptide sequence by Edman degradation.
8	Single band by Western blot, if antibodies are available.
9	Single MW form by Maldi TOF (matrix assisted laser desorption time-of-flight mass spectrometry).

Table 4. Criteria of purity

4. Acknowledgment

The author would like to acknowledge Ms. Sujata Charuvu for her technical assistance.

5. References

Blinks, J. R., Prendergast, F. G., and Allen, D. G. (1976) Photoproteins as Biological Calcium Indicators. *Pharmacol. Rev. 28*: 1-93.

Bokman, S. H. and Ward, W. W. (1981) Renaturation of *Aequorea* Green-Fluorescent Protein. *Biochem. Biophys. Res. Commun. 101*: 1372-1380.

Cheung, W. Y. (1971) Cyclic 3', 5'-Nucleotide Phosophdiesterase-Evidence for and Properties of a Protein Activator. *J. Biol. Chem. 246*: 2859-2869.

Dennison, C. and Lovrien, R. (1997) Three Phase Partitioning: Concentration and Purification of Proteins. *Protein Expr. Purif. 11*:149-161.

Johnson, F. H. and Shimomura, O. (1972) Preparation and Use of Aequorin for Rapid Microdetermination of Ca^{+2} in Biological Systems, *Nature New Biology. 237*:287-288.

Karkhanis, Y. D. and Cormier, M. J. (1971) Isolation and Properties of *Renilla reniformis* Luciferase, a Low Molecular Weight Energy Conversion Enzyme. *Biochemistry 10*: 317-326.

Matthews, J. C., Hori, K. and Cormier, M. J. (1977) Purification and Properties of *Renilla reniformis* Luciferase. *Biochemistry 16*: 85-91.

Morise, H., Shimomura, O., Johnson, F. H. and Winant, J. (1974) Intermolecular Energy Transfer in the Bioluminescence System of *Aequorea. Biochemistry 13*: 2656-2662.

Prendergast, F. G. and Mann, K. G. (1978) Chemical and Physical Properties of Aequorin and the Green-Fluorescent Protein Isolated from *Aequorea forskalea. Biochemistry 17*: 3448-3453.

Roth, A. F. (1985) Purification and Protease Susceptibility of the Green-Fluorescent Protein of *Aequorea aequorea* with a Note on Halistaura. MS Thesis Rutgers University, New Brunswick, NJ.

Scopes, R. K. (1994) Protein Purification, Principles and Practice, 3rd ed. Springer Advanced Texts in Chemistry, Springer-Verlag, N.Y.

Wampler, J. E., Hori, K., Lee, J., and Cormier, M. J. (1971) Structured Bioluminescence. Two Emitters During Both the *In Vitro* and the *In Vivo* Bioluminescence of *Renilla. Biochemistry 10*: 2903-2910.

Ward, W. W. (2005) Biochemical and Physical Properties of Green-Fluorescent Proteins, *In*: Green Fluorescent Proteins: Properties, Applications, and Protocols, 2nd Ed. M. Chalfie and S. Kain (eds) Wiley-Interscience, Inc. pp 45-75.

Ward, W. W. (2012) Novel Purification of Soybean Peroxidase (Glycine max) and SBP Applications in Detecting Bisphenol A. *Cambridge Healthtech Institutes's Pep Talk.* Jan 10, San Diego, CA.

Ward, W. W. and Bokman, S. H. (1982) Reversible Denaturation of *Aequorea* Green-Fluorescent Protein: Physical Separation and Characterization of the Renatured Protein. *Biochemistry 21*: 4535-4540.

Ward, W. W. and Cormier, M. J. (1978) Energy Transfer Via Protein-Protein Interaction in *Renilla* Bioluminescence. *Photochem. Photobiol.* 27:389-396.

Ward, W. W. and Cormier, M. J. (1979) An Energy Transfer Protein in Coelenterate Bioluminescence. Characterization of the *Renilla* Green-Fluorescent Protein (GFP). *J. Biol. Chem. 254*: 781-788.

Ward, W. W., Prentice, H. J., Roth, A. F., Cody, C. W. and Reeves, S. C. (1982) Spectral Perturbations of the *Aequorea* Green-Fluorescent Protein. *Photochem. Photobiol. 35*: 803-808.

Ward, W. W. and Swiatek, G. (2009) Protein Purification. *Current Analytical Chemistry. 5*: 85-105

3

Protein Purification by Affinity Chromatography

Luana C. B. B. Coelho[1], Andréa F. S. Santos[2], Thiago H. Napoleão[1],
Maria T. S. Correia[1] and Patrícia M. G. Paiva[1]

[1]*Universidade Federal de Pernambuco, Centro de Ciências Biológicas,
Departamento de Bioquímica, Av, Prof. Moraes Rego, Recife-PE,*
[2]*University of Minho, IBB-Institute for Biotechnology and Bioengineering,
Centre of Biological Engineering, Campus de Gualtar, Braga,*
[1]*Brazil*
[2]*Portugal*

1. Introduction

Affinity chromatography is a method which depends essentially on the interaction between the molecule to be purified and a solid phase that will allow the separation of contaminants. Lectins are carbohydrate-binding proteins which can be purified by affinity chromatography; also, the presence of multiple molecular forms of lectins in a preparation can be separated. Immobilized lectins have been useful to affinity protein purification. In immunoaffinity chromatography an antibody or an antigen is immobilized on a support so as to purify the protein against which the antibody was developed. Monoclonal antibodies are extremely useful as immunosorbents for purification of antigen. Immobilization of monoclonal antibody on a suitable material to the column produces a support that will bind with high selectivity to protein against which the antibody was developed. Affinity chromatography containing DNA is a highly specific and important technique for the purification of DNA-binding proteins involved in the transcription, replication and recombination. The success of affinity chromatography depends on the conditions used in each chromatographic step. So, the optimization of protocol is essential to achieve optimal protein purification with maximum recovery.

2. Nomenclature and basic concepts

The term "affinity chromatography", first used by Cuatrecasas et al. (1968), refers to a purification technique which depends essentially on the highly specific interaction between the molecule to be purified and the solid phase that will allow the separation of contaminants. This method has several other terms such as "bioselective adsorption", which was appropriately used to denominate an adsorption chromatography that uses a very special kind of affinity between the desired biological product and a biomolecule (Porath, 1973). For example, the biological affinity between an enzyme (protein with catalytic activity) and its substrate and/or other small ligand – usually in the active or allosteric site of the enzyme – results from a selective interaction.

The adsorption corresponds to the fixation of the molecules of a substance (the adsorbate) on the surface of another substance (the adsorbent), which may be immobilized to an insoluble support. The adsorbate and the adsorbent can be referred as bioligands. The bioligands may be specific or may not have absolute specificity of interaction. Many bioligands (e.g. NAD^+, ATP, coenzyme A) may bind different enzymes, being then called group-specific ligands. In the same manner, chitin (a polysaccharide composed by N-acetyl-D-glucosamine units) may be an adsorbent for several different molecules if they possess a group (e.g. a binding or catalytic site) able to interact with chitin.

Affinity chromatography is a powerful tool for the purification of substances in a complex biological mixture. It can also serve to separate denatured and native forms of the same substance. Thus, biomolecules which are difficult to purify have been obtained using bioselective adsorbents, e.g. immobilized metal ions (Ni^{2+} and Zn^{2+}) used to purify proteins containing zinc finger domains with natural affinity to divalent ions (Voráčková et al., 2011). The relative specificity degree of the affinity chromatography is due to the exploitation of biochemical properties inherent in certain molecules, instead of using small differences in physicochemical properties (such as size, form and ionic charge, which are employed by other chromatographic methods).

Affinity chromatography may be used with different final objectives. If the aim is a rapid purification of a macromolecule with high yield, many controls and careful attention are necessary to establish the best conditions for a high bioselectivity of the system; the researcher must be prepared to adjust the chromatographic conditions and to circumvent possible absence of bioselectivity or low yields. If the objective is to first demonstrate a bioselectivity for further purification, the choice of the bioselective adsorbent is dependent on the physiological interaction between the bioselective component and the macromolecule to be purified. In this case, the researcher must spend a lot of time establishing the bioselectivity before starting the isolation experiments.

A good bioselectivity means that the affinity of the molecule by the ligand exceeds all factors of non-specific adsorption that are present in the system. Also, the affinity should not be so strong, since the biomolecule must be removed from the column. A well-designed affinity method should consider the selection of the ligand molecule or the insoluble support to be used; they must have specific and reversible binding affinity for the molecule being purified.

After defining the protocol, purification by affinity chromatography is a rapid method, compared with others less specific. The technique also enables the concentration of the molecule of interest resulting in a small volume of a concentrated product.

Standard procedures of protein purification result in obtainment of homogeneous protein. However, a considerable cost of supplies and hours of work is often required and a low yield is obtained after several steps. The power of affinity chromatography is often larger than other chromatographic techniques, resulting in several hundred or thousand-fold purification factors in a single step.

3. Supports for affinity chromatography

A good support for affinity chromatography should be chemically inert or have minimal interaction with other molecules, having high porosity and large number of functional groups capable of forming covalent bonds with the molecule to be immobilized. Many materials are available (Table 1). A variety of supports with immobilized ligands, or stable media for the immobilization of ligands through different functional groups are commercially available. The ligand molecule to be used should contain a group capable of being chemically modified, often an amino group, which will allow connection with the matrix without destroying its capacity to bind to the molecule of interest.

Supports	References
Affi-gel blue gel	Wong et al., 2006
α-casein-Agarose	Kocabiyik & Ozdemir, 2006
Chitin	Sá et al., 2008; Coelho et al., 2009; Santana et al., 2009; Napoleão et al., 2011a
Fetuin-fractogel	Guzmán-Partida et al., 2004
Fetuin-Sepharose CL-4B	Bhowal et al., 2005
Ferromagnetic levan composite	Angeli et al., 2009
GalNac-Sepharose CL-4B	Gade et al., 1981
Galactosyl-Sepharose	Franco- Fraguas et al., 2003
Glutathione reduced (GSH)-Sepharose	Hamed et al., 2011
Guar gel	Coelho & Silva, 2000; Santos et al., 2009; Nunes et al., 2011; Souza et al., 2011
IMAC (immobilized metal ion affinity chromatography)-Sepharose	Voráčková et al., 2011
Sephadex G25	Santana et al., 2009
Sephadex G50	Fenton-Navarro et al., 2003
Sephadex G75	Correia & Coelho, 1995
Sepharose-manose gel	Latha et al., 2006
Lectin-Sepharose CL-4B	Paiva et al., 2003; Silva et al., 2011
Trypsin-Agarose	Leite et al., 2011

Table 1. Supports for affinity chromatography.

One example is the agarose, a polysaccharide obtained from agar, which provides numerous free hydroxyl groups and is the most widely used (Chung et al., 2009). The ligand may be covalently bound to it through a two step process. In the first step, the agarose reacts with cyanogen bromide to form an "activated" intermediate which is stable and commercially available. In the second step, the molecule to be immobilized reacts with agarose to form the covalently bound product (Voet & Voet, 1995). A support containing trypsin immobilized on agarose was used to purify trypsin inhibitor from liver of *Oreochromis niloticus* (Leite et al., 2011). Chromatography on α-casein-Agarose was useful for purification of an intracellular chymotrypsin-like serine protease from *Thermoplasma volcanium* (Kocabiyik & Ozdemir, 2006).

Sepharose (a tradename of a registered product of GE Healthcare) is a beaded form of agarose cross-linked through lysine side chains. It is a common support for chromatographic separations of biomolecules and can also be activated with cyanogen bromide. For example, glutathione S-transferases from Down syndrome and normal children erythrocytes were purified by chromatography on matrix containing glutathione reduced (GSH) immobilized on Sepharose (Hamed et al., 2011).

Insoluble polysaccharide matrices – such as chitin, guar gel and Sephadex – have been used to purify lectins (carbohydrate-binding proteins) and will be discussed later.

4. Extraction and purification of proteins by affinity chromatography

To obtain a pure protein is essential for structural characterization and exploration of its function in nature. These proteins should be free of contaminants if they will be used for biotechnological purposes, such as the evaluation of their potentiality to purify and characterize other molecules, as well as for studies on the ability to recognize receptors and induce different cellular responses.

Proteins are dependent of environmental conditions to maintain their stability and for this reason some parameters are crucial in all steps of the purification protocol: pH, ionic strength, temperature and dielectric constant. The balance of these parameters, characteristic for each protein, is essential for obtainment of the pure molecule in its native form. The protein activity is due to the maintenance of protein structure that may be stabilized by strong bonds, like disulfide bridges, and weak bonds, like hydrophobic interactions and hydrogen, electrostatic or saline bonds.

In the purification processes of a protein, the following parameters should be considered: the selection of the procedure for protein extraction from the biological source, the assays for monitoring protein concentration in each step, the methods of solubilization, and the environmental conditions for stabilization. The prior separation is based on differences in solubility and usually corresponds to the preparation of a homogenate or extract. After extraction and centrifugation, the separation can be based on molecular mass, electric charge and protein affinity for other molecules.

Many proteins have the ability to bind strongly (but not covalently) to specific molecules and thus can be purified by affinity chromatography. Figure 1 shows the steps of an affinity chromatography for isolation of a protein. Initially, the affinity support must be equilibrated with a binding buffer to achieve adequate conditions for affinity interaction between the protein and the immobilized molecule (step 1). When an impure solution (crude extract or a partially purified preparation) is passed through the affinity support, the protein of interest interacts with the ligand (adsorption) and the other contaminants (other proteins or molecules) are washed from the column with the binding buffer (step 2). The desired molecule can be obtained highly purified by changing the elution conditions to release the protein from the support (step 3). For example, the elution may be performed changing the conditions of pH, ionic strength or temperature (non-bioselective desorption), or with a solution containing a high concentration of free ligand that will compete for the binding-sites of the protein (a bioselective desorption).

A crude extract can be directly applied in an affinity chromatography column. The application of crude extract has the advantage of avoiding other steps that lengthen the process. However, substances that may interfere in this process, like other proteins, nucleic acids and lipids are present in higher concentrations in crude extracts. In general, before the chromatography, one or more steps for partial separation of undesirable constituents are incorporated into the purification protocol.

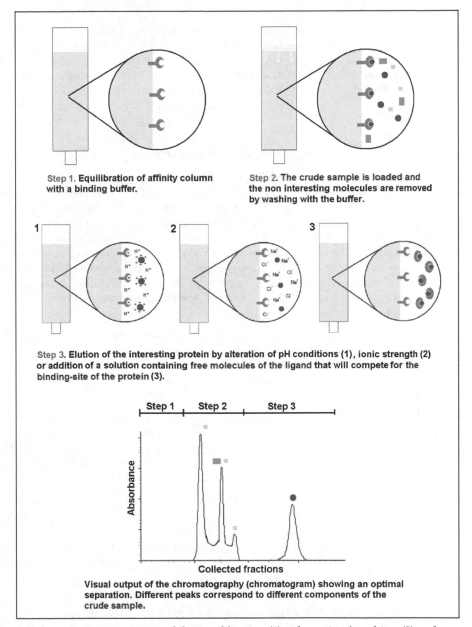

Step 1. **Equilibration of affinity column with a binding buffer.**

Step 2. **The crude sample is loaded and the non interesting molecules are removed by washing with the buffer.**

Step 3. **Elution of the interesting protein by alteration of pH conditions (1), ionic strength (2) or addition of a solution containing free molecules of the ligand that will compete for the binding-site of the protein (3).**

Visual output of the chromatography (chromatogram) showing an optimal separation. Different peaks correspond to different components of the crude sample.

Fig. 1. Schematic representation of the equilibration (1), adsorption/washing (2) and desorption (3) steps of an affinity chromatography for protein purification.

Among the parameters used to evaluate if a preparation is pure can be cited electrophoresis, immunological and chromatographic methods. The homogeneity of a protein preparation should not be judged by isolated parameters. The indication of protein purity is obtained by analysis of various speculations.

Affinity chromatography is a useful tool in proteomics studies; this method plays an essential role in the isolation of protein complexes and in the identification of protein–protein interaction networks. In glycoproteomics, serial lectin affinity chromatography was applied in the process for identification of over thirty proteins from the human blood with O-glycosylation sites (Durham & Regnier, 2006). Affinity chromatography is also required for quantification of protein expression by using isotope-coded affinity tags (Azarkan et al., 2007).

5. Forces that stabilize proteins and affinity interactions

The protein structures are maintained by hydrophobic effects and interactions between polar residues and other types of connections (Voet et al., 2008). For enzymes, the active sites are constituted by amino acid residues in direct contact with the substrate and those amino acid residues indirectly involved in substrate binding through a water molecule as intermediate or by the side chain of an amino acid. Many of the mentioned residues may be in contact with a single substrate; the connection can occur through various combinations of hydrophobic interactions, ionic bonds, hydrogen bonds and charge transfer. The enzyme specificity for a particular substrate depends mainly on the steric positioning of each amino acid in the active site. Substrates or inhibitors can be accommodated in the active site; some are adjusted better than others.

The ideal conditions for affinity chromatography correspond to those in which the adsorbate-adsorbent interaction resembles an enzyme-substrate binding. However, in general, the adsorbent support can interact with proteins applied to the column by ionic interactions, hydrogen bonds, hydrophobic interactions, or other binding sites present on the surface of the protein.

In affinity chromatography occur bioselective and non-bioselective interactions; the contribution of these interactions is dependent of the medium used and the physico-chemical characteristics of the preparation containing the protein to be purified. The bioselective adsorption constitutes one of the most effective and complex methods of protein separation.

In affinity chromatography the bioselective elution (desorption) should be attempted not only to prove that a particular purification was possible due to a bioselective adsorption, but also because the bioselective elution often provides high levels of purification. Large numbers of reversible interactions (hydrophobic attraction and hydrogen or electrostatic bonds) are involved in recognition of the free ligand (in elution solution) by the protein which was adsorbed on the matrix (Scouten, 1981).

6. Lectins: Prototypes in protein purification by affinity chromatography

The term lectin (from Latin *lectus*, past participle of *legere*, which means "to select") was introduced by Boyd (1954) and describes a protein heterogeneous group of non-immune origin, containing two or more binding sites for mono or oligosaccharides. These molecules have the ability to agglutinate cells such as erythrocytes (hemagglutination), lymphocytes, fibroblasts and bacteria, being also able to precipitate glycoconjugates (Goldstein et al., 1980; Barondes, 1988; Kennedy et al., 1995; Correia et al., 2008; Sá et al., 2009a).

To be considered a lectin, the hemagglutinating activity should be inhibited by a carbohydrate; when addition of mono or oligosaccharides neutralizes the agglutination phenomenon, the protein is considered a potential lectin.

The lectin purification may be performed by conventional or high resolution techniques. However, in most of the purification processes, the affinity chromatography is used. The lectins are real models of protein purification exploring affinity interactions.

The lectin extracted from *Canavalia ensiformis* seeds (jack bean) named Concanavalin A (Con A) was the first lectin to be crystallized. Since then, an increasing number of lectins with similar or different specificities have been obtained.

Lectins have been purified from *Cratylia mollis* seeds using Sephadex (cross-linked dextran gel) matrices allying the gel filtration property of this support and the ability of the lectins to bind glucose (Paiva & Coelho, 1992; Correia & Coelho, 1995).

Guar gel beads produced by cross-linking of refined guar gum (a polysaccharide composed of glucose and mannose) with epichlorohydrin in a mixture of water and 2-propanol (Gupta et al., 1979) have been used to purify galactose-specific lectins (Santos et al., 2009; Nunes et al., 2011).

Chitin-binding lectins can be isolated by affinity chromatography on columns containing powder of chitin from crab shells hydrated with the equilibrating solution. This is a cheap, efficient, and rapid technique to purify these lectins, which have great potential as insecticidal and antimicrobial agents (Sá et al., 2009a; Sá et al; 2009b; Santana et al., 2009; Coelho et al., 2009; Ferreira et al., 2011; Napoleão et al., 2011a; Napoleão et al., 2011b).

A ferromagnetic levan (a homopolysaccharide composed of D-fructofuranosyl) composite was developed and efficiently used in purification of *C. mollis* lectin (Angeli et al., 2009). Egg glycoproteins were immobilized and the affinity matrix was efficient to purify lectins from extracts of *Phaseolus vulgaris*, *Lens culinaris*, and *Triticum vulgaris* (Zocatelli et al., 2003).

Lectins can be used for observation of the most diverse phenomena and the study of these proteins allows the evaluation of different cell surfaces. It is known that all cells have a membrane containing carbohydrates, consisting mainly of glycoproteins and glycolipids, that are different for each cell and which may constitute the lectin receptors. In the same cell, the surface structure can change characteristically due to normal development course or cases of illness. The lectins have been used very successfully in histochemistry (Beltrão et al., 1998, Lima et al., 2010) and electrochemistry (Souza et al., 2003, Oliveira et al., 2008, 2011b) with diagnostic purposes.

6.1 Purification of lectins from autochthonous and introduced species at Northeastern Brazil by affinity chromatography

The motivation to search lectins in autochthonous and introduced species from a particular region of a country is primordially due to the perspectives to develop a biotechnological leading edge. In the Laboratory of Glycoproteins from the Department of Biochemistry of the *Universidade Federal de Pernambuco* (Brazil), the first plant tissues evaluated in order to identify hemagglutinating activity (indicating the presence of lectins) were the seeds of *C. mollis* (Paiva & Coelho, 1992; Correia & Coelho, 1995). This legume, also known as camaratu bean, is important as human food and as native forage in the Semi-Arid Region from the State of Pernambuco, northeastern Brazil. Since then, many other lectins have been purified. Examples of lectins purified by affinity chromatography in the Laboratory of Glycoproteins are shown in Table 2.

Lectin	Plant (tissue)	Affinity support used	References
BmoLL	*Bauhinia monandra* (leaf)	Guar gel	Coelho & Silva (2000)
BmoRoL	*B. monandra* (root)	Guar gel	Souza et al. (2011b)
Cramoll	*Cratylia mollis* (seeds)	Sephadex	Paiva & Coelho (1992); Correia & Coelho (1995)
cMoL	*Moringa oleifera* (seeds)	Guar gel	Santos et al. (2009)
WSMoL	*M. oleifera* (seeds)	Chitin	Coelho et al. (2009)
MuBL	*Myracrodruon urundeuva* (bark)	Chitin	Sá et al. (2009b)
MuHL	*M. urundeuva* (heartwood)	Chitin	Sá et al. (2008)
MuLL	*M. urundeuva* (leaf)	Chitin	Napoleão et al. (2011)
PpeL	*Parkia pendula* (seed)	Sephadex	Lombardi et al. (1998)
PpyLL	*Phthirusa pyrifolia* (leaf)	Sephadex	Costa et al. (2010)
OfiL	*Opuntia ficus indica* (cladodes)	Chitin	Santana et al. (2009)

Table 2. Lectins purified by affinity chromatography from different tissues of autochthonous and introduced plants from northeastern Brazil.

Saline extract (0.15 M NaCl) from *C. mollis* seeds showed hemagglutinating activity on erythrocytes from humans and other animals. The lectin activity was inhibited by glucose and mannose. The extract was treated with ammonium sulfate (0-40% and 40-60%), producing three fractions (F): 0-40F and 40-60F (precipitate fractions) and 40-60SF (supernatant fraction) with hemagglutinating activity. The hemagglutinating activity was concentrated (94%) in 40-60F, and a lectin (Cramoll 1) was purified by affinity chromatography on Sephadex G-75 followed by ion exchange chromatography on CM-cellulose (Correia & Coelho, 1995). Additionally, two other molecular forms were obtained from 0-40F (Cramoll 3) and 40-60FS (Cramoll 2) through affinity chromatography on Sephadex G-75, ion exchange using CM-Cellulose column, and molecular exclusion using Bio-Gel P (Paiva & Coelho, 1992). The characterization of the isoforms was performed by electrophoresis and immunological methods. Cramoll 1 was crystallized by Tavares et al. (1996). *C. mollis* lectins showed several biological activities such as mitogenic effect on human lymphocytes (Maciel et al., 2004), antitumor activity on Sarcoma 180 when encapsulated into liposomes (Andrade et al., 2004), potential anti-helminthic against *Schistosoma mansoni* (Melo et al., 2011a), healing activity on cutaneous wounds in healthy and immunocompromised mices (Melo et al., 2011b), and induction of death on epimastigotes of *Trypanosoma cruzi* (Fernandes et al., 2010).

Moringa oleifera is a multipurpose tree with great importance in industry and medicine. Lectins have been found in extracts from distinct tissues of *M. oleifera* (Santos et al., 2009). Seeds from moringa are used to treat water for human consumption and different lectins were detected in this tissue (Santos et al., 2005; Katre et al., 2008; Santos et al., 2009; Coelho et al., 2009). Santos et al. (2005) found a water-soluble *M. oleifera* lectin (WSMoL) that is the unique *M. oleifera* lectin inhibited by fructose. WSMoL was isolated through affinity chromatography on chitin column and showed larvicidal activity against fourth-stage larvae of *Aedes aegypti* (Coelho et al., 2009). This lectin is also a potential natural biocoagulant for water, reducing turbidity, suspended solids and bacteria (Ferreira et al., 2011). Genotoxicity assessment of WSMoL showed that it was not mutagenic and was not able to promote breaks in DNA structure (Rolim et al., 2011).

Santos et al. (2009) purified a lectin with coagulant properties from *M. oleifera* seeds (cMoL) by affinity chromatography on guar gel. cMoL agglutinated erythrocytes from rabbit and human, was insecticidal for *Anagasta kuehniella* and, when immobilized, served as an affinity support able to interact with humic acids (Oliveira et al., 2011a; Santos et al., 2011).

Coelho & Silva (2000) purified a galactose-specific lectin (BmoLL) from the fresh leaves of *Bauhinia monandra*. Also, other galactose-specific lectin was purified from *B. monandra* secondary roots, BmoRoL (Souza et al., 2011). These lectins were purified in milligram quantities by affinity chromatography on guar gel. BmoLL showed insecticidal activity on *Callosobruchus maculatus*, *Anagasta kuehniella* and *Zabrotes subfasciatus* (Macedo et al., 2007) while BmoRoL showed antifungal and termiticidal activities (Souza et al., 2011); thus, these lectins have biotechnological potential for application in control of agricultural pests.

In our studies, the presence of lectin isoforms has been revealed. The exploration and knowledgement of multiple molecular forms of lectins in extracts or in early stages of fractionation is very important. A substantial proportion of proteins have been described with multiple molecular forms having or not defined genetic origin.

The *Parkia pendula* (visgueiro) is a majestic tree from the Brazilian Atlantic Forest that stands out by their generous production of vegetables. Extracts of its seeds showed hemagglutinating activity with erythrocytes from humans and various animal species. The best monosaccharide inhibitors of the hemagglutinating activity from *P. pendula* were α-methyl D-mannoside, D (+)-mannose and D (+)-glucose, in descending order. To purify the lectin, a seed extract in 0.15 M NaCl, was fractionated with ammonium sulfate (40%). The 0-40F recovered 97% of total hemagglutinating activity. The dialyzed preparation was chromatographed by affinity on Sephadex G-75, and eluted with 0.3 M glucose. The purity of the obtained preparation allowed the crystallization of the lectin (Lombardi et al., 1998).

Other supports for purification of *P. pendula* lectin by affinity chromatography were also exploited for its purification. Although the lectin was not inhibited by N-acetyl-D-glucosamine, the support chitin was used to purify two molecular forms of lectin (Souza, 1989). The absence of inhibitory effect of carbohydrate on hemagglutinating activity does not imply in an inability of lectin to adsorb on an affinity support containing this carbohydrate (Lis & Sharon, 1981).

Myracrodruon urundeuva (aroeira-do-sertão) is a plant with importance in traditional medicine and its heartwood is resistant to fungi and termite attack. Lectins were isolated from *M. urundeuva* bark (MuBL), heartwood (MuHL) and leaf (MuLL) by affinity chromatography on chitin columns. Similarly to *P. pendula* lectin, the hemagglutinating activity of MuLL is not inhibited by N-acetyl-D-glucosamine but the lectin bind to chitin. The affinity interaction between MuLL and this monosaccharide was demonstrated by affinity chromatography on N-acetyl-D-glucosamine-Agarose column (Napoleão et al., 2011a).

MuHL showed antimicrobial activity inhibiting the growth of bacteria and fungi (Sá et al., 2009a). The three lectins showed termiticidal activity against *Nasutitermes corniger* and insecticidal effect on fourth-stage larvae of *A. aegypti* (Sá et al., 2008; Sá et al., 2009b; Napoleão et al., 2011a; Napoleão et al., 2011b).

6.2 Applications: Immobilized lectins as affinity supports for protein purification

Various applications of lectins have been developed from the binding of these versatile molecules with free carbohydrates or glycoconjugates present in cell surfaces. The lectin applications have emerged in parallel to their discovery in 1888, with the description of the hemagglutination phenomenon, previously mentioned. Lectins have been applied for different purposes.

An immobilized lectin, covalently attached to a support, can separate glycoproteins or proteoglycans containing specific carbohydrate groups from a crude preparation. The elution of adsorbed material can be performed by treatment of support with a solution containing a competitive glycoside. The elution is usually performed near neutral pH, with minimal deleterious effects to the glycoprotein.

The interaction of a glycoprotein with an immobilized lectin can be used as a suitable technique to obtain preliminary information about the covalently linked carbohydrates to the glycoconjugate in the study. Lectins with different carbohydrate specificities, immobilized on Sepharose, have been applied as an analytical tool to assess and compare the carbohydrate residues.

Coelho (1982), using columns containing lectins with different specificity, detected microheterogeneities in human liver glycosidases. Con A revealed microheterogeneity in type A and B isoenzymes of beta-N-acetylhexosaminidase purified from human placenta.

A preparation of lectin from *C. mollis* containing Cramoll 1,4 isoforms was immobilized on inert support and used as an affinity matrix for purification of glycoproteins from human plasma, including the lecithin cholesterol acyl transferase (Lima et al., 1997). *C. mollis* seed lectins immobilized on cyanogen bromide-activated Sepharose 4B were used to purify a trypsin inhibitor from *Echinodorus paniculatus* seeds (Paiva et al., 2003) and a soybean seed protein with platelet antiaggregation and anticoagulant activities (Silva et al., 2011).

Immobilized *Euonymus europaeus* lectin was an efficient affinity ligand used in the capture step for purification of human influenza A viruses derived from MDCK cells; the main targets were two viral glycoproteins (Opitz et al., 2007).

Lectin affinity chromatography is a powerful fractionation technique in the identification of glycobiomarkers. Immobilized Con A was successfully used in the glycoproteomic analysis of pluripotent murine embryonic stem cells; differential patterns of binding to lectin allowed the identification of stage-specific glycopeptides (Alvarez-Manilla et al., 2010).

7. Immunoaffinity chromatography

The immunoaffinity chromatography consists of an antibody (immunoglobulin) immobilized on a support to purify the protein against which the antibody was developed. Antibodies specific for the protein of interest are produced by the immune system when the exogenous protein is inoculated in the animal; after, polyclonal antibodies are extracted from the blood serum of the animal, isolated and immobilized to constitute a matrix for purification of the protein of interest (Figure 2). Since the polyclonal antibodies are products from many different cells of the immune system, they are heterogeneous, differing in binding affinity for the protein inoculated on the animal.

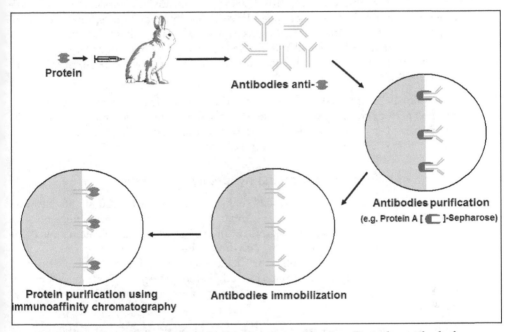

Fig. 2. Immunoaffinity chromatography for protein purification. First, the antibody for a specific protein is developed immunizing an animal. Next, the antibodies produced are purified and immobilized for use as immunoaffinity support.

Polyclonal antibodies (IgY) were developed against *Plasmodium falciparum* proteins and immobilized for use in immunoaffinity chromatography columns; the technique was more efficient than conventional chromatography (using two anion exchange columns) in purification of chimeric proteins expressed in *Escherichia coli* that are candidates for use as vaccine to prevent malaria (Qu et al., 2011).

The native Cramoll 1 was used to develop a serum anti-Cramoll 1 produced by rabbits. The anti-Cramoll 1 immunoglobulin (IgG anti-Cramoll 1) was obtained by affinity chromatography on Protein A-Sepharose. The antibody was conjugated to peroxidase (IgG anti-Cramoll 1-Per) and the conjugate was used to evaluate the structural assessment of the lectin (Correia & Coelho, 1995).

Monoclonal antibodies can be developed by collecting lymphocytes B (producing the desired antibody) from the spleen of the immunized animal and fusing them with a myeloma (a tumor of lymphocytes B). The resulting cells (hybridoma) have an unlimited capacity for division. When developed in culture they will produce large amounts of monoclonal antibodies.

Monoclonal antibodies are extremely useful as immunosorbents for purification of antigens. Immobilization of monoclonal antibody produces a support that can achieve a 10,000-fold purification in a single step. Polyol-responsive monoclonal antibody that recognize a highly conserved sequence in the β-subunit of bacterial RNA polymerase were used to purify RNA polymerase from five species of bacteria in one immunoaffinity chromatography step (Stalder et al., 2011). Nakamura et al. (2010) developed a sensitive and specific monoclonal antibody against a soluble lectin-like oxidized low-density lipoprotein receptor-1 (sLOX-1), expressed prominently in atherosclerotic lesions as a specific biomarker to diagnostic acute coronary syndrome at an early stage; this immunoassay was developed in order to establish a more sensitive assay that may also be useful in predicting cardiovascular disease risk in disease-free subjects.

The antigen-antibody complex often has a strong affinity and it is necessary that the release of the protein is performed in drastic conditions of pH and/or ionic strength. The elution is preferably performed in the reverse direction to sample application. To avoid denaturation of the molecule, the conditions are adjusted immediately after the elution.

8. Purification of DNA-binding proteins by affinity chromatography

DNA-binding proteins can be purified using different techniques of affinity chromatography. One of them is the affinity chromatography containing immobilized DNA, which is a highly specific technique for the purification of DNA-binding proteins involved in the transcription, replication and recombination. The affinity columns are usually generated by immobilization of synthetic oligonucleotides consisting of tandem repeated units or multiple copies of the same sequence (Gadgil et al., 2001). Purification factors can reach 10,000-fold. Figure 3 shows a schematic representation of DNA-affinity chromatography.

DNA-affinity chromatography is a powerful method with broad applicability; this technology has been extended for purifying transcription factors, polymerases, and nucleases (Chockalingam et al., 2001). Since affinity chromatography is based in the specific interaction between molecules, it is highly selective and offers high yield and purity (Gadgil et al., 2001). DNA affinity columns can be constructed depending on the DNA-binding properties of a protein (non-specific, specific, double- or single-stranded DNA). Apart from the purification of DNA-binding proteins, DNA affinity columns can also be used for the purification of nucleic acids, such as RNA and DNA. Golovina et al. (2010) described a fast and simple purification method for the 30S ribosomal subunits carrying lethal mutations using DNA-affinity chromatography. Kerrigan & Kadonaga (2001) developed a DNA affinity resin using agarose activated with cyanogen bromide.

Another technique used for isolation of DNA-binding proteins is ion metal affinity chromatography. The Mvo10b is a DNA-binding protein member of the Sac10b family from the mesophilic archaeon *Methanococcus voltae*, which may play an important role in the organization and accessibility of genetic information in Archaea. This protein was purified by polyethyleneimine precipitation followed by nickel affinity chromatography; this protocol has potential application in the production of other thermophilic and mesophilic proteins in the Sac10b family (Xuan et al., 2009). A telomeric DNA-binding protein (Stn1p) from *Saccharomyces cerevisae* was purified by interaction with nickel-NTA resin followed by chromatography on Superdex 200 column (Qian et al., 2010).

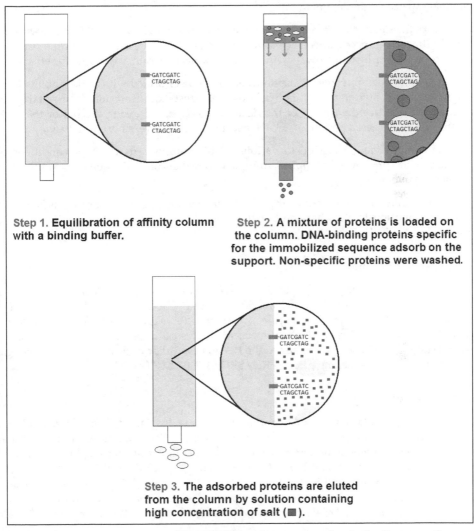

Step 1. Equilibration of affinity column with a binding buffer.

Step 2. A mixture of proteins is loaded on the column. DNA-binding proteins specific for the immobilized sequence adsorb on the support. Non-specific proteins were washed.

Step 3. The adsorbed proteins are eluted from the column by solution containing high concentration of salt (■).

Fig. 3. Schematic representation of DNA-affinity chromatography. In the first step, the support containing a synthetic oligonucleotide consisting of a tandem repeated unit is washed equilibrated. Next, the mixture containing proteins is loaded and the DNA-binding protein that recognizes the oligonucleotide adsorb on the support. The elution is performed by solutions containing a very high concentration of salt.

9. Pitfalls in affinity chromatography

There are numerous problems in affinity chromatography, similar to any other techniques. One of the most common ways to immobilize ligands with free amino groups is by a reaction with Sepharose activated with cyanogen bromide. This method promotes the formation of ionic groups that may cause non-selective electrostatic adsorption. The immobilization of small ligands may create steric impediment problems that limit the functional capacity of the columns.

To minimize the steric interference between the support and substances interacting with the ligand, hydrocarbon spacer arms (such as in Sepharose CL-4B) are often interposed between the substrate and the ligand. However, these spacer arms may cause hydrophobic effects.

In affinity chromatography, the adsorption step is often performed with buffers of low ionic strength and the interference of non-selective electrostatic adsorption is inevitable. In addition, non-selective desorption of the desired molecule, increasing the pH or the ionic strength leads to at least a partial loss of activity, many times reversible but not always.

Special attention should be given to various affinity supports used under conditions of saturation of the protein to be applied to matrix; possible contaminations can result from non-selective interactions.

In immunoaffinity chromatography the disadvantages include the technical difficulty of producing monoclonal antibodies and the drastic conditions that are often required to elute the strongly adsorbed protein.

10. Acknowledgments

The authors express their gratitude to the *Conselho Nacional de Desenvolvimento Científico e Tecnológico* (CNPq) for research grants and fellowships (LCBBC, MTSC and PMGP). We are also grateful to the *Fundação de Amparo à Ciência e Tecnologia do Estado de Pernambuco* (FACEPE) and the *Coordenação de Aperfeiçoamento de Pessoal de Nível Superior* (CAPES) for financial support. It is also acknowledged the Portuguese *Fundação para Ciência e a Tecnologia* (FCT) through the Post-doctoral grant SFRH/BPD/37349/2007 (AFSS).

11. References

Alvarez-Manilla, G.; Warren, N.L.; Atwood III, J.; Orlando, R.; Dalton, S. & Pierce, M. (2010). Glycoproteomic analysis of embryonic stem cells: Identification of potential glycobiomarkers using lectin affinity chromatography of glycopeptides. *Journal of Proteome Research*, Vol. 9, No. 5, (May 2010), pp. 2062-2075, ISSN 1535-3893

Andrade, C.A.S.; Correia, M.T.S.; Coelho, L.C.B.B.; Nascimento, S.C. & Santos-Magalhães, N.S. (2004). Antitumor activity of *Cratylia mollis* lectin encapsulated into liposomes. *International Journal of Pharmaceutics*, Vol. 278, No. 2, (July 2004), pp. 435-445, ISSN 0378-5173

Angeli, R.; Paz, N.V.N.; Maciel, J.C.; Araújo, F.F.B.; Paiva, P.M.G.; Calazans, G.M.T.; Valente, A.P.; Almeida, F.C.L.; Coelho, L.C.B.B.; Carvalho Jr., L.B.; Silva, M.P.C. & Correia, M.T.S. (2009). Ferromagnetic levan composite: an affinity matrix to purify lectin.

Journal of Biomedicine and Biotechnology, Vol. 2009, Article ID 179106, pp. 1-6, ISSN 1110-7243

Azarkan, M.; Huet, J.; Baeyens-Volant, B.; Looze, Y. & Vandenbussche, G. (2007). Affinity chromatography: A useful tool in proteomics studies. *Journal of Chromatography B*, Vol. 849, No. 1-2, (April 2007), pp. 81-90, ISSN 1570-0232

Barondes, S.H. (1988). Bifunctional properties of lectins: lectins redefined. *Trends in Biochemical Sciences*, Vol. 13, No. 12, (December 1988), pp. 480-482, ISSN 0968-0004

Beltrão, E.I.C.; Correia, M.T.S.; Figueredo-Silva, J. & Coelho, L.C.B.B. (1998). Binding evaluation of isoform 1 from *Cratylia mollis* lectin to human mammary tissues. *Applied Biochemistry and Biotechnology*. Vol. 74, No. 3, (September 1998), pp. 125-34, ISSN 0273-2289

Bhowal, J.; Guha, A.K. & Chatterjee, B.P. (2005). Purification and molecular characterization of a sialic acid specific lectin from the phytopathogenic fungus *Macrophomina phaseolina*. *Carbohydrate Research*, Vol. 340, No. 12, (September 2005), pp. 1973-1982, ISSN 0008-6215

Boyd, W.C. (1954) The proteins of immune reactions, In: *The Proteins: Chemistry, Biological Activity and Methods*, vol. 2, Neurath, H. & Bailey, K. (eds.), pp. 765-844, Academic Press, New York, US

Chung, J.A.; Wollack, J.W.; Hovlid, M.L.; Okesli, A.; Chen, Y.; Mueller, J.D.; Distefano, M.D. & Taton, T.A. (2009). Purification of prenylated proteins by affinity chromatography on cyclodextrin-modified agarose. *Analytical Biochemistry*, Vol. 386, No. 1, (March 2009), pp. 1-8, ISSN 0003-2697

Coelho, L.C.B.B. (1982) *Purification and structural studies of human β-N-acetilhexosaminidases*. Ph.D. Thesis. University of London, London, England.

Coelho, L.C.B.B. & Silva, M.B.R. (2000). Simple method to purify milligram quantities of the galactose-specific lectin from the leaves of *Bauhinia monandra*. *Phytochemical Analysis*, Vol. 11, No. 5, (September/October 2000), pp. 295-300, ISSN 1099-1565

Coelho, J.S.; Santos, N.D.L.; Napoleão, T.H.; Gomes, F.S.; Ferreira, R.S.; Zingali, R.B.; Coelho, L.C.B.B.; Leite, S.P.; Navarro, D.M.A.F. & Paiva, P.M.G. (2009). Effect of *Moringa oleifera* lectin on development and mortality of *Aedes aegypti* larvae. *Chemosphere*, Vol. 77, No. 7, (November 2009), pp. 934-938, ISSN 0045-6535

Correia, M.T.S. & Coelho, L.C.B.B. (1995). Purification of glucose/mannose specific lectin, isoform1, from seeds of *Cratylia mollis* Mart. (camaratu bean). *Applied Biochemistry and Biotechnology*, Vol. 55, No. 3, (December 1995), pp. 261-273, ISSN 0273-2289

Correia, M.T.S.; Coelho, L.C.B.B. & Paiva, P.M.G. (2008). Lectins, carbohydrate recognition molecules: Are they toxic?, In: *Recent Trends in Toxicology*, vol. 37, Siddique, Y.H., Vol. 37, pp. 47-59, Transworld Research Network, ISBN 978-81-7895-384-7, Kerala, India

Costa, R.M.P.B.; Vaz, A.F.M.; Oliva, M.L.V.; Coelho, L.C.B.B.; Correia, M.T.S. & Carneiro-da-Cunha, M.G. (2010). A new mistletoe *Phthirusa pyrifolia* leaf lectin with antimicrobial properties. *Process Biochemistry*, Vol. 45, No. 4, (April 2010), pp. 526-533, ISSN 1359-5113

Cuatrecasas, P.; Wilchek, M. & Anfinsen, C.B. (1968). Selective enzyme purification by affinity chromatography. *Proceedings of the National Academy of Sciences of the United States of America*, Vol. 61, No. 2, (October 1968), pp. 636-643, ISSN 0027-8424

Durham, M. & Regnier, F.E. (2006). Targeted glycoproteomics: Serial lectin affinity chromatography in the selection of *O*-glycosylation sites on proteins from the human blood proteome. *Journal of Chromatography A*, Vol. 1132, No. 1-2, (November 2006), pp. 165-173, ISSN 0021-9673

Fenton-Navarro, B.; Arreguín-L, B.; Garcíía-Hernández, E.; Heimer, E.; Aguilar, M.B.; Rodríguez-A, C. & Arreguín-Espinosa, R. (2003). Purification and structural characterization of lectins from the cnidarian *Bunodeopsis antillienis*. *Toxicon*, Vol. 42, No. 5, (October 2003), pp. 525-532, ISSN 0041-0101

Fernandes, M.P.; Inada, N.M.; Chiaratti, M.R.; Araújo, F.F.B.; Meirelles, F.V.; Correia, M.T.S.; Coelho, L.C.B.B.; Alves, M.J.M.; Gadelha, F.R. & Vercesi, A.E. (2010). Mechanism of *Trypanosoma cruzi* death induced by *Cratylia mollis* seed lectin. *Journal of Bioenergetics and Biomembranes*, Vol. 42, No. 1, (February 2010), pp. 69-78, ISSN 0145-479X

Ferreira, R.S.; Napoleão, T.H.; Santos, A.F.S.; Sá, R.A.; Carneiro-da-Cunha, M.G.; Morais, M.M.C.; Silva-Lucca, R.A.; Oliva, M.L.V.; Coelho, L.C.B.B. & Paiva, P.M.G. (2011) Coagulant and antibacterial activities of the water-soluble seed lectin from *Moringa oleifera*. *Letters in Applied Microbiology*, Vol. 53, No. 2, (August 2011), pp. 186-192, ISSN 1472-765X

Franco-Fraguas, L.; Plá, A.; Ferreira, F.; Massaldi, H.; Suárez, N. & Batista-Viera, F. (2003). Preparative purification of soybean agglutinin by affinity chromatography and its immobilization for polysaccharide isolation. *Journal of Chromatography B*, Vol. 790, No. 1-2, (June 2003), pp. 365-372, ISSN 1570-0232

Gade, W.; Jack, M.A.; Dahll, J.B.; Schmidt, E.L. & Wold, F. (1981). The isolation and characterization of a root lectin from Soybean (*Glycine max* (L), Cultivar Chippewa). *The Journal of Biological Chemistry*, Vol. 256, No. 24, (December 1981), pp. 12905-12910, ISSN 0021-9258

Gadgil, H.; Oak, S.A. & Jarrett, H.W. (2001). Affinity purification of DNA-binding proteins. *Journal of Biochemical and Biophysical Methods*, Vol. 49, No. 1-3, (October 2001), pp. 607-624, ISSN 0165-022X

Goldstein, I.J.; Hughes, R.C.; Monsigny, M.; Osawa, T. & Sharon, N. (1980). What should be called a lectin? *Nature*, Vol. 285, (May 1980), pp. 66-68, ISSN 0028-0836

Golovina, A.Y., Bogdanov, A.A., Dontsova, O.A. & Sergiev, P.V. (2010). Purification of 30S ribosomal subunit by streptavidin affinity chromatography. *Biochimie*, Vol. 92, No. 7, (July 2010), pp. 914-917, ISSN 0300-9084

Gupta, K.C.; Sahni, M.K.; Rathaur, B.S.; Narang, C.K. & Mathur, N.K. (1979). Gel filtration medium derived from guar gum. *Journal of Chromatography A*, Vol. 169, (February 1979), pp. 183-190, ISSN 0021-6973

Guzmán-Partida, A.M.; Robles-Burgueño, M.R.; Ortega-Nieblas, M. & Vázquez-Moreno, I. (2004). Purification and characterization of complex carbohydrate specific isolectins from wild legume seeds: *Acacia constricta* is (vinorama) highly homologous to *Phaseolus vulgaris* lectins. *Biochimie*, Vol. 86, No. 4-5, (April-May 2004), pp. 335-342, ISSN 0300-9084

Hamed, R.R.; Maharem, T.M.; Abdel-Meguid, N.; Sabry, G.M.; Abdalla, A-M. & Guneidy, R.A. (2011). Purification and biochemical characterization of glutathione S-transferase from Down syndrome and normal children erythrocytes: A comparative study. *Research in Developmental Disabilities*, Vol. 32, No. 5, (September/October 2011), pp. 1470-1482, ISSN 0891-4222

Katre, U.V.; Suresh, C.G.; Khan, M.I. & Gaikwad, S.M. (2008). Structure-activity relationship of a hemagglutinin from *Moringa oleifera* seeds. *International Journal of Biological Macromolecules*, Vol. 42, No. 2, (March 2008), pp. 203-207, ISSN 0141-8130

Kerrigan, L.A. & Kadonaga, J.T. (2001). Purification of sequence-specific DNA-binding proteins by affinity chromatography, In: *Current Protocols in Protein Science*, Taylor, G., Chapter 9, unit 9.6, Wiley, ISSN 1934-3655, Hoboken, US

Kennedy, J.F.; Paiva, P.M.G.; Correia, M.T.S.; Cavalcanti, M.S.M. & Coelho, L.C.B.B. (1995). Lectins, versatile proteins of recognition: a review. *Carbohydrate Polymers*, Vol. 26, No. 3, pp. 219-230, ISSN 0144-8617

Kocabiyik, S; Ozdemir, I. (2006). Purification and characterization of an intracellular chymotrypsin-like serine protease from *Thermoplasma volcanium*. *Bioscience, Biotechnology, and Biochemistry*, Vol. 70, No. 1, pp. 126-134, ISSN 0916-8451

Latha, V.L.; Rao, R.N. & Nadimpalli, S.K. (2006). Affinity purification, physicochemical and immunological characterization of a galactose-specific lectin from the seeds of *Dolichos lablab* (Indian lablab beans). *Protein Expression and Purification*, Vol. 45, No. 2, (February 2006), pp. 296-306, ISSN 1046-5928

Leite, K.M.; Pontual, E.V.; Napoleão, T.H.; Gomes, F.S.; Carvalho, E.V.M.M.; Paiva, P.M.G.; Coelho, L.C.B.B. (2011). Trypsin inhibitor and antibacterial activities from liver of tilapia fish (*Oreochromis niloticus*), In: *Advances in Environmental Research, vol. 20*, Daniels, J.A. (ed.), Nova Science Publishers Inc., ISBN 978-1-61324-869-0, New York, US

Lima, A.L.R.; Cavalcanti, C.C.B.; Silva, M.C.C.; Paiva, P.M.G.; Coelho, L.C.B.B.; Beltrão, E.I.C. & Correia, M.T.S. (2010). Histochemical evaluation of human prostatic tissues with *Cratylia mollis* seed lectin. *Journal of Biomedicine and Biotechnology*, Vol. 2010, Article ID 179817, pp. 1-6, ISSN 1110-7243

Lima, V.L.M.; Correia, M.T.S.; Cechinel, Y.M.N.; Sampaio, C.A.M.; Owen, J.S. & Coelho, L.C.B.B. (1997). Immobilized *Cratylia mollis* lectin as a potential matrix to isolate plasma glycoproteins, including lecithin-cholesterol acyltransferase, *Carbohydate Polymers*, Vol. 33, No. 1, (May 1997), pp. 27-32, ISSN 0144-8617

Lis, H. & Sharon, N. (1981) Lectins in higher plants, In: *The Biochemistry of Plants vol. 6: A Comprehensive Treatise. Proteins and Nucleic Acids*, Marcus, A. (ed.), Academic Press, ISBN 978-0126754063, New York, US

Lombardi, F.R.; Fontes, M.R.M.; Souza, G.M.O.; Coelho, L.C.B.B.; Arni, R.K. & Azevedo, W.F. (1998). Crystallization and preliminary X-ray analysis of *Parkia pendula* lectin. *Protein and Peptide Letters*, Vol. 5, No. 2, pp. 117-120, ISSN 0929-8665

Macedo, M.L.R.; Freire, M.G.M.; Silva, M.B.R. & Coelho, L.C.B.B. (2007). Insecticidal action of *Bauhinia monandra* leaf lectin (BmoLL) against *Anagasta kuehniella* (Lepidoptera: Pyralidae), *Zabrotes subfasciatus* and *Callosobruchus maculatus* (Coleoptera: Bruchidae). *Comparative Biochemistry and Physiology Part A, Molecular and Integrative Physiology*, Vol. 146, No. 4, (April 2007), pp. 486-498, ISSN 1095-6433

Maciel, E.V.M.; Araújo-Filho, V.S.; Nakazawa, M.; Gomes, Y.M.; Coelho, L.C.B.B. & Correia, M.T.S. Mitogenic activity of *Cratylia mollis* lectin on human lymphocytes. *Biologicals*, Vol. 32, No. 1, (March 2004), pp. 57-60, ISSN 1045-1056

Melo, C.M.L.; Lima, A.L.R.; Beltrão, E.I.C.; Cavalcanti, C.C.B.; Melo-Júnior, M.R.; Montenegro, S.M.L.; Coelho, L.C.B.B.; Correia, M.T.S.; Carneiro-Leão, A.M.A. (2011a). Potential effects of Cramoll 1,4 lectin on murine *Schistosomiasis mansoni*. *Acta Tropica*, Vol. 118, No. 2, (May 2011), pp. 152-158, ISSN 0001-706X

Melo, C.M.L.; Porto, C.S.; Melo-Júnior, M.R.; Mendes, C.M.; Cavalcanti, C.C.B.; Coelho, L.C.B.B.; Porto, A.L.F.; Leão, A.M.A.C.; Correia, M.T.S. (2011b). Healing activity induced by Cramoll 1,4 lectin in healthy and immunocompromised mice. *International Journal of Pharmaceutics*, Vol. 408, No. 1-2, (April 2011), pp. 113-119, ISSN 0378-5173

Nakamura, M.; Ohta, H.; Kume, N.; Hayashida, K.; Tanaka, M.; Mitsuoka, H.; Kaneshige, T.; Misaki, S.; Imagawa, K.; Shimosako, K.; Ogawa, N.; Kita T. & Kominami, G. (2010). Generation of monoclonal antibodies against a soluble form of lectin-like oxidized low-density lipoprotein receptor-1 and development of a sensitive chemiluminescent enzyme immunoassay. *Journal of Pharmaceutical and Biomedical Analysis*, Vol. 51, No. 1, (January 2010), pp. 158-163, ISSN 0731-7085

Napoleão, T.H.; Gomes, F.S.; Lima, T.A.; Santos, N.D.L.; Sá, R.A.; Albuquerque, A.C.; Coelho, L.C.B.B. & Paiva, P.M.G. (2011a). Termiticidal activity of lectins from *Myracrodruon urundeuva* against *Nasutitermes corniger* and its mechanisms. *International Biodeterioration & Biodegradation*, Vol. 65, No. 1, (January 2011), pp. 52-59, ISSN 0964-8305

Napoleão, T.H.; Pontual, E.V.; Lima, T.A.; Santos, N.D.L.; Sá, R.A.; Coelho, L.C.B.B.; Navarro, D.M.A.F.; Paiva, P.M.G. (2011b). Effect of *Myracrodruon urundeuva* leaf lectin on survival and digestive enzymes of *Aedes aegypti* larvae. *Parasitology Research*, doi: 10.1007/s00436-011-2529-7, ISSN 0932-0113

Nunes, E.S.; Souza, M.A.A.; Vaz, A.F.M.; Santana, G.M.S.; Gomes, F.S.; Coelho, L.C.B.B.; Paiva, P.M.G.; Silva, R.M.L.; Silva-Lucca, R.A.; Oliva, M.L.V.; Guarnieri, M.C. & Correia, M.T.S. (2011). Purification of a lectin with antibacterial activity from *Bothrops leucurus* snake venom. *Comparative Biochemistry and Physiology Part B: Biochemistry and Molecular Biology*, Vol. 159, No. 1, (May 2011), pp. 57-63, ISSN 1096-4959

Oliveira, M.D.L.; Correia, M.T.S.; Coelho, L.C.B.B. & Diniz, F.B. (2008). Electrochemical evaluation of lectin–sugar interaction on gold electrode modified with colloidal gold and polyvinyl butyral. *Colloids and Surfaces B: Biointerfaces*, Vol. 66, No. 1, (October 2008), pp. 13-19, ISSN 0927-7765

Oliveira, C.F.R.; Luz, L.A.; Paiva, P.M.G.; Coelho, L.C.B.B.; Marangoni, S. & Macedo, M.L.R. (2011a). Evaluation of seed coagulant *Moringa oleifera* lectin (cMoL) as a bioinsecticidal tool with potential for the control of insects. *Process Biochemistry*, Vol. 46, No. 2, (February 2011), pp. 498-504, ISSN 1359-5113

Oliveira, M.D.L.; Nogueira, M.L.; Correia, M.T.S.; Coelho, L.C.B.B. & Andrade, C.A.S. (2011b). Detection of dengue virus serotypes on the surface of gold electrode based on *Cratylia mollis* lectin affinity. *Sensors and Actuators B: Chemical*, Vol. 155, No. 2, (July 2011), pp. 789-795, ISSN 0925-4005

Opitz, L.; Salaklang, J.; Büttner, H.; Reichl, U.; Wolff, M.W. (2007). Lectin-affinity chromatography for downstream processing of MDCK cell culture derived human influenza A viruses. *Vaccine*, Vol. 25, No. 5, (January 2007), pp. 939-947, ISSN 0264-410X

Paiva, P.M.G. & Coelho, L.C.B.B. (1992). Purification and partial characterization of two lectins isoforms from *Cratylia mollis* Mart. (Camaratu Bean). *Applied Biochemistry and Biotechnology*, Vol. 36, No. 2, (August 1992), pp. 113-118, ISSN 0273-2289

Paiva, P.M.G.; Souza, A.F.; Oliva, M.L.V.; Kennedy, J.F.; Cavalcanti, M.S.M.; Coelho, L.C.B.B. & Sampaio, C.A.M. (2003). Isolation of a trypsin inhibitor from *Echinodorus paniculatus* seeds by affinity chromatography on immobilized *Cratylia mollis* isolectins. *Bioresource Technology*, Vol. 88, No. 1, (May 2003), pp. 75-79, ISSN 0960-8524

Porath, J. (1973). Conditions for biospecific adsorption. *Biochimie*, Vol. 55, No. 8, (October 1973), pp. 943-951, ISSN 0300-9084

Qian, W.; Fu, X.; Zhou, J. (2010). Purification and characterization of Stn1p, a single-stranded telomeric DNA binding protein. *Protein Expression and Purification*, Vol. 73, No. 2, (October 2010), pp. 107-112, ISSN 1046-5928

Qu, J.; Lin, Y.; Ma, R.; Wang, H. (2011). Immunoaffinity purification of polyepitope proteins against *Plasmodium falciparum* with chicken IgY specific to their C-terminal epitope tag. *Protein Expression and Purification*, Vol. 75, No. 2, (February 2011), pp. 225-229, ISSN 1046-5928

Rolim, L.A.D.M.M.; Macedo, M.F.S.; Sisenando, H.A.; Napoleão, T.H.; Felzenswalb, I.; Aiub, C.A.F.; Coelho, L.C.B.B.; Medeiros, S.R.B. & Paiva, P.M.G. (2011). Genotoxicity evaluation of *Moringa oleifera* seed extract and lectin. *Journal of Food Science*, Vol. 76, No. 2, (March 2011), pp. T53-T58, ISSN 1750-3841

Sá, R.A.; Gomes, F.S.; Napoleão, T.H.; Santos, N.D.L.; Melo, C.M.L.; Gusmão, N.B.; Coelho, L.C.B.B.; Paiva, P.M.G. & Bieber, L.W. (2009a). Antibacterial and antifungal activities of *Myracrodruon urundeuva* heartwood. *Wood Science and Technology*, Vol. 43, No. 1-2, (February 2009), pp. 85-95, ISSN 1432-5225

Sá, R.A.; Napoleão, T.H.; Santos, N.D.L.; Gomes, F.S.; Albuquerque, A.C.; Xavier, H.S.; Coelho, L.C.B.B.; Bieber, L.W. & Paiva, P.M.G. (2008). Induction of mortality on *Nasutitermes corniger* (Isoptera, Termitidae) by *Myracrodruon urundeuva* heartwood lectin. *International Biodeterioration & Biodegradation*, Vol. 62, No. 4, (December 2008), pp. 460-464, ISSN 0964-8305

Sá, R.A.; Santos, N.D.L.; da Silva, C.S.B.; Napoleão, T.H.; Gomes, F.S.; Cavada, B.S.; Coelho, L.C.B.B.; Navarro, D.M.A.F.; Bieber, L.W. & Paiva, P.M.G. (2009b). Larvicidal activity of lectins from *Myracrodruon urundeuva* on *Aedes aegypti*. *Comparative Biochemistry and Physiology Part C, Toxicology and Pharmacology*, Vol. 149, No. 3, (April 2009), pp. 300-306, ISSN 1532-0456

Santana, G.M.S.; Albuquerque, L.P.; Simões, D.A.; Gusmão, N.B.; Coelho, L.C.B. B. & Paiva, P.M.G. (2009). Isolation of lectin from *Opuntia ficus* indica cladodes. *Acta Horticulturae*, Vol. 811, (February 2009), pp. 281-286, ISSN 0567-7572

Santos, A.F.S.; Argolo, A.C.C.; Coelho, L.C.B.B.; Paiva, P.M.G. (2005). Detection of water soluble lectin and antioxidant component from *Moringa oleifera* seeds. *Water Research*, Vol. 39, No. 6, (March 2005), pp. 975-980, ISSN 0043-1354

Santos, A.F.S.; Carneiro-da-Cunha, M.G.; Teixeira, J.A.; Paiva, P.M.G.; Coelho, L.C.B.B. & Nogueira, R.M.O.B. (2011a). Interaction of *Moringa oleifera* seed lectin with humic acid. *Chemical Papers*, Vol. 65, No. 4, (August 2011), pp. 406-411, ISSN 1336-9075

Santos, A.F.S.; Luz, L.A.; Argolo, A.C.C.; Teixeira, J.A.; Paiva, P.M.G. & Coelho, L.C.B.B. (2009) Isolation of a seed coagulant *Moringa oleifera* lectin. *Process Biochemistry*, Vol. 44, No. 4, (April 2009), pp. 504–508, ISSN 1359-5113

Scouten, W.H. (1981), *Affinity Chromatography: Bioselective adsorption on inert matrices*. John Wiley & Sons Inc., ISBN 978-0471026495, US

Silva, M.C.C.; Santana, L.A.; Silva-Lucca, R.A.; Lima, A.L.R.; Ferreira, J.G.; Paiva, P.M.G.; Coelho, L.C.B.B.; Oliva, M.L.V.; Zingali, R.B. & Correia, M.T.S. (2011) Immobilized *Cratylia mollis* lectin: An affinity matrix to purify a soybean (*Glycine max*) seed protein with in vitro platelet antiaggregation and anticoagulant activities. *Process Biochemistry*, Vol. 46, No. 1, (January 2011), pp. 74–80. ISSN 1359-5113

Souza, J.D.; Silva, M.B.R.; Argolo, A.C.C.; Napoleão, T.H.; Sá, R.A.; Correia, M.T.S.; Paiva, P.M.G.; Silva, M.D.C. & Coelho, L.C.B.B. (2011b). A new *Bauhinia monandra* galactose-specific lectin purified in milligram quantities from secondary roots with antifungal and termiticidal activities. *International Biodeterioration and Biodegradation*, Vol. 65, No. 5, (August 2011), pp. 696-702, ISSN 0964-8305

Souza, S.R.; Dutra, R.F.; Correia, M.T.S.; Pessoa, M.M.A., Lima-Filho, J.L. & Coelho, L.C.B.B. (2003). Electrochemical potential of free and immobilized *Cratylia mollis* seed lectin. *Bioresource Technology*, Vol. 88, No. 3, (July 2003), pp. 255-258. ISSN 09608524

Souza, G.M.O. (1989) *Estudos cromatográficos de preparações da lectina de Parkia pendula L. (Visgueiro)*. Mastering Dissertation. Universidade Federal de Pernambuco, Recife, Brazil.

Stadler, E.S.; Nagy, L.H.; Batalla, P.; Arthur, T.M.; Thompson, N.E.; Burgess, R.R. (2010). The epitope for the polyol-responsive monoclonal antibody 8RB13 is in the flap-domain of the beta-subunit of bacterial RNA polymerase and can be used as an epitope tag for immunoaffinity chromatography. *Protein Expression and Purification*, Vol. 77, No. 1, (May 2011), pp. 26-33, ISSN 1046-5928

Tavares, G.A.; Caracelli, I.; Burger, R.; Correia, M.T.S.; Coelho, L.C.B.B. & Oliva, G. (1996). Crystallization and preliminary X-ray studies on the lectin from the seeds of *Cratylia mollis*. *Acta Crystallographica. Section D, Biological Crystallography*, Vol. D52, No. 5, (September 1996), pp. 1046-1047, ISSN 1399-0047

Wong, J.H.; Wong, C.C.T. & Ng, T.B. (2006). Purification and characterization of a galactose-specific lectin with mitogenic activity from pinto beans. *Biochimica et Biophysica Acta – General Subjects*, Vol. 1760, No. 5, (May 2006), pp. 808-813, ISSN 0006-3002

Voráčková, I.; Suchanová, Š.; Ulbrich, P.; Diehl, W.E. & Ruml, T. (2011). Purification of proteins containing zinc finger domains using immobilized metal ion affinity chromatography. *Protein Expression and Purification*, Vol. 79, No. 1, (September 2011), pp. 88-95, ISSN 1046-5928

Voet, D. & Voet, J. G. (1995). *Biochemistry*. John Wiley & Sons, Inc., ISBN 9780471586517, US.

Voet, D., Voet, J. G. & Pratt, C. W. (2008). *Fundamentals of Biochemistry: Life at Molecular Level*. John Wiley & Sons, Inc., ISBN 978-0-470-12930-2, US

Xuan, J.; Yao, H.; Feng, Y. & Wang, J. (2009). Cloning, expression and purification of DNA-binding protein Mvo10b from *Methanococcus voltae*. *Protein Expression and Purification*, Vol. 64, No. 2, (April 2009), pp. 162-166, ISSN 1046-5928

Zocatelli, G.; Pellegrina, C.D.; Vincenzi, S.; Rizzi, C.; Chignola, R. & Peruffo, A.D.B. (2003) Egg-matrix for large-scale single-step affinity purification of plant lectins with different carbohydrate specificities. *Protein Expression and Purification*, Vol. 27, No. 1, (January 2003), pp. 182-185, ISSN 1046-5928

4

Purification Systems Based on Bacterial Surface Proteins

Tove Boström*, Johan Nilvebrant* and Sophia Hober
Royal Institute of Technology, Stockholm,
Sweden

1. Introduction

Affinity purification is based on the selective and reversible interaction between two binding partners, of which one is bound to a chromatography matrix and the other may be either a native target protein or a recombinant protein fused with an affinity tag (Cuatrecasas et al. 1968). Recombinant DNA-technology allows straightforward construction of gene fusions to provide fusion proteins with two or more functions. The main intention is to facilitate downstream purification; however gene fusions may also improve solubility and proteolytic stability and assist in refolding (Waugh 2005). There are many fusion partners for which commercially available purification systems exist, ranging in size from a few amino acids to whole proteins (Flaschel & Friehs 1993; Terpe 2003). A commonly used purification handle is the poly-histidine (His) tag, enabling purification of the recombinant protein on a column with immobilized metal ions (Hochuli et al. 1988). Other commonly used tags include the FLAG peptide (binding to anti-FLAG monoclonal antibodies), the strep-tag (binding to streptavidin), glutathione S-transferase (binding to glutathione) and maltose binding protein (binding to amylose) (Terpe 2003). Many affinity chromatography strategies also exist for the purification of native proteins, however these are slightly less specific and generally purify classes of proteins, as individual proteins each need a specific ligand. Today, many different ligands are available that can separate specific groups of proteins, for example phosphorylated, glycosylated or ubiquitinylated proteins (Azarkan et al. 2007).

Several bacterial surface proteins that show high affinity against different host proteins as immunoglobulins (Ig:s) and serum albumin, but also other host serum proteins, have been identified, see table 1 for examples. These proteins have different specificities regarding species and immunoglobulin classes and also bind to different parts of the immunoglobulin molecules. Therefore they have proven to be highly suitable for applications within protein purification. Many such proteins are expressed by pathogenic strains of the *Staphylococci* and *Streptococci* genera, and one biological function of these surface proteins is to help the bacteria evade the immune system of the host by covering the bacterium with host proteins (Achari et al. 1992; Sauer-Eriksson et al. 1995; Starovasnik et al. 1996). A significant property of serum albumin is the capability to bind other molecules and act as a transporter in the

---

* Authors Contributed Equally

blood. Bacteria able to bind albumin may therefore also benefit by scavenging albumin-bound nutrients (de Chateau et al. 1996). One of the most studied immunoglobulin-binding proteins is the surface-exposed protein A of *Staphylococcus aureus*. Several animal models have demonstrated a decreased virulence for mutants of *S. aureus* that lack Staphylococcal protein A (SPA) on their surface (Foster 2005). Another staphylococcal surface protein, *S. aureus* binder of IgG (Sbi), has also been described (Atkins et al. 2008; Zhang et al. 1998). Several cell surface proteins binding immunoglobulins and other host proteins have also been discovered in *Streptococcus* strains. Streptococcal protein G (SPG), which binds both to immunoglobulins and serum albumin of different species (Kronvall 1973), is the most investigated. Proteins M, H and Arp (short for IgA receptor protein) are expressed by the human-specific pathogen group A streptococci and have different specificities (Akerstrom et al. 1991; Akesson et al. 1990; Fischetti 1989; Lindahl & Akerstrom 1989; Smeesters et al. 2010). Protein L is expressed by the anaerobic bacterial species *Finegoldia magna* (formerly known as *Peptostreptococcus magnus*). It has been shown that this protein binds to the light chains of human IgG molecules (Bjorck 1988). Another protein expressed by *F. magna* is the peptostreptococcal albumin-binding protein (PAB), which displays high sequence similarity with the albumin-binding parts of SPG. However, the species specificity differs somewhat and PAB binds mainly to albumin from primates (Lejon et al. 2004). Protein B, which is expressed by group B streptococci, binds exclusively to human IgA of both subclasses as well as its secretory form (Faulmann et al. 1991).

Among the identified staphylococcal and streptococcal immunoglobulin-binding surface proteins, SPA (Grov et al. 1964; Oeding et al. 1964; Verwey 1940) and SPG (Bjorck & Kronvall 1984) have been subjects for substantial research and have found several applications in the field of biotechnology. SPA exists in different forms in various strains of *S. aureus*, either as a cell wall component, or as a secreted form (Guss et al. 1985; Lofdahl et al. 1983). This indicates that the function of SPA stretches beyond only immune system evasion and SPA has for example been shown to activate TNFR1, a receptor for tumor necrosis factor-α (TNF-α), with pneumonia as a possible outcome (Gomez et al. 2004). SPA includes five homologous immunoglobulin-binding domains that share high sequence identity (Moks et al. 1986). SPG contains, apart from two or three regions binding to IgG, also two or three homologous domains binding serum albumin, depending on the strain (Kronvall et al. 1979). Although they differ somewhat regarding sequence length, there is great homology between the variants (Olsson et al. 1987). The IgG-binding domains of SPG differ from their counterparts in SPA, regarding subclass and species specificity as well as structure (Bjorck & Kronvall 1984; Gouda et al. 1992; Gronenborn et al. 1991; Kronvall et al. 1979). Today, SPA and SPG are widely used in different biotechnological areas, the most widespread being affinity purification of antibodies and proteins fused with the fragment crystallizable (Fc) antibody region. Other applications are for example depletion of IgG or albumin from serum and plasma samples (Fu et al. 2005; Hober et al. 2007). The selective affinity of SPA and SPG for different immunoglobulin types enables efficient isolation of specific antibody subclasses from an immunoglobulin mixture. SPA and SPG bind both to the Fc- and fragment antigen-binding (Fab)-portions of the antibody, the latter enabling purification also of antibody fragments (Akerstrom et al. 1985; Erntell et al. 1988; Jansson et al. 1998). The history behind these proteins, along with their structural and binding properties will be discussed in section 2. In this section we will also cover some applications of SPA and SPG in protein purification and related areas. As both proteins consist of repeated homologous domains, a natural development has been to investigate the utility of them individually. In section 3 we introduce how these domains have been generated and how they have found applicability in the protein purification field. With the recombinant DNA technology, it has become more feasible to create proteins with new properties and several improvements have been made to the domains of SPA and SPG regarding for example stability and binding specificity using rational design or combinatorial engineering. Modified domain variants have proven to be very useful as ligands in affinity purification of antibodies and as fusion partners for purification of target proteins. The engineered proteins have been used in a wide range of applications, including affinity chromatography and depletion. These efforts are presented in section 4, where we also discuss possible future developments.

Protein	Origin	Binding specificity
Staphylococcal protein A (SPA)	*S. aureus*	IgG, IgM, IgA of different species
S. aureus binder of IgG (Sbi)	*S. aureus*	IgG of different species (weak binding to IgM)
Streptococcal protein G (SPG)	Group C and G streptococci	IgG and albumin of different species
Protein M	Group A streptococci	Human IgG, IgA, albumin among others
Protein H	Group A streptococci	Human IgG
Protein Arp	Group A streptococci	Human IgA (weak binding to IgG)
Protein B	Group B streptococci	Human IgA
Protein L	*F. magna*	Human IgG
Peptostreptococcal albumin-binding protein (PAB)	*F. magna*	Human albumin

Table 1. Overview of some staphylococcal and streptococcal surface proteins that bind different immunoglobulin classes, albumin and other host serum proteins.

2. Protein A and protein G applied in protein purification

SPA and SPG represent the best-characterized bacterial surface proteins. Several structures of their immunoglobulin- and albumin-binding, in the case of protein G, domains have been solved. Species specificities and affinities of the full-length proteins as well as individual domains have been determined. Based on the interesting properties and accumulated knowledge regarding these proteins, they have found many different applications in the field of biotechnology. In this section, we will first present some background information on the proteins, before describing some examples of where the proteins have been utilized in different applications related to protein expression and purification.

2.1 Staphylococcal protein A

The interaction between SPA and IgG has been widely studied and SPA has for a long time been used as a tool in many biotechnological applications (Langone 1982). The molecule was discovered already in 1940, when extraction of cells of the J13 strain of *S. aureus* yielded an antigenic fraction, which was found to consist of proteins (Verwey 1940) and the protein received its name in 1964 (Oeding et al. 1964). It was observed in 1958 that SPA stimulated an immune response in rabbits, wherefore it was believed that SPA participated in an antigen-antibody interaction. However, it was later shown that the observed interaction between SPA and the immunoglobulin did not involve the antigen-binding site, but rather the constant Fc-region and the interaction was therefore denoted a "pseudo-immune" reaction (Forsgren & Sjoquist 1966). This interaction causes many immunological effects similar to an antigen-antibody interaction, including complement activation and hypersensitivity reactions (Martin et al. 1967; Sjoquist & Stalenheim 1969).

The gene for SPA was sequenced in 1984 (Uhlen et al. 1984) and the corresponding protein was shown to be a surface protein of about 58 kDa consisting of a single polypeptide chain. The protein can be divided into three regions with different functions. The N-terminal part consists of a signal peptide (Ss) followed by five homologous IgG-binding domains (E, D, A, B and C) and the C-terminal region (X and M) anchors the protein to the bacterial cell wall (Abrahmsen et al. 1985; Guss et al. 1984; Lofdahl et al. 1983; Moks et al. 1986; Schneewind et al. 1995; Uhlen et al. 1984), see figure 1.

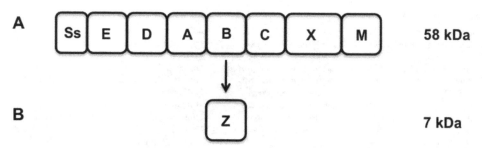

Fig. 1. (A) Organization of the different regions of SPA; An N-terminal signal sequence (Ss), which localizes the protein to the cell surface, five homologous IgG-binding domains (E, D, A, B and C) and two domains for anchoring of the protein to the cell wall (X and M). (B) The IgG-binding Z-domain, which is an engineered version of the B-domain, discussed in section 3.

SPA is produced by many strains of *S. aureus* and most of them typically produce a cell wall-bound variant. Usually about 85% of the protein is anchored to the cell wall whereas 15% exists as a soluble protein in the cytoplasm, however some strains produce the soluble variant exclusively (Movitz 1976). SPA is produced in the form of a precursor protein that contains a 36 amino acid N-terminal signal sequence, which directs the protein to the cell wall before it is cleaved off. There is a high sequence identity between the five IgG-binding domains. A "homology gradient" along the protein sequence has been established as two regions lying next to each other show a higher degree of sequence identity than two domains situated further apart. This indicates that the IgG-binding domains have evolved through step-wise gene duplications. The gene sequence of SPA reveals an unusually large number of changed nucleotides compared to changed amino acids, indicating that an evolutionary pressure has aimed to preserve the primary amino acid sequence (Sjodahl 1977; Uhlen et al. 1984). The sequence similarity between the five domains varies between 65-90% (Starovasnik et al. 1996), see figure 2. The E- and C-domains, situated closest to the N- and C-terminus, respectively, exhibit higher sequence dissimilarity when compared to the other domains. The C-domain seems to have diverged more to the cell wall anchoring part X, however without affecting the IgG-binding affinity (Jansson et al. 1998; Sjodahl 1977). Region X anchors the protein to the bacterial cell wall by binding to peptidoglycan with the N-terminus, thereby exposing the IgG-binding regions to the extracellular space (Schneewind et al. 1995; Sjodahl 1977; Ton-That et al. 1997).

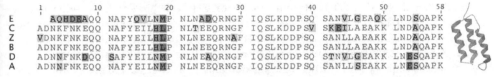

```
      1        10         20         30         40         50        58
E      AQHDEAQQ  NAFYQVLNMP  NLNADQRNGF  IQSLKDDPSQ  SANVLGEAQK  LNDSQAPK
C    ADNKFNKEQQ  NAFYEILHLP  NLTEEQRNGF  IQSLKDDPSV  SKEILAEAKK  LNDAQAPK
Z    VDNKFNKEQQ  NAFYEILHLP  NLNEEQRNAF  IQSLKDDPSQ  SANLLAEAKK  LNDAQAPK
B    ADNKFNKEQQ  NAFYEILHLP  NLNEEQRNGF  IQSLKDDPSQ  SANLLAEAKK  LNDAQAPK
D    ADNNFNKDQQ  SAFYEILNMP  NLNEAQRNGF  IQSLKDDPSQ  STNVLGEAKK  LNESQAPK
A    ADNNFNKEQQ  NAFYEILNMP  NLNEEQRNGF  IQSLKDDPSQ  SANLLSEAKK  LNESQAPK
```

Fig. 2. Sequence alignment of the five immunoglobulin-binding domains of SPA and the synthetic Z-domain. Differences are highlighted (Nilsson, B. et al. 1987). The structure of the domains is also shown. (Reconstructed from PDB structure 1Q2N) (Zheng et al. 2004).

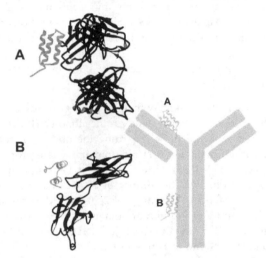

Fig. 3. Overview of binding sites of domains from SPA to (A) Fab and (B) Fc (Reconstructed from PDB structures 1DEE and 1FC2) (Deisenhofer 1981; Derrick et al. 1999; Graille et al. 2000; Lejon et al. 2004; Sauer-Eriksson et al. 1995).

The first crystallographic structure of the B-domain in complex with Fc showed a structure of two helices and the part corresponding to the third helix being irregularly folded (Deisenhofer 1981). However, further studies have been performed on several of the IgG-binding domains and a triple helix conformation has been determined in both the bound and unbound state. No significant difference in structure between the domains has been observed (Deisenhofer 1981; Gouda et al. 1992; Graille et al. 2000; Starovasnik et al. 1996) and only a minor side chain rearrangement of a phenylalanine occurs upon binding, as observed in the E-domain (Starovasnik et al. 1996). Each IgG-binding domain of SPA recognizes two separate binding sites on the immunoglobulin molecule located on the Fc and Fab parts, respectively (figure 3).

Eleven amino acids have been suggested to be important for the Fc interaction. All of them are situated in the first two helices of each IgG-binding domain and they are highly conserved within the five homologous regions (Deisenhofer 1981; Moks et al. 1986; Uhlen et al. 1984). Even though the third helix is not involved in Fc binding, it has been shown that it has structural importance. For example, truncated mutants of domain B where residues corresponding to the third helix are deleted have decreased affinity to Fc due to an overall loss in stability (Gouda et al. 1992). A region at the CH2 and CH3 interface of Fc interacts with SPA (Graille et al. 2000), but neither CH2 or CH3 can bind to SPA independently (Haake et al. 1982). Region E, which is situated closest to the signal peptide, differs the most from the other domains in this region (Moks et al. 1986). When comparing all five domains, domain B shows the lowest number of substitutions and its sequence can therefore be seen as a consensus sequence for all IgG-binding domains (Uhlen et al. 1984). Several binding studies have shown an approximate affinity to Fc of 5 nM for SPA and between 10-100 nM for the individual domains as determined by Surface Plasmon Resonance (SPR) (Karlsson et al. 1995; Nilsson, J. et al. 1994; Roben et al. 1995). A decreased affinity to Fc has been observed for the E-domain compared to the other four domains, which demonstrate similar affinities (Jansson et al. 1998; Moks et al. 1986). However, even though individual domains demonstrate similar affinities to Fc, a greatly increased apparent affinity is observed when several domains are combined (Ljungberg et al. 1993). Despite the fact that SPA contains five IgG-binding domains, it only has the capacity to bind on average 2.5 IgG molecules simultaneously (Ghose et al. 2007).

Besides IgG, SPA also interacts weakly with IgM and IgA. However, those low affinity interactions involve the Fab part of the antibody rather than Fc (Inganas 1981; Ljungberg et al. 1993). There have been contradictions regarding the ability of individual IgG-binding domains to bind to Fab, and initially only certain domains where proposed to be responsible for the binding (Ljungberg et al. 1993). However, it was later shown that all five domains bind both Fc and Fab individually. The affinity of SPA to Fab has been determined and dissociation equilibrium constants in the range of 2-200 nM, depending on the VH3 genes, have been reported. The affinity of individual domains is lower and reported numbers lay in the range 100-500 nM as determined by SPR (Jansson et al. 1998; Roben et al. 1995). The crystal structure of domain D bound to Fab of human IgM has been solved and the binding site is non-overlapping with the binding site to Fc. The binding site on the SPA domains involves highly conserved residues in the second and third helix and the loop in-between (Graille et al. 2000). The binding site on Fab involves residues situated on four β-strands in the VH region, see figure 3B. The interactions involve mainly polar side chains, as opposed to the binding between SPA and Fc, where the binding site is mainly hydrophobic (Graille et al. 2000). Only one position in each domain of SPA participates in both the interaction with Fc and Fab and its contribution is small in both cases. A single domain of SPA can bind to Fc and Fab simultaneously, as has been shown for domain D in an enzyme-linked immunosorbent assay (ELISA) (Roben et al. 1995) and for domain E in competition assays using affinity chromatography and calorimetry (Starovasnik et al. 1999).

IgG origin	SPA binding	SPG binding
Human	Yes*	Yes
Mouse	Yes**	Yes
Rat	No	Yes
Rabbit	Yes	Yes
Cow	Yes	Yes
Goat	No	Yes

Table 2. Some examples of IgG-binding specificities of SPA and SPG (* IgG1, IgG2, IgG4 and some IgG3, ** IgG2 an IgG3 but not IgG1).

SPA binds to IgG from different species, with varying affinity. In one study, a competition assay was used to analyze the binding to IgG of sera from 80 animals. (^{125}I)SPA was incubated with sera and the fraction of non-bound protein was analyzed by capturing the protein on IgG-coupled beads. The results showed a 10^6-fold variation in affinity between species (Richman et al. 1982). SPA binds to IgG from for example human, mouse and rabbit, but not rat (Reis et al. 1984; Richman et al. 1982), see table 2. The protein is not only species-specific, but also subclass-specific and binds for example to murine IgG2 but not IgG1 and to all human IgG subclasses except IgG3. The interaction between SPA and IgG does however not seem to be entirely subclass-specific. Some allotypes of human IgG3 bind to SPA, whereas some do not and a possible explanation includes an amino acid in a loop of CH3, which is either a histidine or an arginine (Haake et al. 1982; Reis et al. 1984; Scott et al. 1997). The affinity against human IgG1 has been shown to be higher than the binding against IgG2 and IgG4 (Reis et al. 1984) and binding is only observed to Fab parts from the human VH3 family and their homologues in other mammalian species. However, this is a common family from which about 50% of inherited VH genes originate (Graille et al. 2000).

2.2 Streptococcal protein G

SPG was discovered in 1973, when it was found that β-hemolytic streptococci carried IgG-binding proteins on their surfaces (Kronvall 1973). Different groups of streptococcal strains were determined to bind immunoglobulins with different affinities. Group A streptococci infect only humans, while group C and G streptococci also commonly infect animals (Myhre & Kronvall 1977). These observed differences led to the introduction of a classification system, where SPA was classified as a type I Fc-binding protein, SPG from group A streptococci as type II and proteins from human group C, G and L streptococci as type III proteins (Myhre & Kronvall 1977). There is no major difference between the binding characteristics of streptococcal proteins from groups II and III when it comes to human IgG subclasses (Kronvall et al. 1979). Bovine group G streptococci together form the type IV Fc receptor group. They show a limited specificity and bind weakly to human IgG (Myhre & Kronvall 1981). Group C streptococcal strains from the species *S. zooepidemicus* form another group, with properties similar to SPA with regard to human IgG specificity, however this group differs from the type I proteins when it comes to specificity to non-human IgG:s (Myhre & Kronvall 1980). The protein G molecules from groups C and G have the same principal structures and share high sequence similarity (Sjobring et al. 1991). Apart from binding to IgG, SPG can also bind serum albumin (Kronvall et al. 1979). There are many different group C and G streptococcal strains and they have been classified into three different groups based on size and binding patterns. SPG from certain strains have lost their serum albumin-binding capacity and the affinity against IgG differs about ten times between different strains. That protein G from all strains bind IgG, while the affinity against albumin has been lost in some strains, indicates that the evolutionary pressure on keeping the immunoglobulin-binding properties of SPG is greater than keeping the affinity against albumin. Hence, IgG-binding would seem to be essential for the bacteria, while binding to albumin seems less critical (Sjobring et al. 1991). Two strains that have been widely studied are G148, containing three IgG-binding and three serum albumin-binding domains and GX7809, containing two of each (Olsson et al. 1987). Apart from this, there is a very high sequence homology between SPG from the two strains (>99%) and only five mutations (including two silent) have been observed. This indicates a very recent divergence of the two proteins or a deletion of part of the gene (Olsson et al. 1987). In this chapter, we will focus primarily on SPG from group G streptococci and the Fc-binding proteins type III.

Based on the nucleotide sequences of G148 and GX7809 (Fahnestock et al. 1986; Guss et al. 1986), the protein has been divided into several regions. An N-terminal signal sequence (Ss), followed by a serum albumin-binding region (A1-A3) and an IgG-binding region (C1-C3) separated by a spacer region (S). The protein also includes a region that anchors it to the bacterial cell wall (W) (Akerstrom et al. 1987; Olsson et al. 1987), see figure 4. The C-terminal IgG-binding domains are denoted C1-C3 or B1-B2, depending on the strain. However, B1 and B2 are identical to C1 and C3, respectively (Achari et al. 1992; Fahnestock et al. 1986; Sauer-Eriksson et al. 1995), wherefore from now on in this text the domains will be referred to as C1-C3. The IgG-binding domains are very stable, despite the lack of stabilizing disulfide bonds (Achari et al. 1992). They each constitute 55 amino acids and are separated by two 15 amino acid spacers, D1 and D2 (Guss et al. 1986; Lian et al. 1992). There is only a two amino acid difference between C1 and C2, four additional substitutions exist between C1 and C3 and consequently four amino acids differ between C2 and C3 (Achari et al. 1992; Gronenborn & Clore 1993; Sauer-Eriksson et al. 1995), see figure 5. Despite the high sequence similarity, the affinity against IgG is different for the different domains (Lian et al. 1992). Similarly to SPA, there is a homology gradient between the domains, which indicates that they have arisen through gene duplications (Guss et al. 1986). It seems as if initially, one IgG-binding domain was split into the C1 and C3 parts with a spacer in-between. The C2-domain seems to originate from both of these domains; the N-terminal end from C1 and the C-terminal end from C3. This is further indicated as the two spacers D1 and D2 are identical and probably diverged relatively recently (Guss et al. 1986; Olsson et al. 1987). With this evolutionary explanation, one is inclined to believe that protein G from strain GX7809, which contains only two IgG-binding domains, represents a variant that stayed at the intermediate stage (Olsson et al. 1987). Comparisons made to the IgG-binding domains of SPA reveal no sequence homology even though the proteins compete for the same binding site on IgG and the two different IgG-binding domains are therefore an excellent example of

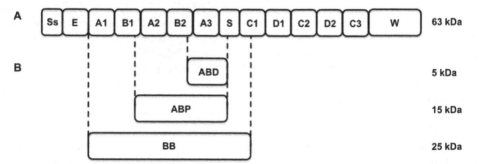

Fig. 4. (A) Organization of the different regions of SPG; An N-terminal signal sequence (Ss), which localizes the protein to the cell surface, a serum albumin-binding region (A1-A3) and an IgG-binding region (C1-C3) separated by a spacer region (S) and a part for anchoring of the protein to the cell wall (W). (B) Derivatives of the serum albumin-binding region that have been used in biotechnological applications.

```
      1        10          20          30          40          50     55
C2   TYKLVINGKT LKGETTTEAV DAATAEKVFK QYANDNGVDG EWTYDDATKT FTVTE
C1   TYKLILNGKT LKGETTTEAV DAATAEKVFK QYANDNGVDG EWTYDDATKT FTVTE
B1   TYKLILNGKT LKGETTTEAV DAATAEKVFK QYANDNGVDG EWTYDDATKT FTVTE
C3   TYKLVINGKT LKGETTTKAV DABTAEKAFK QYANDNGVDG VWTYDDATKT FTVTE
B2   TYKLVINGKT LKGETTTKAV DABTAEKAFK QYANDNGVDG VWTYDDATKT FTVTE
```

Fig. 5. Sequence alignment of the immunoglobulin-binding domains of SPG. Differences are highlighted (Fahnestock et al. 1986; Guss et al. 1986). The structure of the domains is also shown (Reconstructed from PDB structure 1FCC) (Sauer-Eriksson et al. 1995).

convergent evolution (Frick et al. 1992; Olsson et al. 1987). This is further strengthened by the notion that an eleven amino acid long peptide binding to Fc can inhibit binding of both SPA and SPG (Frick et al. 1992). The structure shared by the IgG-binding domains of SPG is different from that of the domains of SPA. Several studies have revealed it to be a four-stranded β-sheet of two β-hairpins connected by an α-helix and short loop regions (Achari et al. 1992; Gronenborn et al. 1991; Lian et al. 1992; Lian et al. 1991; Sauer-Eriksson et al. 1995).

SPG from strain G148 has three serum albumin-binding domains, whereas SPG from GX7809 has only two, each of about 46 amino acids. The differing number of repeats can be explained similarly as for the IgG-binding domains (Kraulis et al. 1996; Kronvall et al. 1979; Olsson et al. 1987). The binding sites of IgG and serum albumin on SPG are situated on opposite sides of the molecule and IgG cannot inhibit the binding of SPG to serum albumin (Bjorck et al. 1987). The structure of the albumin-binding unit is very similar to the structure of the IgG-binding domains of SPA, although the helices differ somewhat in length (Gouda et al. 1992; Kraulis et al. 1996). This interesting observation suggests a possible evolutionary relationship despite the lack of sequence homology (Kraulis et al. 1996). No structural data exists on the complex between human serum albumin (HSA) and the albumin-binding domains of SPG, however a complex between albumin and a highly sequence similar albumin-binding domain, the second GA (G-related albumin-binding)-module of PAB (sometimes referred to as ALB8-GA) has been determined using both NMR and crystallography. No significant structural change was observed upon binding for either of the two proteins (Cramer et al. 2007; Johansson et al. 1997; Lejon et al. 2004). The GA-module shows almost 60% sequence identity to the albumin-binding domains of SPG and it is even likely that the GA-module originates from these domains, wherefore this structural data is likely to also correspond to the albumin-binding domains of SPG (de Chateau & Bjorck 1994; de Chateau et al. 1996), see figure 6. At least 16 homologous albumin-binding domains from four different bacterial species have been identified (Johansson et al. 2002a). Taken together, this indicates that the fold, stability and mode of interaction of the homologs are very similar (Johansson et al. 1997; Lejon et al. 2004).

The interaction surface between Fc and the SPG-domain (C1-C3) can be divided into three centers; region I is a network of hydrogen bonds and consists of two residues in the center of the α-helix separated by three residues so that they are pointing in the same direction, region II also contains amino acids in the α-helix, very close to the two from region I and also separated by three residues. Region III includes one residue in the C-terminal end of the α-helix, two residues in the N-terminus of the third β-strand and two in the loop connecting

```
         1          10          20          30            40      45
         |          |           |           |             |       |
ALB8-GA  LKNAKEDAIA ELKKAGITSD  FYFNAINKAK  TV EEVNALKN   EILKA
ABD3     LAEAKVLANR ELDKYGV SD  YYKNLINNAK  TV EGVKALID   EILAA
ABD1     LAKAKADALK EFNKYGV SD  YYKNLINNAK  TV EGVKDLQA   QVVES
ABD2     LAEAKVLANR ELDKYGV SD  YHKNLINNAK  TV EGVKDLQA   QVVES
```

Fig. 6. Sequence alignment of the albumin-binding domains of SPA together with the second GA-module derived from *F. magna* (ALB8-GA). Differences are highlighted and only the 44-45 amino acid motifs that are most highly conserved are displayed. The structure of the domain is also shown (Reconstructed from PDB structure 1GJT) (Johansson et al. 2002a).

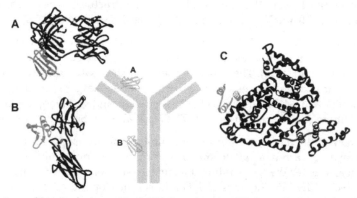

Fig. 7. Overview of binding sites of IgG-binding domain C2 from SPG to (A) Fab and (B) Fc. The binding of the GA-module of PAB to HSA is shown in (C) (Reconstructed from PDB structures 1QKZ, 1FCC and 1TF0) .

them (Sauer-Eriksson et al. 1995). SPG-domains bind to the cleft between CH2 and CH3 (figure 7A), as opposed to the domains of SPA, which bind more on the CH2 side of Fc. However, in the complex, the third strand of the SPG-domain is situated approximately in the same region as the first Fc-binding helix of the SPA-domain. Consequently, SPA and SPG cannot simultaneously bind the same Fc-molecule (Stone et al. 1989). Furthermore, the interactions between SPG and Fc involve many charged and polar residues, forming hydrogen bonds and salt bridges while the binding between SPA and Fc involve mostly hydrophobic interactions (Sauer-Eriksson et al. 1995). The strength of the binding to Fc has been determined using SPR to around 20-100 nM for the C2-domain and low nanomolar values for the whole SPG molecule have been reported (Akerstrom & Bjorck 1986; Gulich et al. 2002; Sagawa et al. 2005). The binding of SPA and SPG to Fc is pH-dependent. SPG binds most efficiently to IgG at pH 4-5 and the binding is weakened with increased pH. SPA on the other hand, binds with highest affinity at pH 8 (Akerstrom & Bjorck 1986). This difference is due to the composition of the interaction interfaces. C3 has higher affinity to Fc than C1 and C2, all differences in binding affinity have been explained based on structural data (Lian et al. 1992; Sauer-Eriksson et al. 1995). The difference in binding can partly be explained by the existence of a carboxylic acid in the binding site of SPG-domain C3, a position in the α-helix, which is an alanine in C1 and C2 (Achari et al. 1992; Sauer-Eriksson et al. 1995). It has also been speculated that the substitution Glu/Val located on the third β-strand in C1/C3 would improve binding to Fc, this has however been argued (Gronenborn & Clore 1993; Sauer-Eriksson et al. 1995).

Similarly to SPA, SPG also has the capability to bind Fab although the binding strength is about ten times weaker than the affinity to Fc (Bjorck & Kronvall 1984; Erntell et al. 1983; Erntell et al. 1988; Sagawa et al. 2005). The affinity constant for domain C2 has been determined to around 150 nM (Sagawa et al. 2005). As for SPA, the binding sites to Fc and Fab are not overlapping. The α-helix does not participate in the Fab-binding, but instead the second β-strand forms an extended β-sheet structure with the last β-strand of the CH1-domain of Fab, see figure 7B, and the interactions are mainly between main chain atoms (Sauer-Eriksson et al. 1995). The third IgG-binding domain of SPG has been analyzed in a crystal complex with Fab and forms twelve hydrogen bonds with Fab residues. Eight of these are between SPG and main-chain atoms on the Fab CH1-domain, contributing to the broad specificity of SPG to different IgG subclasses and species. The remaining four hydrogen bonds are between SPG and side chain atoms on the Fab CH1-domain, and these amino acids are highly conserved among γ heavy chain subclasses (Derrick & Wigley 1994). SPG binds well to IgG whereas, in contrast to SPA, no binding has been observed to IgA or IgM (Achari et al. 1992; Bjorck & Kronvall 1984; Kronvall et al. 1979). However, SPG has broader subclass specificity than SPA and it binds to all four subclasses of human IgG, whereas SPA shows no affinity against certain human allotypes of IgG3. A histidine residue in a loop on CH3, situated in the binding interface of SPA and IgG, may explain this observation. Histidine possibly blocks the interaction and in allotypes of IgG3 binding to SPA, this amino acid has been substituted for an arginine. In SPG, this residue does not participate in the Fc-binding, which explains why SPG binds well to all IgG3 allotypes (Haake et al. 1982; Sauer-Eriksson et al. 1995; Shimizu et al. 1983). SPG has also been shown to bind more strongly than SPA to IgG from several species including human, mouse, rat, cow, rabbit and goat (Akerstrom & Bjorck 1986; Akerstrom et al. 1985; Fahnestock et al. 1986). However, there have been contradictions to this statement and results showing no significant difference in the binding strengths of SPA and SPG have also been published (Guss et al. 1986; Kronvall et al. 1979). The abilities of SPA and SPG to bind a large number of IgGs from different species were determined using a competitive ELISA setup (GE Healthcare, Antibody Purification Handbook 18-1037-46, 2007). From this study none of the proteins could be denoted the overall superior binder; they have different advantages.

Early biochemical analysis of the interaction between SPG and HSA revealed that the binding site involved mainly the second domain of HSA and engages loops 7-8 (Falkenberg et al. 1992). Isolation of a small fragment of HSA corresponding to these residues inhibited the binding between intact HSA and SPG (Falkenberg et al. 1992). This finding also suggested that all albumin-binding domains of SPG share the same epitope on HSA, which is in concordance with the evolutionary hypothesis that gene duplications gave rise to the multiple homologous domains. Each of the three albumin-binding regions in SPG is approximately 5 kDa in size (Johansson et al. 2002a). The third domain has been most extensively investigated. It consists of 46 amino acids folded into a three helical bundle as determined by nuclear magnetic resonance (NMR) (Kraulis et al. 1996). The sequence is devoid of cysteines and the structure does not depend on any stabilizing factors such as bound ligands, metal ions or disulphide bonds. This is true also for the larger constructs containing two and a half (ABP) or three (BB) sequential domains (Stahl & Nygren 1997), see figure 4. NMR-perturbation studies have indicated that the albumin-binding residues are mainly localized to the second helix and the loop preceding it (Johansson et al. 2002b). Those observations are in agreement with a mutational analysis of the binding site to HSA on the albumin-binding domain from SPG (Linhult et al. 2002). By comparing the binding of different point mutants as well as a few double and one triple mutant, the binding site was deduced to reside mainly in the second helix. The first helix does not take part in the binding and only small parts of the third helix are involved (Linhult et al. 2002). It has also been demonstrated that a variant with a truncated N-terminus has a significantly reduced affinity against HSA (Johansson et al. 2002a). Only five N-terminal amino acids differ somewhat in a crystal structure of the domain compared to the NMR-data. In the crystal structure of the GA module, amino acids are more ordered and extend the first helix by an additional turn. The GA-module mainly binds to the surface of domain II of HSA. The structural data from the complex between the GA-module and HSA shows that the binding interface is of hydrophobic nature with two bordering hydrogen bond networks (Lejon et al. 2004). The binding surface is centered around a tyrosine of the GA-module in a hydrophobic cleft on HSA. This residue, as well as the flexibility of the backbone structure, has been linked to the broader host specificity of the albumin-binding domain derived from SPG compared to the GA-module from *F. magna* (He et al. 2006; Johansson et al. 2002b; Lejon et al. 2004). Bacteria that express the GA-module have only been isolated from humans, whereas streptococci expressing SPG are known to infect all mammalian species (Johansson et al. 2002b). This may also explain the higher affinity of the GA-module against HSA as compared to the domains derived from SPG (Johansson et al. 2002a). The third albumin-binding domain of SPG binds strongly to human, mouse, rabbit and rat serum albumin, among others, with a low nanomolar affinity as determined by SPR. It binds less efficiently to hen and horse serum albumin and only weak or no binding is observed to albumins of bovine origin (Falkenberg et al. 1992; Johansson et al. 2002a; Linhult et al. 2002; Nygren et al. 1990; Raeder et al. 1991). When comparing affinities against IgG and albumin from different species, it is common that strong binding to one of the molecules means weaker binding to the other. The only exception is man, as SPG binds well to both human IgG and human albumin (Nygren et al. 1990).

2.3 Utilization of full-length SPA and SPG in protein purification

SPA and SPG are today used in a number of applications concerning protein purification. The most widespread application is the use of SPA and SPG coupled to chromatography resins for the purification of antibodies and Fc-tagged recombinant proteins (Lindmark et al. 1983; Ohlson et al. 1988). Even though both SPA and SPG bind to antibodies, as mentioned above they bind with different specificities to immunoglobulins of various species and subclasses. This makes the proteins suitable for slightly different applications. Furthermore, SPA has a higher stability than SPG, making it more suitable for large-scale, industrial applications (Boyle 1990; Hober et al. 2007). SPG is usually the protein of choice when isolation of the total IgG fraction of a sample is desired. Due to its broader specificity, SPG would generate a higher yield of antibody from a sample containing for example human antibodies of different subclasses. On the other hand, SPA can be a better choice for isolation of specific subclasses of IgG. SPA has been reported to separate mouse IgG1, IgG2a and IgG2b in pure fractions. Surprisingly, in this study, mouse IgG1 was found to bind to the SPA columns under certain conditions with very high salt concentrations. The binding of SPA to IgG is pH sensitive and different pH were used to separately elute IgG1, IgG2a and IgG2b from the column (Ey et al. 1978). Affinity membranes with SPA or SPG have also been used to purify human and murine IgG (Dancette et al. 1999). An important issue in chromatography is the possibility to clean the column after purification to be able to reuse it. This is most efficiently done with high concentrations of NaOH, but unfortunately this is a problem in affinity chromatography since it often results in denaturation of proteinacious ligands. However, SPA has been shown to cope well with high NaOH concentrations with only a small decrease in binding capability (Hale et al. 1994). Use of SPA has been evaluated in therapeutic applications as well, for example in the treatment of autoimmune disorders. Patients with an autoimmune disease produce autoantibodies, which can be removed from the blood by plasma exchange. A simpler way to remove the antibodies, without unwanted removal of all other serum components, is immunoadsorption. IgG is selectively removed from serum by immobilized SPA, however whether or not this is the reason for the success of the treatment is controversial. Two products are currently available on the market for immunoadsorption; the ProSorba® column and the ImmunoSorba® column, which are both accepted by the food and drug administration (Matic et al. 2001; Poullin et al. 2005; Silverman et al. 2005).

An early study demonstrated the successful use of SPG in a western blot setup, where an iodine-labeled variant of SPG was used as a secondary detection reagent. A mixture of antigens was separated on a sodium dodecyl sulfate polyacrylamide gel and transferred to a nitrocellulose membrane. Primary antigen-specific antibodies were bound to the membrane before the iodine-labeled SPG was added for detection (Akerstrom et al. 1985). SPA and SPG have also been used in immunocapture, with the aim to capture target proteins from a complex sample. Antigen-specific antibodies are cross-linked to an SPA or SPG matrix, which ensures correct orientation and hence no blocking of antigen binding sites, leading to a higher yield of immunocaptured material (Kaboord & Perr 2008; Podlaski & Stern 2000; Sisson & Castor 1990).

When analyzing serum samples, the problem arises that there is a huge difference in abundance of different proteins. A concentration range spanning ten orders of magnitude from the least to the most abundant proteins, represented by IgG and albumin, has been reported. This makes it hard to analyze the low abundant proteins, which are often of interest in for example biomarker analysis or plasma profiling (Anderson & Anderson 2002). Depletion of IgG using SPG (Faulkner et al. 2011; Fu et al. 2005) prior to further analysis is common to decrease the complexity of a sample. Different types of matrices are commonly used, such as porous particles, monoliths and affinity membranes (Urbas et al. 2009). There are several products on the market using natural, recombinant or even stabilized derivatives of SPA and SPG on affinity media. The media are mostly agarose, sepharose or acrylamide (Grodzki & Berenstein 2010; Hober et al. 2007). A different approach has been evaluated for depletion of serum proteins using antibody fragments, in combination with SPA and SPG. Specific antibody fragments bind serum proteins and the complexes are subsequently captured using a combined SPA and SPG resin (Ettorre et al. 2006).

SPA has been used as a fusion partner to simplify production and purification of recombinant proteins (Nilsson, B. et al. 1985). SPA is a stable protein and generates functional fusion proteins when produced in different bacterial hosts (Abrahmsen et al. 1985; Nilsson, B. et al. 1985). Different variants of SPA have been used; often the X region has been deleted to hinder the protein from being incorporated into the bacterial cell wall (Nilsson, B. & Abrahmsen 1990; Uhlen et al. 1983). Several properties of SPA make it a good fusion partner; the acidic properties of SPA can help stabilize basic proteins at neutral pH

(Nilsson, B. & Abrahmsen 1990). SPA does not contain any cysteines, which could otherwise interfere with the target protein through formation of disulphide bridges (Uhlen et al. 1983). The secretion signal associated with SPA enables secretion of the fusion protein if the membrane anchoring region of SPA is deleted (Abrahmsen et al. 1985). Introduction of a cleavage site between SPA and the target protein enables cleavage of SPA after purification (Nilsson, B. & Abrahmsen 1990). SPA has also been used as a fusion partner to antigens in the production of antibodies by immunization of animals. In this context the protein acts as an adjuvant to increase the immune response towards the antigen (Lowenadler et al. 1986).

3. Domains of bacterial surface proteins

Both SPA and SPG are multi-domain proteins with several domains filling the same function. A protein domain is defined as the smallest structural unit that alone possess characteristics that are associated with the whole protein (Holland et al. 2006). It can fold independently, and should have the same conformation as when included in the whole protein. Furthermore, a protein domain should be able to function on its own. All IgG- and serum albumin-binding domains of SPA and SPG have these properties and can therefore be used individually. Protein domains have several advantages compared to their full-length ancestors, wherefore a natural development has been to utilize single or multiple IgG- or albumin-binding domains from SPA and SPG as replacements for the full-length proteins. One advantage is their small size, which both decreases the protein production cost and simplifies the production procedure, for example due to a more straightforward folding process. Protein domains also have the advantage of easy characterization, for example structural and binding studies are more easily performed with smaller proteins. The isolated binding property contained in a small domain enables efficient use as capture ligands on columns for affinity chromatography. Alternatively, protein domain(s) may be fused to a recombinant target protein to facilitate recovery by affinity purification on easily prepared IgG or albumin media (Nygren et al. 1988). However, commercial matrices are not widely available, perhaps due to the harsh elution conditions required. The smaller domains have many additional advantages, which make them favorable as fusion partners. (I) Domains of SPA or SPG have surface-exposed termini, assuring that the fusion tag will not interfere with the structure of the fusion protein. (II) They do not contain any cysteines that can form disulphides with the fusion protein and interfere with the folding process. (III) The domains are highly soluble and refold easily after treatment with denaturants, which can aid the refolding of the fusion partner. (IV) Fusion proteins can be produced at high levels in *Escherichia coli* and still remain soluble. (V) It is generally easy to insert cleavage sites for proteases between the fusion tag and the target protein, which enables recovery of native protein. Different hosts have successfully been applied for production of fusion proteins with domains of SPA and SPG, including gram-positive and –negative bacteria, yeast, plant, mammalian and insect cells (Stahl et al. 1997). Protein domains from SPA and SPG, and commonly slightly modified versions of these, are also frequently used as affinity ligands for purification of antibodies, antibody fragments and Fc-fused proteins, which is a common strategy to express proteins in mammalian hosts (Hober et al. 2007; Ljungquist et al. 1989; Lo et al. 1998). In the case of SPG, another advantage with using single domains instead of the full-length protein is the fact that SPG is a dual affinity protein, binding to both IgG and albumin. This dual affinity is a drawback for example in antibody purification, as antibodies are commonly purified from serum in which serum albumin is present at high abundance. Several examples where advantageous properties of domains of SPA and SPG are exploited are given below.

In this section we focus on different approaches that utilize mono- or multi-domain derivatives of the IgG-binding domains from SPA or the albumin-binding domains from SPG. Those domains have been most widely explored in the context of protein expression and purification. The most thoroughly studied and used IgG-binding domain from SPA is the B-domain; from which the synthetic stabilized Z-domain has been designed. Although the Z-domain is by definition a synthetic domain, its widespread use in a large range of applications makes it a natural focus of this section. In the brief part that follows, the emphasis is on different uses of the immunoglobulin-binding domains of SPG, which have found their main application areas as models in studies of protein folding and dynamics rather than within the field of protein purification. Of the albumin-binding domains of SPG, the third domain has been investigated and utilized the most. Therefore, the use of this domain is the main topic for the section covering domains of SPG.

3.1 Domains of protein A

As mentioned above, each IgG-binding domain of SPA independently folds into a three-helix bundle that can bind to the Fc or Fab region of an antibody. All five IgG-binding domains of SPA have high sequence identity (figure 2), although when comparing them one by one, the IgG-binding domain B was found to contain the least substitutions and it may therefore be seen as a consensus sequence of the IgG-binding domains (Uhlen et al. 1984). A pair of modifications has been introduced into the B-domain, with the aim to increase its stability and potential as a fusion partner. The modified variant of the B-domain has been given the name Z (Nilsson, B. et al. 1987).

3.1.1 The Z-domain derived from the B-domain of protein A

The B-domain of SPA is the most thoroughly studied of the five IgG-binding domains and has been subject to rational improvements yielding the synthetic Z-domain. Two amino acids have been changed, mainly to increase the chemical stability of the protein. An Asn28-Gly29 dipeptide has been changed to Asn28-Ala29 to ensure resistance to hydroxylamine (Nilsson, B. et al. 1987). This facilitates efficient removal of Z after purification, by the introduction of a hydroxylamine cleavage site in the joint between Z and the fusion protein. As the asparagine in the dipeptide cleavage site is believed to be involved in Fc binding, the glycine was instead mutated. The Z-domain also lacks methionines, which makes it stable against proteolytic cleavage with cyanogen bromide (Nilsson, B. et al. 1987). To facilitate the cloning procedure, an *AccI* cleavage site was introduced by exchanging Ala1 to Val1, situated outside the first helical region (Nilsson, B. et al. 1987). All of the five native IgG-binding domains of SPA exhibit binding to both Fc and Fab (Jansson et al. 1998). However, the modifications incorporated in the B-domain to produce Z resulted in a loss of affinity to the Fab-part of the antibody, although the Z-domain retains its Fc-binding capacity along with high stability and solubility. Binding studies show that neither Z, ZZ or a pentameric variant of it bind Fab (Ljungberg et al. 1993). The Fab-interaction involves the second and the third helix of the IgG-binding domains of SPA, and position Gly29 has been shown to be important (Graille et al. 2000). In the Z-domain, this position is mutated to an alanine, which could explain the loss of binding. As many as ten repeat domains of Z in succession have been expressed in bacteria, the long construct was however susceptible to some homologous recombination even in a RecA-negative host (Nilsson, B. et al. 1987). This observation may explain the high frequency of silent mutations found in the native SPA-gene and suggests a selection pressure to avoid homologous recombination of the regions encoding the domains (Nilsson, B. et al. 1987).

As the original B-domain, the Z-domain consists of 58 amino acids. Despite the substitutions in Z compared to the B-domain of SPA, the structures are very similar and the Z-domain has also been determined to be a three-helical bundle (Jendeberg et al. 1996; Tashiro et al. 1997). Helices two and three are situated in an anti-parallel fashion and the first helix is anti-parallel to the second helix but slightly tilted. NMR has been used to evaluate the conformation of Z and the dimer ZZ in solution and circular dichroism (CD) spectroscopy has been used to investigate structural changes upon binding to Fc. Only minor structural changes were observed in both the monomer and the dimer during complex formation. In addition, both the bound and unbound states were shown to contain a structured third helix (Jendeberg et al. 1996), as opposed to the original crystal complex where the third helix is not well resolved (Deisenhofer 1981). In several studies, the dimer ZZ has been used instead of the monomeric Z-domain. ZZ has been shown to bind more strongly to Fc than does the monomeric Z, due to a lower off-rate achieved through the avidity effect (Nilsson, J. et al. 1994). ZZ has been suggested to be a preferred arrangement for many applications, which yields strong Fc binding in combination with efficient secretion and small overall size (Ljungquist et al. 1989; Nilsson, B. et al. 1987).

The interaction between the Z-domain and human IgG1 has been further investigated, for example by the construction of four single amino acid mutants. Amino acids were chosen that were thought to be important in the binding surface, based on structural data from the crystal complex (Deisenhofer 1981). The mutants were evaluated in a competition assay where radioactively labeled Z was used as a tracer. All four mutants were found to have a decreased affinity against IgG1 compared to Z, which led to the conclusion that positions Ile31, Lys35, Leu17 and Asn28 are important for Fc-binding (Cedergren et al. 1993). Those results also confirmed the importance of Asn28 for binding; this position is found in the hydroxylamine site that was altered as part of the development of Z. It was later shown that the kinetics of the interaction of Fc with the B- or Z-domain were indeed identical (Jendeberg et al. 1995; Starovasnik et al. 1996).

The Z-domain has found use in several ways in the field of protein purification, mostly as a fusion tag for efficient production and purification of recombinant target proteins. Usage of the Z-domain as a fusion tag enables production and purification of recombinant proteins with very high yields. The majority of proteins that have been purified fused to the Z-domain have been produced as soluble proteins and several examples exist where ZZ-fusions have facilitated the recovery of proteins secreted into the periplasmic space or to the culture medium (Hansson et al. 1994; Uhlen et al. 1992). For example, human insulin-like growth factor II was produced as a secreted fusion to ZZ and affinity purified on IgG Sepharose (Wadensten et al. 1991). In a similar strategy, a secreted protein built up from a repeat-structure of a malaria antigen (M5) tagged with a dimeric Z-tag could be recovered from the culture medium by expanded bed adsorption and ion exchange chromatography. The initial capture was followed by a polishing step by affinity purification facilitated by the

IgG-binding fusion (Hansson et al. 1994). Z displays fast kinetics, enabling the use of high flow rates and columns with immobilized IgG can be reused many times (Uhlen & Moks 1990). Fusion proteins including the Z-domain are easy to detect by immunoblotting after purification, as Z binds to antibodies normally used in these setups (Stahl et al. 1997). The Z-domain has been used as a solubilizing fusion partner as it folds easily and may therefore aid the *in vitro* folding process of proteins with complex folding patterns. It has also been used as a fusion partner to insulin-like growth factor I (IGF-I), a protein with a complicated folding pattern that involves formation of three disulfide bonds. The fusion tag was shown to confer a higher overall solubility to IGF-I, which was shown to be at least 120 times more soluble when fused to either Z or the dimer ZZ. In addition, Z also decreased the degree of multimerization of IGF-I (Samuelsson et al. 1994; Samuelsson & Uhlen 1996; Samuelsson et al. 1991). The Z-domain has also been used in the production of very insoluble proteins in the form of inclusion bodies. As IgG-sepharose columns are resistant to 0.5 M guanidine hydrochloride, it is possible to perform the purification step in the presence of a chaotropic agent, which keeps the target proteins in a soluble state (Stahl et al. 1997). The D-domain of SPA has also recently been shown to function as a solubility and stability enhancing tag (Heel et al. 2010).

Competitive elution protocols have been developed as a milder alternative to the strategies normally used. This concept has been proven effective for Z-fusion proteins eluted with bivalent ZZ, which has a roughly 10-fold higher apparent affinity as a result of avidity effects (Nilsson, J. et al. 1994). The feasibility of the competitive elution strategy was demonstrated for a Z-fusion to the Klenow fragment of DNA polymerase I expressed in *E. coli* (Nilsson, J. et al. 1994). The competitor in this study was also tagged by a dimeric albumin-binding domain to facilitate effective removal after elution from the IgG column by capture of the competitor on an HSA column, without interfering with the tag still present on the final product. This approach should in principle be applicable to fusions to an albumin-binding domain as well, provided the purification steps are used in the reverse order for a protein tagged with a monovalent albumin-binding tag. It is possible to recombinantly introduce a proteolytic cleavage site in the joint between the Z-domain and the recombinant protein to enable removal of the fusion tag after purification. Several efficient cleavage agents have been identified for removal of Z (Forsberg et al. 1992). In one study ZZ was fused to proinsulin and three different short linkers containing trypsin cleavage sites were introduced between the tag and the target protein (Jonasson et al. 1996).

The Z-domain can also be utilized for purification of antibodies or Fc-fused target proteins, similarly to the full-length SPA or SPG. However, the IgG-binding domains may more easily be engineered to facilitate site-directed immobilization on a solid support. For example thiol-directed immobilization has been employed, where a C-terminal cysteine was recombinantly introduced to enable immobilization of Z, ZZ or pentameric Z. The C-terminal residue had little impact on the binding capacity for Fc, determined by measuring the amount of protein eluted from the column. This strategy is advantageous since the ligands are correctly oriented and no ligands are truncated since an intact C-terminus is required for coupling to the column (Ljungquist et al. 1989). Furthermore, the Z-domain has been used as a means for site-directed immobilization of antibodies on cells or viruses. For example, yeast cells have been engineered to express a dimeric form of the Z-domain on the cell surface (Nakamura, Y. et al. 2001). The engineered cells were applied as renewable immunosorbents for affinity purification of antibodies from serum. In addition, cells expressing Z were used for detection of antigens, after a primary incubation of the sample with target-specific immunoglobulins. Other examples include the capture of antibodies on phage (Mazor et al. 2010), and display of Z on baculovirus (Ojala et al. 2004) or *E. coli* (Mazor et al. 2008; Mazor et al. 2007).

3.2 Domains of streptococcal protein G

The immunoglobulin-binding domains C1-C3 have been expressed, purified and studied independently (Akerstrom et al. 1985; Akerstrom et al. 1987). Whereas the immunoglobulin-binding domains of SPA have been the subjects of a large number of studies, the domains of SPG conferring the same binding activity have however not been as extensively investigated in the context of bioseparation. They have been utilized for antibody purification, mostly due to the broader subclass specificity compared to SPA. However, the immunoglobulin-binding domains of SPG have been best characterized and utilized as models to deepen the understanding of protein folding and dynamics. Regions responsible for albumin binding have also been isolated (Nygren et al. 1988). Fragments spanning two and a half (BB) or three (ABP, albumin-binding protein) of the albumin-binding motifs of SPG have been expressed and characterized (Larsson et al. 1996; Nygren et al. 1988). A smaller albumin-binding segment that has been widely studied alone comprises the third albumin-binding repeat flanked by a few amino acids from the B2- and S-regions, respectively (Nygren et al. 1990). This molecule is referred to as the albumin-binding domain (ABD) in the text.

3.2.1 The IgG-binding domains derived from SPG

SPG contains three homologous IgG-binding domains, referred to as C1-C3. The immunoglobulin-binding domains each consist of 55 amino acids and fold into a four-stranded β-sheet connected by an α-helix and short loops (Akerstrom et al. 1985; Lian et al. 1992). In analogy to the albumin-binding domains of SPG and the IgG-binding domains of SPA, C1-C3 are unusually stable to harsh thermal or chemical treatment and can be effectively refolded after denaturation (Alexander, P. et al. 1992). Each domain comprises non-overlapping binding-sites for both the Fc- and the Fab-regions of IgG from several subclasses (Erntell et al. 1988; Lian et al. 1992). The subclass specificity of SPG is broader than for SPA since the immunoglobulin-binding domains also bind IgG3 (Bjorck & Kronvall 1984). SPG has been widely used for purification of immunoglobulins or antibody fragments (Akerstrom et al. 1985; Cassulis et al. 1991; Hober et al. 2007).

The immunoglobulin-binding domains of SPG have not been as extensively used as gene fusions or ligands for affinity capture as the domains of SPA, perhaps as a result of the later identification of SPG and lower tolerance to alkaline conditions of the immunoglobulin-binding domains compared to SPA. However, SPG is widely used for purification of antibody fragments and the inherent tolerance for chaotropic agents facilitates rigorous cleaning (Winter et al. 1994). A few diverse examples of fusions to C-domains are exemplified here to illustrate some additional applications. The C1-domain has been used to increase the expression levels and aid in refolding of small recombinant proteins or peptides (Cheng & Patel 2004; Nadaud et al. 2010; Pazehoski et al. 2011). A repeat of the C3-domain has in another study been combined with luciferase to form a fusion protein with ability to detect antibodies bound to bacteria through a light-emitting reaction (Nakamura, M. et al. 2011). Another fusion strategy produced an adherent protein able to capture antibodies in microwells when a hydrophobic domain of elastin was combined with an immunoglobulin-binding domain from SPG (Tanaka et al. 2006). As mentioned above, the use of the C1-C3 domains of SPG has been more focused around basic biophysical questions. Since the initial structural characterization of the C1-domain (Gronenborn et al. 1991) all three IgG-binding domains of SPG have become popular model systems for studies on protein stability, folding, structure and dynamics (Alexander, P. et al. 1992; Clore & Schwieters 2004; Derrick & Wigley 1994; Franks et al. 2005; Hall & Fushman 2003; Ulmer et al. 2003). The vast number of studies within those fields have been reviewed elsewhere and lie out of the scope of this chapter. The surprising structural similarity between the albumin-binding domains of SPG and the immunoglobulin-binding domains of SPA (Falkenberg et al. 1992) has motivated several studies where the folding patterns and sequence-structure relationships have been experimentally dissected. Interestingly, it was recently demonstrated that a domain with the same immunoglobulin-binding fold as found in C1-C3 could be transformed into a three-helix bundle domain, similar to the albumin-binding domains of SPG, with acquired affinity against albumin through a defined mutational pathway (Alexander, P. A. et al. 2009; He et al. 2005).

3.2.2 The albumin-binding domains derived from SPG

Different regions of the albumin-binding part of SPG have been affinity purified by an effective one-step HSA-chromatography protocol (Nygren et al. 1988). This method has also been applied to a wide range of proteins fused to different albumin-binding fragments of SPG. Those albumin-binding affinity tags have, in analogy to SPA-based tags, been shown to be proteolytically stable, highly soluble and possible to produce in high yields (Larsson et al. 1996; Nilsson, J. et al. 1997b; Nygren et al. 1988; Stahl et al. 1989). Due to the harsh conditions required to elute tightly bound proteins from HSA columns, different approaches using milder routines have been evaluated. The low pH most often applied for elution may be harmful for the fusion partner of interest. For ABP-fusion proteins, other elution strategies including heat (Nilsson, J. et al. 1997a), high pH (Makrides et al. 1996) and lithium diiodosalicylate (Lorca et al. 1992) have been successfully investigated. Furthermore, the different binding affinities measured for albumin from different species has also been proposed as a means to achieve milder elution conditions by for example using albumin from mouse as the affinity ligand instead of the human equivalent (Nygren et al. 1990). The albumin-binding fragments of SPG that have been studied are easily refolded and retain activity after harsh treatment (Oberg 1994, as cited in Kraulis et al. 1996). This property has been utilized to facilitate recovery and refolding of fusion proteins from inclusion bodies (Murby 1994, as cited in Murby et al. 1996; Stahl & Nygren 1997). It can sometimes be an advantage to produce proteins as inclusion bodies since high production yields can be achieved and the insoluble proteins are protected from proteolysis (Murby et al. 1996). An ABP-fusion has been combined with hydrophobicity engineering to express and recover a slightly modified variant of a very insoluble and easily degraded fragment of the human respiratory syncytial virus (RSV) major glycoprotein G (Murby et al. 1995). In another study (Murby 1994, as cited in Murby et al. 1996), efficient recovery was demonstrated in the presence of chaotropic agents (0.5 M guanidine hydrochloride) for precipitation prone fragments of the fusion glycoprotein F from the same virus expressed as ZZ- or BB-tagged fusions.

Sometimes removal of the fusion tag is necessary to obtain a product of desired quality. Several chemical and enzymatic methods for tag removal have been devised (Arnau et al. 2006; LaVallie et al. 2001). Chemical methods are often scalable and relatively inexpensive, they may however produce side-chain modifications, denaturation of the target protein or be too unspecific to be generally applicable to many larger proteins (Parks et al. 1994). In general, more specific agents, such as proteases, are required to avoid unwanted cleavage in the coupled target protein. Strategies where the protease carries the same tag as it cleaves off from the protein of interest have been described. This facilitates simultaneous capture of enzyme and cleaved tag in a single step that leaves pure target protein. Such systems are exemplified by 3C protease in the PreScission system (Walker et al. 1994) and a similar approach was developed for use with the ABP-tag fused to both the protease and the target protein (Graslund, T. et al. 1997). Recent developments include the Profinity Exact system where tagged proteins are captured by a modified subtilisin that is subsequently activated to specifically cleave off, but retain, the tag and release the pure target protein (Bio-Rad Laboratories, Hercules, CA).

A maintained tag can sometimes be a way to achieve directed immobilization or detection. This has been demonstrated for several formats involving different parts of the albumin-binding regions of SPG (Baumann et al. 1998; Konig & Skerra 1998; Stahl et al. 1989). Albumin-binding fusions have also found interesting applications *in vivo* for delivery of subunit vaccines or protein therapeutics (Nilsson, J. et al. 1997b; Sjolander et al. 1997; Stahl & Nygren 1997). Fusion of the BB-fragment of SPG to CD4 resulted in stabilization of the protein *in vivo* in mice as well as in macaques (Nygren 1991, as cited in Sjolander et al. 1997). The same concept has been evaluated in rats for the human soluble complement receptor type 1 (Makrides et al. 1996). Those early attempts suggested that the minimal binding motif ABD might also be useful to improve the *in vivo* stability of proteins. Several recent studies have indeed demonstrated that fusions to albumin-binding domains efficiently prolong the half-life of the fused protein *in vivo* (Andersen et al. 2011; Hopp et al. 2010; Nilsson, F. Y. & Tolmachev 2007; Stork et al. 2009). Albumin-binding proteins may also have immunopotentiating properties when used as carriers for a fused immunogen. This concept was originally evaluated for malaria antigens by Sjölander (Sjolander et al. 1995; Sjolander et al. 1997; Sjolander et al. 1993) and has been observed for other antigens as well, for example an antigen derived from the syncytical virus subgroup A (Libon et al. 1999). However, it is not clearly elucidated whether those effects result from prolonged half-life, occurrence of T-cell epitopes or a combination of both (Stahl & Nygren 1997). Some B- and T-cell epitopes have been identified in the albumin-binding region of SPG (Goetsch et al. 2003) and an immunogenicity mapping of the albumin-binding protein has been undertaken (Steen et al., manuscript 2011). Related work has shown that the Z-domain can stimulate B-cells and therefore act as an adjuvant. This has been demonstrated using ZZ as a fusion partner for immunization (Lowenadler et al. 1987; Stahl et al. 1989). A dual expression system for immunogens expressed as either a fusion to ZZ or BB was also devised (Stahl et al. 1989). This strategy facilitated immunization with an immunogen tagged in one way followed by evaluation of the antibody response using a differently tagged antigen, which eliminates the background response raised against the fusion partner.

Gene fusions are not limited to only one fusion partner or one end of the gene of interest. Several studies exist where two or several tags have been used in combination. A dual affinity fusion strategy where the gene of interest is expressed flanked by immunoglobulin and albumin-binding regions was shown to be able to provide active protein after two sequential affinity purification steps, as exemplified with human insulin-growth factor II (Hammarberg et al. 1989). The cell lysate was initially passed through an IgG column and secondly through an HSA column. This system will not prevent truncated forms of the protein from being produced, but they will not be collected in both purification steps. Interestingly, it was observed that the proteolytic stability of the recombinant target protein was increased when expressed between two tags compared to an N-terminal ZZ-tag alone (Hammarberg et al. 1989). This indicates that the C-terminal affinity tag confers increased overall stability to the fusion protein. Those findings motivated another study where this phenomenon was investigated in more detail (Murby et al. 1991). Here, the dual-affinity fusion approach was used to assess the degradation of the albumin-binding region of SPG in E. coli, when fused to different recombinant partners with an N-terminal ZZ-protein (Murby et al. 1991). The tagging made independent recovery of the C- and N-terminal regions possible, thereby providing a means for characterization of proteolytic events. Proteins were captured on an HSA column and proteolysis products in the flow through were recovered by a subsequent IgG-affinity chromatography step. In addition to demonstrating that susceptibility to proteolysis can be addressed by gene fusion strategies, this study also showed that small fragments around 6 kDa with functional HSA-binding activity could be recovered. A related system has been used to stabilize various mammalian proteins from degradation when expressed in E. coli (Murby et al. 1991). A similar approach utilized dual tagging with albumin- and metal-binding gene fusions in the termini to recover unstable derivatives of an IgG-binding domain from SPA (Jansson et al. 1990). Taken one step further, a tri-functional Bio-His-ABP tag has been evaluated (Nilsson, J. et al. 1996). This multi-functional tag combines properties of different tags to facilitate detection or immobilization (by streptavidin binding to the in vivo biotinylated bio-tag), refolding and solubility enhancing properties of ABP and possibility to purify under both native and denaturing conditions by immobilized metal ion affinity chromatography through the hexahistidine tag. In the same study, an alternative format for multiple tagging; including a FLAG-epitope tag, a His-tag, a strep-tag and an IgG-binding Z-domain; was evaluated. The main advantage with the former version is the lack of reactivity against antibodies, which broadens its applicability (Nilsson, J. et al. 1996). Tags based on SPA or SPG have also been used in tandem affinity purification strategies to efficiently recover low abundant target proteins and acquire very high purity (Burckstummer et al. 2006; Rigaut et al. 1999).

4. Engineered protein domains derived from SPA and SPG

The robustness of the individual domains of SPA and SPG, together with the knowledge from previous studies of their various properties and applications, has motivated novel protein engineering efforts. Using the small and stable domains as starting points, a wide range of new proteins has been developed for various purposes. Several rational and combinatorial approaches have been attempted to provide small proteins with novel or improved properties that advance the field of protein purification. In this section we summarize some strategies for stabilization, miniaturization, surface engineering and combination of various domains or modified variants of domains derived from SPA and

SPG. The focus lies on modification of the Z-domain derived from SPA and the third albumin-binding domain derived from SPG (figure 8), some associated examples based on related domains are also discussed.

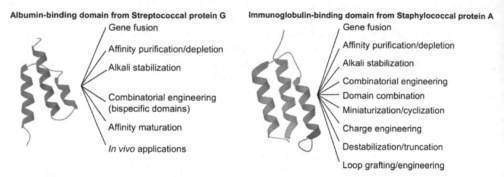

Fig. 8. Engineered protein domains. Several strategies have been devised to engineer new or modified properties into protein domains to facilitate for example affinity purification, increase or decrease stability or introduce novel binding sites.

4.1 Protein engineering to address stability and its implications on affinity

Several innovative strategies have been devised for stabilization of domains derived from the bacterial surface proteins and some examples are reviewed below. The Z-domain represents one of the first rational engineering efforts of a protein domain for increased stability, however at the expense of Fab-binding ability (Nilsson, B. et al. 1987). The C1-domain of SPG has also been engineered for increased stability by an approach that first identified candidate positions for stabilizing mutations and then selected variants from a combinatorial library encompassing combinations of potentially stabilizing mutations (Wunderlich & Schmid 2006). An ingenious system that links the stability of a particular protein variant to the infectivity of a phage particle carrying the same mutant was devised for this purpose. Earlier attempts to stabilize C1 using computational design tools (Dahiyat & Mayo 1997; Malakauskas & Mayo 1998) only resulted in modest improvements compared to this strategy. Stability improvements of C1 have also been achieved by screening of a small combinatorial library built on the domain with a fragment complementation system (Lindman et al. 2010). In this method, fluorescence functions as a reporter for increased affinity between split fragments of green fluorescent protein, which in turn is linked to stability of the C1-fusion. Other applications of the C1-domain include stabilization of a fusion partner to facilitate for example NMR-based studies (Cheng & Patel 2004). Contrarily, to decrease the stability of C1, a screen for destabilized but active variants of the same domain by phage display has been employed (O'Neil et al. 1995). Engineered variants of the C1-domain with a pH-sensitive binding to Fc have also been described recently (Watanabe et al. 2009). Common for those strategies is that they have used C1, which is one of the best-studied models for protein folding, stability and dynamics and therefore a good model to evaluate stabilization approaches (Magliery & Regan 2004). In general, combinatorial library approaches have provided new tools to address the relationship between sequence and folding, stability and function (Forrer et al. 1999; Hoess 2001; Kotz et al. 2004; Magliery & Regan 2004).

Native SPA and SPG are remarkably stable to challenging chemical or physical conditions (Boyle 1990; Girot et al. 1990; Hale et al. 1994). The alkaline tolerance of the immunoglobulin-binding domains of SPG is not as good as for SPA, however resistance of proteins to environmental challenges can be substantially improved by protein engineering. In a strategy that was first proven effective for the albumin-binding domain of SPG, all four asparagine residues susceptible to base catalyzed deamidation were replaced with other residues (Gulich et al. 2000a). This approach resulted in a molecule that could withstand repeated cycles of cleaning with high concentrations of sodium hydroxide, without any considerable loss in performance when linked to a chromatographic resin. An increased thermal stability was also achieved without significant alteration in structure or function (Gulich et al. 2000a). To improve the performance of this ligand even further, different linker regions and multimeric formats have been evaluated for tolerance to extended time periods of alkali exposure (Linhult et al. 2003). The alkali stabilized albumin-binding domain provides a robust affinity ligand for purification or depletion of albumin from for example serum to facilitate detection of low abundant biomarkers (Eriksson et al. 2010; Linhult et al. 2003). Furthermore, the strategy for increasing tolerance to alkali has proven effective also when applied to the Z-domain (Linhult et al. 2004) and the C2-domain (Gulich et al. 2002). The strategies employed to those domains evaluated, apart from asparagine residues, also replacements of glutamine residues, and showed that all replacements have to be evaluated on a case-by-case basis. Some residues may be crucial for binding or stability and thereby impossible to alter, and one can also take homologous sequences into account to decide on suitable replacements. When applying the stabilization strategy to the originally very stable Z-domain, a destabilized point mutant where a core phenylalanine was substituted for an alanine (Cedergren et al. 1993), was used as a model to assess the influence of various mutations (Linhult et al. 2004). Finally, the improvements suggested by the bypass mutagenesis approach were grafted back onto the original domain to verify the result (Linhult et al. 2004). A multimeric variant of this alkali stabilized Z-domain is now utilized as a ligand on a commercially available resin for purification of antibodies or Fc-fused proteins (MabSelect SuRe, GE Healthcare). Furthermore, use of only this domain for purification of antibodies instead of native SPA results in negligible binding to the variable region and therefore a more homogenous pH elution profile (Ghose et al. 2005). A truncated variant of the C2-domain with increased alkaline stability has been described (Goward et al. 1990), however the mutational strategy described above yielded additional tolerance (Gulich et al. 2002). The alkaline stability of SPG was further evaluated and improved in a recent study (Palmer et al. 2008). The mutational strategy was expanded to consider thermodynamic stabilization that retains the tertiary structure and modification of surface electrostatics as well. These examples illustrate the potential of both a relatively straightforward approach to improve the tolerance to alkaline conditions of proteins and a strategy based on more detailed structural understanding. In theory, combinatorial approaches should also be feasible to address this issue, provided that a sensitive enough selection protocol, which can distinguish between differences in tolerance to a specific challenge, can be set up.

Strong interactions between the native domains and their target molecules often require harsh conditions for elution. This may potentially be harmful for the target protein, for example recombinant proteins fused to an affinity tag. To address this problem, modified variants of the B- or Z-domain of SPA have been developed. Variants of the B-domain of

SPA with different C-terminal truncations have been used for affinity chromatographic purification of human IgG (Bottomley et al. 1995). Several variants with decreased affinities were produced and shown to enable elution at elevated pH-values compared to the full-length protein A. The objective of another approach was to replace or extend the loop linking the second and third helix of the Z-domain, thereby yielding destabilized variants of the molecule that facilitate milder elution (Gulich et al. 2000b). Destabilized Z-variants that dissociate from IgG more rapidly were coupled to columns and shown to elute bound immunoglobulin at pH 4.5, which is more than one unit higher than what in commonly used (Gulich et al. 2000b). Potential use of such destabilized immunoglobulin-binding molecules includes utilization as affinity tags for pH sensitive proteins or, when used on a solid support, purification of antibodies using milder conditions. The C1-domain has also served as a more general model to understand how introduction of various loops can affect stability (Zhou et al. 1996). The modified variants have however not been used in chromatographic systems.

4.2 Miniaturization and protein mimicking

Several attempts have aimed at miniaturizing the binding units of protein A, and a few studies have investigated smaller variants of SPG-derived domains. Smaller proteins can more easily be produced and modified by peptide synthesis, they may be useful as starting points for design of smaller organic mimetics or aid in the understanding of protein stability and structure (Braisted & Wells 1996). Analogues of the B-domain of SPA that were truncated in both termini, which are not directly involved in the domain core packing (Deisenhofer 1981), have been produced (Huston et al. 1992). The modified proteins were shown to have lowered affinity against IgG. This was also observed for C-terminally truncated variants of the B-domain that were used as affinity ligands for the removal of immune complexes from blood (Bottomley et al. 1995). Similar results indicating that constructs with a truncated C-terminus are less stable have been reported for dual-tagged variants of the Z-domain (Jansson et al. 1990). Other studies have also addressed folding and stability of mutant variants of the B-domain (Bottomley et al. 1994; Popplewell et al. 1991; Sato et al. 2006). A more systematic approach including iterative structure based design and phage display selections resulted in a miniaturized Z-domain that only comprises 38 residues (Braisted & Wells 1996). This was achieved by stabilizing the first two helices so that IgG-binding would be retained even without the third helix (Braisted & Wells 1996). Structural evaluation of the Z38-variant by NMR (Starovasnik et al. 1997) further strengthened the conclusion that the conformational paratope responsible for IgG-binding was shared with the original Z-domain. Guided by the structural data, a disulphide bonded 34 amino acid truncated variant was designed, synthesized and structurally characterized (Starovasnik et al. 1997). Both this variant and the 38 residue version formed two anti-parallel helices similar to the topology of the corresponding parts in the Z-domain and both demonstrated retained binding to IgG with affinities comparable to the original scaffold (Starovasnik et al. 1997). In addition, an X-ray structure of the miniaturized cyclic Z-domain in complex with Fc demonstrated that the fold and fundamental interactions were preserved in the interaction (Wells et al. 2002). This cyclization strategy has since then been modified to a backbone cyclization through a native chemical ligation reaction (Jarver et al. 2011). The resulting miniaturized molecule was shown to be able to capture human polyclonal IgG when immobilized onto a solid support. In an interesting modification of the original two-helix version of the Z-domain, an elastin sequence was inserted in the inter-helix turn (Reiersen & Rees 1999). This modification dramatically altered the helical structure of the resulting protein. However, in contrast to the starting molecule, the elastin-turn mutant exhibited a more than 20-fold improvement of Fc-binding affinity when the temperature was increased. This effect is hypothesized to arise through a temperature- or salt-induced formation of a ß-turn that stabilizes the alignment of the Fc-binding helices and represents a modular switch to alter structure and activity (Reiersen & Rees 1999, 2000). For small protein domains, synthesis provides a straightforward means for site-specific labeling, chemical cross-linking or introduction of non-natural building blocks to make novel variants available for different applications. Deeper understanding of interaction interfaces between proteins may also facilitate rational design of small molecular weight mimics (Wells et al. 2002); miniaturized proteins only represent intermediates for the challenging task of designing small molecule mimetics.

In rational molecular design, starting from a structurally defined scaffold and a binding surface rather than a sequential stretch of amino acids such as a loop region, usually results in a more defined binding molecule (Stahl & Nygren 1997). For example, the key determinants of the interaction site between the IgG-binding domains of SPA and Fc have stood model for the generation of several small protein mimetic organic molecules. This concept was beautifully demonstrated for the interaction between the B-domain of SPA and the Fc-part of IgG (Li et al. 1998). Using the hydrophobic core dipeptide Phe132-Tyr133 as a starting point, a novel triazine mimetic was rationally designed, synthesized and utilized for purification of antibodies (Li et al. 1998). Since then, mimetics have also been developed for other antibody-binding proteins using modified synthetic molecular scaffolds and chemistries (Haigh et al. 2009; Lowe 2001; Roque et al. 2005). The SPA mimetic peptide PAM is another example of a protein A mimetic ligand. PAM was selected from a combinatorial peptide library, and to further increase the stability of this molecule D-amino acids have been used to hinder degradation of the molecule by proteases (Verdoliva et al. 2002). A synthetic protein called MAbsorbent®A2P (ProMetic BioSciences), which binds all subclasses of human IgG, has also been described (Newcombe et al. 2005). One can also use thiophilic ligands for antibody purification, the most common is called "T-gel", which carries linear ligands with two sulfur atoms and displays good selectivity for antibodies in the presence of high concentrations of lyotropic salts (Boschetti 2001). These small molecule imitations may provide a competitive, robust, scalable and chemically resistant alternative to SPA, SPG or domains thereof for purification of antibodies or Fc-fused proteins. They may achieve increased stability compared to proteinacious ligands, but may however be limited to lower flow-rates since the binding is normally not as fast as for the protein-based ligands.

4.3 Engineering and improving new binding surfaces

Protein engineering may also be applied to modify or evaluate larger binding areas (Sidhu & Koide 2007). Surface exposed amino acids of the Z-domain have been replaced with charged amino acids to generate modified variants of the molecule that carry an excess of positive or negative charge (Graslund, T. et al. 2000; Hedhammar et al. 2004). These molecules, Z_{basic} and Z_{acid}, have efficiently been employed as affinity fusion tags for the purification of recombinant target proteins by cation- or anion-exchange chromatography.

Target protein capture through the Z_{basic}-tag has also been exploited for solid-phase refolding of denatured proteins purified from solubilized inclusion bodies (Hedhammar et al. 2006), capture of fusion proteins by cation-exchange chromatography in an expanded bed adsorption mode (Graslund, T. et al. 2002b) and for high-throughput protein expression and purification (Alm et al. 2007). Those examples illustrate that compact, stable protein domains may be extensively engineered and still retain the beneficial characteristics of the original domain.

Another engineering approach related to the concept of affecting stability through modification of loops has been reported. Here, a biologically active peptide that was selected by phage display to inhibit cathepsin L, was grafted into the loop between the second and third helix of the Z-scaffold (Bratkovic et al. 2006). Loop grafting, and thereby transfer of a novel biological function, could be achieved without loss of structure, as evaluated by CD spectroscopy. Moreover, all constructs also retained their IgG-binding ability (Bratkovic et al. 2006). Consequently, the Z-domain could be utilized as a stable carrier for a new functional entity without loosing its structure or inherent Fc-binding capability.

Combinatorial approaches using robust protein domains can be a valuable tool for the development of tailored purification strategies for native biomolecules (Jonasson et al. 2002; Nygren & Uhlen 1997). Engineering protein surfaces to accommodate novel binding regions provides a means to produce proteins with new functions. On the Z-domain, 13 discontinuous surface-exposed amino acids on the same two helices that mediate the interaction with Fc have been targeted for randomization (Nord et al. 1995). The amino acids involved in the Fc-binding, as identified in the crystal complex of the B-domain and Fc (Deisenhofer 1981), are situated on the outer surfaces of the first and second helix and are not involved in the packing of the core. The Fc-binding surface covers an area of roughly 600 $\mathring{A}^2$, which is comparable to interfaces observed in antibody-antigen interactions (Lo Conte et al. 1999; Nygren 2008). This targeted randomization approach provides a combinatorial library from which so called Affibody molecules with novel binding specificities may be selected (Nord et al. 1997; Nord et al. 1995). To enable selection of variants with desired specificities, the combinatorial library was fused to the gene encoding phage coat protein III and fusions were expressed on filamentous phage. Post selection output was subsequently expressed as fusions to an albumin-binding domain to facilitate evaluation (Nord et al. 1997; Nord et al. 1995). Currently, a large number of alternative display and selection systems are available, many of which have been utilized for selection of Affibody molecules as well as other scaffold proteins (Binz et al. 2005; Lofblom et al. 2010; Nygren 2008; Nygren & Skerra 2004). Early targets for selection of Z-based binding molecules include Taq DNA polymerase, human insulin and a human apolipoprotein variant (Nord et al. 1997) and as of today Affibody molecules have been selected against a large number of targets for use in a variety of applications (Lofblom et al. 2010; Nygren 2008). Several variants have found use within protein purification applications. The before mentioned molecules specific for Taq DNA polymerase or human apolipoprotein A were, in the form of dimers, successfully utilized as affinity ligands for the capture of their respective targets from *E. coli* lysates (Nord 2000). Repeated cycles were performed with elution at low pH, without any observed loss in capacity or selectivity of the Affibody-coupled columns. Furthermore, *in situ* sanitation of columns with 0.5 M NaOH did not result in any significant loss of performance

(Nord et al. 2000). Affibody-mediated capture has also been demonstrated for many proteins, including for example human Factor VIII produced in Chinese hamster ovary cells (Nord et al. 2001), depletion of transferrin (Gronwall et al. 2007b), human IgA (Ronnmark et al. 2002a), amyloid-ß-peptide (Gronwall et al. 2007a), human IgG (Eriksson et al. 2010) or combinations of proteins (Ramstrom et al. 2009). Affibody molecules are available, together with several other capture agents, in commercial multiple affinity removal systems (MARS) (MARS-7, MARS-14 columns, Agilent Technologies). Those kits and utilization of non-antibody based capture proteins have been shown to have advantages compared to utilization of native SPA or SPG for depletion (Coyle et al. 2006; Echan et al. 2005; Eriksson et al. 2010). In addition to protein capture on columns, binding molecules based on the Z-domain have also been utilized for capture in protein microarray applications (Renberg et al. 2007; Renberg et al. 2005).

Another recent example of how novel specificity may be incorporated in small protein domains is illustrated by selection of Affibody molecules with increased affinity to mouse IgG1 (Grimm et al. 2011). The original Z-domain has practically non-existing affinity against mouse IgG, which represents the most widely used within biotechnology. The new specificity possessed by the mouse IgG1-specific binding molecule facilitates specific recovery of monoclonal mouse antibodies from hybridoma supernatants rich in bovine immunoglobulin that may cross-react with alternative capture agents (Grimm et al. 2011). Furthermore, anti-ideotypic Affibody molecules have been generated using other affinity ligands or SPA itself as the target in the selections (Eklund et al. 2002; Wallberg et al. 2011). One such molecule was recently used to facilitate the recovery of untagged Affibody molecules, aimed for imaging studies of human epidermal growth factor receptor 2 over-expressing tumor xenografts, from *E. coli* lysates (Wallberg et al. 2011). An interesting related approach utilized an Affibody molecule specific for SPA as affinity fusion for purification of fusion proteins on readily available protein A media (Graslund, S. et al. 2002a). Similarly, purification of Fc-fused Affibody molecules in an artificial antibody format on protein A Sepharose has been described (Ronnmark et al. 2002b). Together, those examples demonstrate the usefulness of custom-made affinity molecules in various applications. Several structures of Affibody molecules alone or in complex with their targets have been solved, which further expands the understanding of structure- and function-relationships in engineered binding molecules and provides detailed insights for the interactions (Eigenbrot et al. 2010; Hogbom et al. 2003; Hoyer et al. 2008; Lendel et al. 2006; Nygren 2008; Wahlberg et al. 2003). Some applications however demand higher binding affinities than is normally achieved by a single selection from a naïve library. Different approaches to affinity mature Affibody molecules have been devised. For example helix shuffling, error-prone PCR (Grimm et al. unpublished results) or construction of targeted libraries with more focused diversification based on first generation binding molecules have been developed (Gunneriusson et al. 1999; Nord et al. 2001; Orlova et al. 2006). Alternatively, multimeric formats may provide a sufficient gain in apparent affinity for more demanding applications (Nord et al. 1997).

The same miniaturizing strategies that were originally applied to the Z-domain have now also been demonstrated on Affibody molecules with novel binding specificities (Ren et al. 2009; Webster et al. 2009). Those studies demonstrated that the two-helix format could provide a starting template for the design of miniaturized binding molecules, nonetheless some specific optimization may be required to yield a molecule fit for use. Another study has also shown that truncation of a binding molecule based on structural data, here an Affibody molecule specific for the amyloid-ß-peptide, can provide improved variants (Lindgren et al. 2010). This may however require case-by-case optimization and only be applicable when detailed structural data is available. The prospect of producing binding molecules by solid phase peptide synthesis has also motivated an optimization of the Z-scaffold for synthesis. This has been accomplished by utilizing a well-characterized human epidermal growth factor receptor 2-binding molecule as a template (Feldwisch et al. 2010). In addition, the recent scaffold optimization resulted in increased thermal and chemical stability as well as improved solubility. A successful grafting of binding-surfaces for a selection of molecules with other target specificities onto the new scaffold was also demonstrated (Feldwisch et al. 2010). Taken together, a wide range of technologies are now available for the construction of combinatorial libraries, selection of molecules with desired properties, affinity maturation and even miniaturization to provide novel or improved affinity reagents for bioseparation as well as many other applications (Binz et al. 2005; Nygren 2008; Nygren & Skerra 2004). Several synthetically produced and modified variants have so far been described for the Z-domain and C1-domain (Boutillon et al. 1995; Ekblad et al. 2009; Engfeldt et al. 2005). Robust and tailor-made target-specific affinity ligands provide an interesting approach to recover recombinant or naturally occurring proteins in their native forms and will certainly find even broader use in the future. Recent development of new orthogonal aminoacyl-tRNA synthetase/tRNA pairs, which allows for addition of various unnatural amino acids to recombinantly expressed proteins, may aid the further advancement of this expanding field of protein engineering (Liu & Schultz 2010). The addition of building blocks with novel properties to the 20 amino acids chosen by nature may further expand the fitness landscape in which proteins evolve to fulfill novel or enhanced functions. Recent progress includes phage-based *in vitro* evolution systems that utilize bacteria designed to read a 21 amino acid code (Liu et al. 2008).

In a similar fashion as explored for the Z-domain derived from SPA, the albumin-binding domain of SPG has been used as a scaffold for the design of a combinatorial library (Alm et al. 2010). From this library, bispecific binding molecules with retained binding to albumin and an additional acquired affinity to a novel target molecule have been selected by phage display (Alm et al. 2010). In a proof-of-principle study, target proteins with different characteristics were genetically fused to a bispecific ABD-molecule that had been identified through biopanning against the Z-domain. Following expression in bacterial hosts, the target proteins could efficiently be purified to high homogeneity by a two-step affinity purification protocol utilizing the two binding specificities of the tag for the Z-domain and HSA. Affinity maturation of ABD-based, bispecific molecules have also been demonstrated exploiting a cell-displayed library, designed for targeted randomization based on phage display-selected TNF-α-binding molecules (Nilvebrant et al., manuscript 2011). Furthermore, the affinity of the ABD-molecule itself has been addressed in a combinatorial engineering approach (Jonsson et al. 2008). Through several rounds of affinity maturation and rational design where 15 of the 46 amino acids that constitute the domain were randomized, a molecule with an extremely strong affinity against HSA was achieved. Both this molecule and the original albumin-binding domain have successfully been used as gene fusions with for example antibody fragments (Kontermann 2009) or Affibody molecules (Tolmachev et al. 2009) to provide improved persistence *in vivo*, mediated by the binding to serum albumin. Moreover, a recent protein engineering effort was aimed at de-immunizing the affinity-matured albumin-binding domain described above. Identified T-cell epitopes could be removed without influencing the stability, solubility or high affinity of the protein domain (Affibody AB, unpublished results).

Phage display has also been used in an attempt to evolve albumin-binding domains with different species specificities and gain understanding about their mode of interaction, biophysical properties and structural basis for specificity (He et al. 2007; He et al. 2006). A GA-domain derived from *F. magna* with affinity against two phylogenetically distinct serum albumins was successfully selected (Rozak et al. 2006). The binding mode of the resulting molecule, referred to as phage-selected domain-1, to albumin of different species has been further characterized by chemical shift perturbation measurements (He et al. 2007) and structural evaluation (He et al. 2006). The results demonstrate that increased flexibility is not a requirement for broadened specificity (He et al. 2006) and also indicate that a core mutation stabilizes the backbone in a conformation that more closely resembles the structure found in the complex between the GA-module and HSA (He et al. 2007; Lejon et al. 2004). This core residue, a tyrosine, is therefore the main reason for the broader species specificity of the albumin-binding domain from SPG compared to the GA-module derived from *F. magna*. Those efforts illustrate how homologs of a naturally evolved protein scaffold can be used as a starting point to alter the binding specificities through minor modifications of the binding surface. The *in vitro* recombination technique used in those experiments, offset recombinant polymerase chain reaction (Rozak & Bryan 2005), may also be a useful tool to further evaluate or evolve other homologous small protein domains.

Most of the modifications reported for the C1-C3 domains of SPG relate to structural or biophysical questions that lie outside the scope of this chapter (Gronenborn et al. 1991; Gronenborn et al. 1996; Malakauskas & Mayo 1998). However, one interesting example that relates to engineering of novel binding surfaces is the computational *de novo* design of a protein-protein heterodimer based on the C1-domain (Huang et al. 2007). Through rational design, molecules that spontaneously formed heterodimers could be produced. This demonstrates a step forward, among many other examples, on the path to envision a link between design of a primary sequence and a desired structure and function.

4.4 Generation of hybrid proteins

In order to broaden the class- and subclass specificity of immunoglobulin-binding proteins, several hybrid proteins have been compiled from domains of various bacterial surface proteins. The first hybrid protein was developed as a fusion between domains of protein A and G (Eliasson et al. 1988). Four constructs encoding either five domains from SPA, two domains from SPG, or combinations of domains from both, as well as the synthetic Z-variant instead of the native SPA-domains, were evaluated. It was shown that binding specificities from different immunoglobulin-binding proteins could successfully be combined in the hybrid proteins (Eliasson et al. 1989; Eliasson et al. 1988). In a similar approach, immunoglobulin-binding domains from SPA and SPG were combined and expressed in fusion to β-galactosidase to provide a novel enzymatic tool for immunoassays with broad antibody specificity (Strandberg et al. 1990). Similar concepts have since then been applied to produce hybrid molecules of protein L from *F. magna* and protein G

(Kihlberg et al. 1992) as well as protein L and A (Svensson et al. 1998). Immunoglobulin-binding domains of protein L have a fold that resembles the immunoglobulin-binding domains of SPG and interact with the light chain of many antibodies, which provides potential for broadened specificity of the hybrid proteins (Bjorck 1988; Wikstrom et al. 1994). Protein LG was constructed from four domains of protein L combined with two domains from protein G (Kihlberg et al. 1992). Protein LA was assembled from four domains each of the primary proteins (Svensson et al. 1998). The hybrid protein with the broadest combination of specificities has been further minimized in the form of a fusion of a single domain from protein L with one domain from protein G (Harrison et al. 2008). The fused domains were shown to be able to fold and interact with their respective target proteins in an independent manner. A combinatorial approach has also been described to combine individual domains of protein A, G and L (Yang et al. 2008). Randomly arranged domains were displayed on phage and selected against four different immunoglobulin-baits. Powerful library and selection technologies may provide a means to further improve or fine-tune the available range of hybrid proteins to tailor-make new ligands for specific purification or detection of antibodies, antibody fragments as well as many other target proteins.

5. Conclusions

For a few decades, SPA and SPG have been widely investigated to provide the deep understanding we have today about the evolution of the proteins, the structure of the domains and their binding specificities. This, in turn, has enabled us to find many applications for the proteins in a wide range of areas, the most common being ligands for antibody purification or depletion of abundant proteins from complex samples. As structural studies show that individual domains of SPA and SPG fold individually, it is possible to use single domains of the proteins, which have obtained especially good applicability as fusion proteins for production of recombinant proteins. Recombinant DNA technology enables simple construction of expression vectors where a domain of SPA or SPG is fused to a protein of interest. The domains not only simplify the purification procedure, but may also act as solubilizing and stabilizing agents.

Moreover, protein engineering has been applied to improve or combine properties of the stable domains derived from the bacterial surface proteins. Those efforts have resulted in new refined proteins with wide applicability. Furthermore, those techniques have been demonstrated to provide new insights in protein folding and dynamics as well, using small and stable protein domains as models to deepen the understanding of complicated biophysical processes. In summary, small, stable scaffolds have already proven their value in the biotechnological field in many ways and new, innovative applications are currently being investigated. Those rational and combinatorial engineering concepts have the potential to generate alternatives to antibodies as affinity capture agents in demanding, large-scale applications and thereby expand the applicability of affinity chromatography to a wider range of target proteins.

6. Acknowledgment

The authors would like to acknowledge John Löfblom for critical reading of the text.

7. References

Abrahmsen, L., T. Moks, B. Nilsson, U. Hellman & M. Uhlen (1985). "Analysis of signals for secretion in the staphylococcal protein A gene." *EMBO J* 4(13B): 3901-3906.

Achari, A., S. P. Hale, A. J. Howard, G. M. Clore, A. M. Gronenborn, K. D. Hardman & M. Whitlow (1992). "1.67-A X-ray structure of the B2 immunoglobulin-binding domain of streptococcal protein G and comparison to the NMR structure of the B1 domain." *Biochemistry* 31(43): 10449-10457.

Akerstrom, B. & L. Bjorck (1986). "A physicochemical study of protein G, a molecule with unique immunoglobulin G-binding properties." *J Biol Chem* 261(22): 10240-10247.

Akerstrom, B., T. Brodin, K. Reis & L. Bjorck (1985). "Protein G: a powerful tool for binding and detection of monoclonal and polyclonal antibodies." *J Immunol* 135(4): 2589-2592.

Akerstrom, B., A. Lindqvist & G. Lindahl (1991). "Binding properties of protein Arp, a bacterial IgA-receptor." *Mol Immunol* 28(4-5): 349-357.

Akerstrom, B., E. Nielsen & L. Bjorck (1987). "Definition of IgG- and albumin-binding regions of streptococcal protein G." *J Biol Chem* 262(28): 13388-13391.

Akesson, P., J. Cooney, F. Kishimoto & L. Bjorck (1990). "Protein H--a novel IgG binding bacterial protein." *Mol Immunol* 27(6): 523-531.

Alexander, P., S. Fahnestock, T. Lee, J. Orban & P. Bryan (1992). "Thermodynamic analysis of the folding of the streptococcal protein G IgG-binding domains B1 and B2: why small proteins tend to have high denaturation temperatures." *Biochemistry* 31(14): 3597-3603.

Alexander, P. A., Y. He, Y. Chen, J. Orban & P. N. Bryan (2009). "A minimal sequence code for switching protein structure and function." *Proc Natl Acad Sci U S A* 106(50): 21149-21154.

Alm, T., J. Steen, J. Ottosson & S. Hober (2007). "High-throughput protein purification under denaturing conditions by the use of cation exchange chromatography." *Biotechnol J* 2(6): 709-716.

Alm, T., L. Yderland, J. Nilvebrant, A. Halldin & S. Hober (2010). "A small bispecific protein selected for orthogonal affinity purification." *Biotechnol J* 5(6): 605-617.

Andersen, J. T., R. Pehrson, V. Tolmachev, M. B. Daba, L. Abrahmsen & C. Ekblad (2011). "Extending half-life by indirect targeting of the neonatal Fc receptor (FcRn) using a minimal albumin binding domain." *J Biol Chem* 286(7): 5234-5241.

Anderson, N. L. & N. G. Anderson (2002). "The human plasma proteome: history, character, and diagnostic prospects." *Mol Cell Proteomics* 1(11): 845-867.

Arnau, J., C. Lauritzen, G. E. Petersen & J. Pedersen (2006). "Current strategies for the use of affinity tags and tag removal for the purification of recombinant proteins." *Protein Expr Purif* 48(1): 1-13.

Atkins, K. L., J. D. Burman, E. S. Chamberlain, J. E. Cooper, B. Poutrel, S. Bagby, A. T. Jenkins, E. J. Feil & J. M. van den Elsen (2008). "S. aureus IgG-binding proteins SpA and Sbi: host specificity and mechanisms of immune complex formation." *Mol Immunol* 45(6): 1600-1611.

Azarkan, M., J. Huet, D. Baeyens-Volant, Y. Looze & G. Vandenbussche (2007). "Affinity chromatography: a useful tool in proteomics studies." *J Chromatogr B Analyt Technol Biomed Life Sci* 849(1-2): 81-90.

Baumann, S., P. Grob, F. Stuart, D. Pertlik, M. Ackermann & M. Suter (1998). "Indirect immobilization of recombinant proteins to a solid phase using the albumin binding domain of streptococcal protein G and immobilized albumin." *J Immunol Methods* 221(1-2): 95-106.

Binz, H. K., P. Amstutz & A. Pluckthun (2005). "Engineering novel binding proteins from nonimmunoglobulin domains." *Nat Biotechnol* 23(10): 1257-1268.

Bjorck, L. (1988). "Protein L. A novel bacterial cell wall protein with affinity for Ig L chains." *J Immunol* 140(4): 1194-1197.

Bjorck, L., W. Kastern, G. Lindahl & K. Wideback (1987). "Streptococcal protein G, expressed by streptococci or by Escherichia coli, has separate binding sites for human albumin and IgG." *Mol Immunol* 24(10): 1113-1122.

Bjorck, L. & G. Kronvall (1984). "Purification and some properties of streptococcal protein G, a novel IgG-binding reagent." *J Immunol* 133(2): 969-974.

Boschetti, E. (2001). "The use of thiophilic chromatography for antibody purification: a review." *J Biochem Biophys Methods* 49(1-3): 361-389.

Bottomley, S. P., A. G. Popplewell, M. Scawen, T. Wan, B. J. Sutton & M. G. Gore (1994). "The stability and unfolding of an IgG binding protein based upon the B domain of protein A from Staphylococcus aureus probed by tryptophan substitution and fluorescence spectroscopy." *Protein Eng* 7(12): 1463-1470.

Bottomley, S. P., B. J. Sutton & M. G. Gore (1995). "Elution of human IgG from affinity columns containing immobilised variants of protein A." *J Immunol Methods* 182(2): 185-192.

Boutillon, C., R. Wintjens, G. Lippens, H. Drobecq & A. Tartar (1995). "Synthesis, three-dimensional structure, and specific 15N-labelling of the streptococcal protein G B1-domain." *Eur J Biochem* 231(1): 166-180.

Boyle, M. D. P. (1990). *Bacterial Immunoglobulin Binding Proteins: Applications in immunotechnology*, Academic Press.

Braisted, A. C. & J. A. Wells (1996). "Minimizing a binding domain from protein A." *Proc Natl Acad Sci U S A* 93(12): 5688-5692.

Bratkovic, T., A. Berlec, T. Popovic, M. Lunder, S. Kreft, U. Urleb & B. Strukelj (2006). "Engineered staphylococcal protein A's IgG-binding domain with cathepsin L inhibitory activity." *Biochem Biophys Res Commun* 349(1): 449-453.

Burckstummer, T., K. L. Bennett, A. Preradovic, G. Schutze, O. Hantschel, G. Superti-Furga & A. Bauch (2006). "An efficient tandem affinity purification procedure for interaction proteomics in mammalian cells." *Nat Methods* 3(12): 1013-1019.

Cassulis, P., M. Magasic & V. DeBari (1991). "Ligand affinity chromatographic separation of serum IgG on recombinant protein G-silica." *Clin Chem* 37(6): 882-886.

Cedergren, L., R. Andersson, B. Jansson, M. Uhlen & B. Nilsson (1993). "Mutational analysis of the interaction between staphylococcal protein A and human IgG1." *Protein Eng* 6(4): 441-448.

Cheng, Y. & D. J. Patel (2004). "An efficient system for small protein expression and refolding." *Biochem Biophys Res Commun* 317(2): 401-405.

Clore, G. M. & C. D. Schwieters (2004). "Amplitudes of protein backbone dynamics and correlated motions in a small alpha/beta protein: correspondence of dipolar coupling and heteronuclear relaxation measurements." *Biochemistry* 43(33): 10678-10691.

Coyle, E. M., L. L. Blazer, A. A. White, J. L. Hess & M. D. Boyle (2006). "Practical applications of high-affinity, albumin-binding proteins from a group G streptococcal isolate." *Appl Microbiol Biotechnol* 71(1): 39-45.

Cramer, J. F., P. A. Nordberg, J. Hajdu & S. Lejon (2007). "Crystal structure of a bacterial albumin-binding domain at 1.4 A resolution." *FEBS Lett* 581(17): 3178-3182.

Cuatrecasas, P., M. Wilchek & C. B. Anfinsen (1968). "Selective enzyme purification by affinity chromatography." *Proc Natl Acad Sci U S A* 61(2): 636-643.

Dahiyat, B. I. & S. L. Mayo (1997). "Probing the role of packing specificity in protein design." *Proc Natl Acad Sci U S A* 94(19): 10172-10177.

Dancette, O. P., J. L. Taboureau, E. Tournier, C. Charcosset & P. Blond (1999). "Purification of immunoglobulins G by protein A/G affinity membrane chromatography." *J Chromatogr B Biomed Sci Appl* 723(1-2): 61-68.

de Chateau, M. & L. Bjorck (1994). "Protein PAB, a mosaic albumin-binding bacterial protein representing the first contemporary example of module shuffling." *J Biol Chem* 269(16): 12147-12151.

de Chateau, M., E. Holst & L. Bjorck (1996). "Protein PAB, an albumin-binding bacterial surface protein promoting growth and virulence." *J Biol Chem* 271(43): 26609-26615.

Deisenhofer, J. (1981). "Crystallographic refinement and atomic models of a human Fc fragment and its complex with fragment B of protein A from Staphylococcus aureus at 2.9- and 2.8-A resolution." *Biochemistry* 20(9): 2361-2370.

Derrick, J. P., M. C. Maiden & I. M. Feavers (1999). "Crystal structure of an Fab fragment in complex with a meningococcal serosubtype antigen and a protein G domain." *J Mol Biol* 293(1): 81-91.

Derrick, J. P. & D. B. Wigley (1994). "The third IgG-binding domain from streptococcal protein G. An analysis by X-ray crystallography of the structure alone and in a complex with Fab." *J Mol Biol* 243(5): 906-918.

Echan, L. A., H. Y. Tang, N. Ali-Khan, K. Lee & D. W. Speicher (2005). "Depletion of multiple high-abundance proteins improves protein profiling capacities of human serum and plasma." *Proteomics* 5(13): 3292-3303.

Eigenbrot, C., M. Ultsch, A. Dubnovitsky, L. Abrahmsen & T. Hard (2010). "Structural basis for high-affinity HER2 receptor binding by an engineered protein." *Proc Natl Acad Sci U S A* 107(34): 15039-15044.

Ekblad, T., V. Tolmachev, A. Orlova, C. Lendel, L. Abrahmsen & A. E. Karlstrom (2009). "Synthesis and chemoselective intramolecular crosslinking of a HER2-binding affibody." *Biopolymers* 92(2): 116-123.

Eklund, M., L. Axelsson, M. Uhlen & P. A. Nygren (2002). "Anti-idiotypic protein domains selected from protein A-based affibody libraries." *Proteins* 48(3): 454-462.

Eliasson, M., R. Andersson, A. Olsson, H. Wigzell & M. Uhlen (1989). "Differential IgG-binding characteristics of staphylococcal protein A, streptococcal protein G, and a chimeric protein AG." *J Immunol* 142(2): 575-581.

Eliasson, M., A. Olsson, E. Palmcrantz, K. Wiberg, M. Inganas, B. Guss, M. Lindberg & M. Uhlen (1988). "Chimeric IgG-binding receptors engineered from staphylococcal protein A and streptococcal protein G." *J Biol Chem* 263(9): 4323-4327.

Engfeldt, T., B. Renberg, H. Brumer, P. A. Nygren & A. E. Karlstrom (2005). "Chemical synthesis of triple-labelled three-helix bundle binding proteins for specific fluorescent detection of unlabelled protein." *Chembiochem* 6(6): 1043-1050.

Eriksson, C., J. M. Schwenk, A. Sjoberg & S. Hober (2010). "Affibody molecule-mediated depletion of HSA and IgG using different buffer compositions: a 15 min protocol for parallel processing of 1-48 samples." *Biotechnol Appl Biochem* 56(2): 49-57.

Erntell, M., E. B. Myhre & G. Kronvall (1983). "Alternative non-immune F(ab')2-mediated immunoglobulin binding to group C and G streptococci." *Scand J Immunol* 17(3): 201-209.

Erntell, M., E. B. Myhre, U. Sjobring & L. Bjorck (1988). "Streptococcal protein G has affinity for both Fab- and Fc-fragments of human IgG." *Mol Immunol* 25(2): 121-126.

Ettorre, A., C. Rosli, M. Silacci, S. Brack, G. McCombie, R. Knochenmuss, G. Elia & D. Neri (2006). "Recombinant antibodies for the depletion of abundant proteins from human serum." *Proteomics* 6(16): 4496-4505.

Ey, P. L., S. J. Prowse & C. R. Jenkin (1978). "Isolation of pure IgG1, IgG2a and IgG2b immunoglobulins from mouse serum using protein A-sepharose." *Immunochemistry* 15(7): 429-436.

Fahnestock, S. R., P. Alexander, J. Nagle & D. Filpula (1986). "Gene for an immunoglobulin-binding protein from a group G streptococcus." *J Bacteriol* 167(3): 870-880.

Falkenberg, C., L. Bjorck & B. Akerstrom (1992). "Localization of the binding site for streptococcal protein G on human serum albumin. Identification of a 5.5-kilodalton protein G binding albumin fragment." *Biochemistry* 31(5): 1451-1457.

Faulkner, S., G. Elia, M. Hillard, P. O'Boyle, M. Dunn & D. Morris (2011). "Immunodepletion of albumin and immunoglobulin G from bovine plasma." *Proteomics* 11(11): 2329-2335.

Faulmann, E. L., J. L. Duvall & M. D. Boyle (1991). "Protein B: a versatile bacterial Fc-binding protein selective for human IgA." *Biotechniques* 10(6): 748-755.

Feldwisch, J., V. Tolmachev, C. Lendel, N. Herne, A. Sjoberg, B. Larsson, D. Rosik, E. Lindqvist, G. Fant, I. Hoiden-Guthenberg, J. Galli, P. Jonasson & L. Abrahmsen (2010). "Design of an optimized scaffold for affibody molecules." *J Mol Biol* 398(2): 232-247.

Fischetti, V. A. (1989). "Streptococcal M protein: molecular design and biological behavior." *Clin Microbiol Rev* 2(3): 285-314.

Flaschel, E. & K. Friehs (1993). "Improvement of downstream processing of recombinant proteins by means of genetic engineering methods." *Biotechnol Adv* 11(1): 31-77.

Forrer, P., S. Jung & A. Pluckthun (1999). "Beyond binding: using phage display to select for structure, folding and enzymatic activity in proteins." *Curr Opin Struct Biol* 9(4): 514-520.

Forsberg, G., B. Baastrup, H. Rondahl, E. Holmgren, G. Pohl, M. Hartmanis & M. Lake (1992). "An evaluation of different enzymatic cleavage methods for recombinant fusion proteins, applied on des(1-3)insulin-like growth factor I." *J Protein Chem* 11(2): 201-211.

Forsgren, A. & J. Sjoquist (1966). ""Protein A" from S. aureus. I. Pseudo-immune reaction with human gamma-globulin." *J Immunol* 97(6): 822-827.

Foster, T. J. (2005). "Immune evasion by staphylococci." *Nat Rev Microbiol* 3(12): 948-958.

Franks, W. T., D. H. Zhou, B. J. Wylie, B. G. Money, D. T. Graesser, H. L. Frericks, G. Sahota & C. M. Rienstra (2005). "Magic-angle spinning solid-state NMR spectroscopy of the beta1 immunoglobulin binding domain of protein G (GB1): 15N and 13C chemical shift assignments and conformational analysis." *J Am Chem Soc* 127(35): 12291-12305.

Frick, I. M., M. Wikstrom, S. Forsen, T. Drakenberg, H. Gomi, U. Sjobring & L. Bjorck (1992). "Convergent evolution among immunoglobulin G-binding bacterial proteins." *Proc Natl Acad Sci U S A* 89(18): 8532-8536.

Fu, Q., C. P. Garnham, S. T. Elliott, D. E. Bovenkamp & J. E. Van Eyk (2005). "A robust, streamlined, and reproducible method for proteomic analysis of serum by delipidation, albumin and IgG depletion, and two-dimensional gel electrophoresis." *Proteomics* 5(10): 2656-2664.

Ghose, S., M. Allen, B. Hubbard, C. Brooks & S. M. Cramer (2005). "Antibody variable region interactions with Protein A: implications for the development of generic purification processes." *Biotechnol Bioeng* 92(6): 665-673.

Ghose, S., B. Hubbard & S. M. Cramer (2007). "Binding capacity differences for antibodies and Fc-fusion proteins on protein A chromatographic materials." *Biotechnol Bioeng* 96(4): 768-779.

Girot, P., Y. Moroux, X. P. Duteil, C. Nguyen & E. Boschetti (1990). "Composite affinity sorbents and their cleaning in place." *J Chromatogr* 510: 213-223.

Goetsch, L., J. F. Haeuw, T. Champion, C. Lacheny, T. N'Guyen, A. Beck & N. Corvaia (2003). "Identification of B- and T-cell epitopes of BB, a carrier protein derived from the G protein of Streptococcus strain G148." *Clin Diagn Lab Immunol* 10(1): 125-132.

Gomez, M. I., A. Lee, B. Reddy, A. Muir, G. Soong, A. Pitt, A. Cheung & A. Prince (2004). "Staphylococcus aureus protein A induces airway epithelial inflammatory responses by activating TNFR1." *Nat Med* 10(8): 842-848.

Gouda, H., H. Torigoe, A. Saito, M. Sato, Y. Arata & I. Shimada (1992). "Three-dimensional solution structure of the B domain of staphylococcal protein A: comparisons of the solution and crystal structures." *Biochemistry* 31(40): 9665-9672.

Goward, C. R., J. P. Murphy, T. Atkinson & D. A. Barstow (1990). "Expression and purification of a truncated recombinant streptococcal protein G." *Biochem J* 267(1): 171-177.

Graille, M., E. A. Stura, A. L. Corper, B. J. Sutton, M. J. Taussig, J. B. Charbonnier & G. J. Silverman (2000). "Crystal structure of a Staphylococcus aureus protein A domain complexed with the Fab fragment of a human IgM antibody: structural basis for recognition of B-cell receptors and superantigen activity." *Proc Natl Acad Sci U S A* 97(10): 5399-5404.

Graslund, S., M. Eklund, R. Falk, M. Uhlen, P. A. Nygren & S. Stahl (2002a). "A novel affinity gene fusion system allowing protein A-based recovery of non-immunoglobulin gene products." *J Biotechnol* 99(1): 41-50.

Graslund, T., M. Hedhammar, M. Uhlen, P. A. Nygren & S. Hober (2002b). "Integrated strategy for selective expanded bed ion-exchange adsorption and site-specific protein processing using gene fusion technology." *J Biotechnol* 96(1): 93-102.

Graslund, T., G. Lundin, M. Uhlen, P. A. Nygren & S. Hober (2000). "Charge engineering of a protein domain to allow efficient ion-exchange recovery." *Protein Eng* 13(10): 703-709.

Graslund, T., J. Nilsson, A. M. Lindberg, M. Uhlen & P. A. Nygren (1997). "Production of a thermostable DNA polymerase by site-specific cleavage of a heat-eluted affinity fusion protein." *Protein Expr Purif* 9(1): 125-132.

Grimm, S., F. Yu & P. A. Nygren (2011). "Ribosome Display Selection of a Murine IgG(1) Fab Binding Affibody Molecule Allowing Species Selective Recovery Of Monoclonal Antibodies." *Mol Biotechnol* 48(3): 263-276.

Grodzki, A. C. & E. Berenstein (2010). "Antibody purification: affinity chromatography - protein A and protein G Sepharose." *Methods Mol Biol* 588: 33-41.

Gronenborn, A. M. & G. M. Clore (1993). "Identification of the contact surface of a streptococcal protein G domain complexed with a human Fc fragment." *J Mol Biol* 233(3): 331-335.

Gronenborn, A. M., D. R. Filpula, N. Z. Essig, A. Achari, M. Whitlow, P. T. Wingfield & G. M. Clore (1991). "A novel, highly stable fold of the immunoglobulin binding domain of streptococcal protein G." *Science* 253(5020): 657-661.

Gronenborn, A. M., M. K. Frank & G. M. Clore (1996). "Core mutants of the immunoglobulin binding domain of streptococcal protein G: stability and structural integrity." *FEBS Lett* 398(2-3): 312-316.

Gronwall, C., A. Jonsson, S. Lindstrom, E. Gunneriusson, S. Stahl & N. Herne (2007a). "Selection and characterization of Affibody ligands binding to Alzheimer amyloid beta peptides." *J Biotechnol* 128(1): 162-183.

Gronwall, C., A. Sjoberg, M. Ramstrom, I. Hoiden-Guthenberg, S. Hober, P. Jonasson & S. Stahl (2007b). "Affibody-mediated transferrin depletion for proteomics applications." *Biotechnol J* 2(11): 1389-1398.

Grov, A., B. Myklestad & P. Oeding (1964). "Immunochemical Studies on Antigen Preparations from Staphylococcus Aureus. 1. Isolation and Chemical Characterization of Antigen A." *Acta Pathol Microbiol Scand* 61: 588-596.

Gulich, S., M. Linhult, P. Nygren, M. Uhlen & S. Hober (2000a). "Stability towards alkaline conditions can be engineered into a protein ligand." *J Biotechnol* 80(2): 169-178.

Gulich, S., M. Linhult, S. Stahl & S. Hober (2002). "Engineering streptococcal protein G for increased alkaline stability." *Protein Eng* 15(10): 835-842.

Gulich, S., M. Uhlen & S. Hober (2000b). "Protein engineering of an IgG-binding domain allows milder elution conditions during affinity chromatography." *J Biotechnol* 76(2-3): 233-244.

Gunneriusson, E., K. Nord, M. Uhlen & P. Nygren (1999). "Affinity maturation of a Taq DNA polymerase specific affibody by helix shuffling." *Protein Eng* 12(10): 873-878.

Guss, B., M. Eliasson, A. Olsson, M. Uhlen, A. K. Frej, H. Jornvall, J. I. Flock & M. Lindberg (1986). "Structure of the IgG-binding regions of streptococcal protein G." *EMBO J* 5(7): 1567-1575.

Guss, B., K. Leander, U. Hellman, M. Uhlen, J. Sjoquist & M. Lindberg (1985). "Analysis of protein A encoded by a mutated gene of Staphylococcus aureus Cowan I." *Eur J Biochem* 153(3): 579-585.

Guss, B., M. Uhlen, B. Nilsson, M. Lindberg, J. Sjoquist & J. Sjodahl (1984). "Region X, the cell-wall-attachment part of staphylococcal protein A." *Eur J Biochem* 138(2): 413-420.

Haake, D. A., E. C. Franklin & B. Frangione (1982). "The modification of human immunoglobulin binding to staphylococcal protein A using diethylpyrocarbonate." *J Immunol* 129(1): 190-192.

Haigh, J. M., A. Hussain, M. L. Mimmack & C. R. Lowe (2009). "Affinity ligands for immunoglobulins based on the multicomponent Ugi reaction." *J Chromatogr B Analyt Technol Biomed Life Sci* 877(14-15): 1440-1452.

Hale, G., A. Drumm, P. Harrison & J. Phillips (1994). "Repeated cleaning of protein A affinity column with sodium hydroxide." *J Immunol Methods* 171(1): 15-21.

Hall, J. B. & D. Fushman (2003). "Characterization of the overall and local dynamics of a protein with intermediate rotational anisotropy: Differentiating between conformational exchange and anisotropic diffusion in the B3 domain of protein G." *J Biomol NMR* 27(3): 261-275.

Hammarberg, B., P. A. Nygren, E. Holmgren, A. Elmblad, M. Tally, U. Hellman, T. Moks & M. Uhlen (1989). "Dual affinity fusion approach and its use to express recombinant human insulin-like growth factor II." *Proc Natl Acad Sci U S A* 86(12): 4367-4371.

Hansson, M., S. Stahl, R. Hjorth, M. Uhlen & T. Moks (1994). "Single-step recovery of a secreted recombinant protein by expanded bed adsorption." *Biotechnology (N Y)* 12(3): 285-288.

Harrison, S. L., N. G. Housden, S. P. Bottomley, A. J. Cossins & M. G. Gore (2008). "Generation and expression of a minimal hybrid Ig-receptor formed between single domains from proteins L and G." *Protein Expr Purif* 58(1): 12-22.

He, Y., Y. Chen, D. A. Rozak, P. N. Bryan & J. Orban (2007). "An artificially evolved albumin binding module facilitates chemical shift epitope mapping of GA domain interactions with phylogenetically diverse albumins." *Protein Sci* 16(7): 1490-1494.

He, Y., D. A. Rozak, N. Sari, Y. Chen, P. Bryan & J. Orban (2006). "Structure, dynamics, and stability variation in bacterial albumin binding modules: implications for species specificity." *Biochemistry* 45(33): 10102-10109.

He, Y., D. C. Yeh, P. Alexander, P. N. Bryan & J. Orban (2005). "Solution NMR structures of IgG binding domains with artificially evolved high levels of sequence identity but different folds." *Biochemistry* 44(43): 14055-14061.

Hedhammar, M., T. Alm, T. Graslund & S. Hober (2006). "Single-step recovery and solid-phase refolding of inclusion body proteins using a polycationic purification tag." *Biotechnol J* 1(2): 187-196.

Hedhammar, M., T. Graslund, M. Uhlen & S. Hober (2004). "Negatively charged purification tags for selective anion-exchange recovery." *Protein Eng Des Sel* 17(11): 779-786.

Heel, T., M. Paal, R. Schneider & B. Auer (2010). "Dissection of an old protein reveals a novel application: domain D of Staphylococcus aureus Protein A (sSpAD) as a secretion--tag." *Microb Cell Fact* 9: 92.

Hober, S., K. Nord & M. Linhult (2007). "Protein A chromatography for antibody purification." *J Chromatogr B Analyt Technol Biomed Life Sci* 848(1): 40-47.

Hochuli, E., W. Bannwarth, H. Dobeli, R. Gentz & D. Stuber (1988). "Genetic Approach to Facilitate Purification of Recombinant Proteins with a Novel Metal Chelate Adsorbent." *Bio-Technology* 6(11): 1321-1325.

Hoess, R. H. (2001). "Protein design and phage display." *Chem Rev* 101(10): 3205-3218.

Hogbom, M., M. Eklund, P. A. Nygren & P. Nordlund (2003). "Structural basis for recognition by an in vitro evolved affibody." *Proc Natl Acad Sci U S A* 100(6): 3191-3196.

Holland, T. A., S. Veretnik, I. N. Shindyalov & P. E. Bourne (2006). "Partitioning protein structures into domains: why is it so difficult?" *J Mol Biol* 361(3): 562-590.

Hopp, J., N. Hornig, K. A. Zettlitz, A. Schwarz, N. Fuss, D. Muller & R. E. Kontermann (2010). "The effects of affinity and valency of an albumin-binding domain (ABD) on the half-life of a single-chain diabody-ABD fusion protein." *Protein Eng Des Sel* 23(11): 827-834.

Hoyer, W., C. Gronwall, A. Jonsson, S. Stahl & T. Hard (2008). "Stabilization of a beta-hairpin in monomeric Alzheimer's amyloid-beta peptide inhibits amyloid formation." *Proc Natl Acad Sci U S A* 105(13): 5099-5104.

Huang, P. S., J. J. Love & S. L. Mayo (2007). "A de novo designed protein protein interface." *Protein Sci* 16(12): 2770-2774.

Huston, J. S., C. Cohen, D. Maratea, F. Fields, M. S. Tai, N. Cabral-Denison, R. Juffras, D. C. Rueger, R. J. Ridge, H. Oppermann & et al. (1992). "Multisite association by recombinant proteins can enhance binding selectivity. Preferential removal of immune complexes from serum by immobilized truncated FB analogues of the B domain from staphylococcal protein A." *Biophys J* 62(1): 87-91.

Inganas, M. (1981). "Comparison of mechanisms of interaction between protein A from Staphylococcus aureus and human monoclonal IgG, IgA and IgM in relation to the classical FC gamma and the alternative F(ab')2 epsilon protein A interactions." *Scand J Immunol* 13(4): 343-352.

Jansson, B., C. Palmcrantz, M. Uhlen & B. Nilsson (1990). "A dual-affinity gene fusion system to express small recombinant proteins in a soluble form: expression and characterization of protein A deletion mutants." *Protein Eng* 3(6): 555-561.

Jansson, B., M. Uhlen & P. A. Nygren (1998). "All individual domains of staphylococcal protein A show Fab binding." *FEMS Immunol Med Microbiol* 20(1): 69-78.

Jarver, P., C. Mikaelsson & A. E. Karlstrom (2011). "Chemical synthesis and evaluation of a backbone-cyclized minimized 2-helix Z-domain." *J Pept Sci* 17(6): 463-469.

Jendeberg, L., B. Persson, R. Andersson, R. Karlsson, M. Uhlen & B. Nilsson (1995). "Kinetic analysis of the interaction between protein A domain variants and human Fc using plasmon resonance detection." *J Mol Recognit* 8(4): 270-278.

Jendeberg, L., M. Tashiro, R. Tejero, B. A. Lyons, M. Uhlen, G. T. Montelione & B. Nilsson (1996). "The mechanism of binding staphylococcal protein A to immunoglobin G does not involve helix unwinding." *Biochemistry* 35(1): 22-31.

Johansson, M. U., M. de Chateau, M. Wikstrom, S. Forsen, T. Drakenberg & L. Bjorck (1997). "Solution structure of the albumin-binding GA module: a versatile bacterial protein domain." *J Mol Biol* 266(5): 859-865.

Johansson, M. U., I. M. Frick, H. Nilsson, P. J. Kraulis, S. Hober, P. Jonasson, M. Linhult, P. A. Nygren, M. Uhlen, L. Bjorck, T. Drakenberg, S. Forsen & M. Wikstrom (2002a). "Structure, specificity, and mode of interaction for bacterial albumin-binding modules." *J Biol Chem* 277(10): 8114-8120.

Johansson, M. U., H. Nilsson, J. Evenas, S. Forsen, T. Drakenberg, L. Bjorck & M. Wikstrom (2002b). "Differences in backbone dynamics of two homologous bacterial albumin-binding modules: implications for binding specificity and bacterial adaptation." *J Mol Biol* 316(5): 1083-1099.

Jonasson, P., S. Liljeqvist, P. A. Nygren & S. Stahl (2002). "Genetic design for facilitated production and recovery of recombinant proteins in Escherichia coli." *Biotechnol Appl Biochem* 35(Pt 2): 91-105.

Jonasson, P., J. Nilsson, E. Samuelsson, T. Moks, S. Stahl & M. Uhlen (1996). "Single-step trypsin cleavage of a fusion protein to obtain human insulin and its C peptide." *Eur J Biochem* 236(2): 656-661.

Jonsson, A., J. Dogan, N. Herne, L. Abrahmsen & P. A. Nygren (2008). "Engineering of a femtomolar affinity binding protein to human serum albumin." *Protein Eng Des Sel* 21(8): 515-527.

Kaboord, B. & M. Perr (2008). "Isolation of proteins and protein complexes by immunoprecipitation." *Methods Mol Biol* 424: 349-364.

Karlsson, R., L. Jendeberg, B. Nilsson, J. Nilsson & P. A. Nygren (1995). "Direct and competitive kinetic analysis of the interaction between human IgG1 and a one domain analogue of protein A." *J Immunol Methods* 183(1): 43-49.

Kihlberg, B. M., U. Sjobring, W. Kastern & L. Bjorck (1992). "Protein LG: a hybrid molecule with unique immunoglobulin binding properties." *J Biol Chem* 267(35): 25583-25588.

Konig, T. & A. Skerra (1998). "Use of an albumin-binding domain for the selective immobilisation of recombinant capture antibody fragments on ELISA plates." *J Immunol Methods* 218(1-2): 73-83.

Kontermann, R. E. (2009). "Strategies to extend plasma half-lives of recombinant antibodies." *BioDrugs* 23(2): 93-109.

Kotz, J. D., C. J. Bond & A. G. Cochran (2004). "Phage-display as a tool for quantifying protein stability determinants." *Eur J Biochem* 271(9): 1623-1629.

Kraulis, P. J., P. Jonasson, P. A. Nygren, M. Uhlen, L. Jendeberg, B. Nilsson & J. Kordel (1996). "The serum albumin-binding domain of streptococcal protein G is a three-helical bundle: a heteronuclear NMR study." *FEBS Lett* 378(2): 190-194.

Kronvall, G. (1973). "A surface component in group A, C, and G streptococci with non-immune reactivity for immunoglobulin G." *J Immunol* 111(5): 1401-1406.

Kronvall, G., A. Simmons, E. B. Myhre & S. Jonsson (1979). "Specific absorption of human serum albumin, immunoglobulin A, and immunoglobulin G with selected strains of group A and G streptococci." *Infect Immun* 25(1): 1-10.

Langone, J. J. (1982). "Protein A of Staphylococcus aureus and related immunoglobulin receptors produced by streptococci and pneumonococci." *Adv Immunol* 32: 157-252.

Larsson, M., E. Brundell, L. Nordfors, C. Hoog, M. Uhlen & S. Stahl (1996). "A general bacterial expression system for functional analysis of cDNA-encoded proteins." *Protein Expr Purif* 7(4): 447-457.

LaVallie, E. R., J. M. McCoy, D. B. Smith & P. Riggs (2001). "Enzymatic and chemical cleavage of fusion proteins." *Curr Protoc Mol Biol* Chapter 16: Unit16 14B.

Lejon, S., I. M. Frick, L. Bjorck, M. Wikstrom & S. Svensson (2004). "Crystal structure and biological implications of a bacterial albumin binding module in complex with human serum albumin." *J Biol Chem* 279(41): 42924-42928.

Lendel, C., J. Dogan & T. Hard (2006). "Structural basis for molecular recognition in an affibody:affibody complex." *J Mol Biol* 359(5): 1293-1304.

Li, R., V. Dowd, D. J. Stewart, S. J. Burton & C. R. Lowe (1998). "Design, synthesis, and application of a protein A mimetic." *Nat Biotechnol* 16(2): 190-195.

Lian, L. Y., J. P. Derrick, M. J. Sutcliffe, J. C. Yang & G. C. Roberts (1992). "Determination of the solution structures of domains II and III of protein G from Streptococcus by 1H nuclear magnetic resonance." *J Mol Biol* 228(4): 1219-1234.

Lian, L. Y., J. C. Yang, J. P. Derrick, M. J. Sutcliffe, G. C. Roberts, J. P. Murphy, C. R. Goward & T. Atkinson (1991). "Sequential 1H NMR assignments and secondary structure of an IgG-binding domain from protein G." *Biochemistry* 30(22): 5335-5340.

Libon, C., N. Corvaia, J. F. Haeuw, T. N. Nguyen, S. Stahl, J. Y. Bonnefoy & C. Andreoni (1999). "The serum albumin-binding region of streptococcal protein G (BB) potentiates the immunogenicity of the G130-230 RSV-A protein." *Vaccine* 17(5): 406-414.

Lindahl, G. & B. Akerstrom (1989). "Receptor for IgA in group A streptococci: cloning of the gene and characterization of the protein expressed in Escherichia coli." *Mol Microbiol* 3(2): 239-247.

Lindgren, J., A. Wahlstrom, J. Danielsson, N. Markova, C. Ekblad, A. Graslund, L. Abrahmsen, A. E. Karlstrom & S. K. Warmlander (2010). "N-terminal engineering of amyloid-beta-binding Affibody molecules yields improved chemical synthesis and higher binding affinity." *Protein Sci* 19(12): 2319-2329.

Lindman, S., A. Hernandez-Garcia, O. Szczepankiewicz, B. Frohm & S. Linse (2010). "In vivo protein stabilization based on fragment complementation and a split GFP system." *Proc Natl Acad Sci U S A* 107(46): 19826-19831.

Lindmark, R., K. Thoren-Tolling & J. Sjoquist (1983). "Binding of immunoglobulins to protein A and immunoglobulin levels in mammalian sera." *J Immunol Methods* 62(1): 1-13.

Linhult, M., H. K. Binz, M. Uhlen & S. Hober (2002). "Mutational analysis of the interaction between albumin-binding domain from streptococcal protein G and human serum albumin." *Protein Sci* 11(2): 206-213.

Linhult, M., S. Gulich, T. Graslund, P. A. Nygren & S. Hober (2003). "Evaluation of different linker regions for multimerization and coupling chemistry for immobilization of a proteinaceous affinity ligand." *Protein Eng* 16(12): 1147-1152.

Linhult, M., S. Gulich, T. Graslund, A. Simon, M. Karlsson, A. Sjoberg, K. Nord & S. Hober (2004). "Improving the tolerance of a protein a analogue to repeated alkaline exposures using a bypass mutagenesis approach." *Proteins* 55(2): 407-416.

Liu, C. C., A. V. Mack, M. L. Tsao, J. H. Mills, H. S. Lee, H. Choe, M. Farzan, P. G. Schultz & V. V. Smider (2008). "Protein evolution with an expanded genetic code." *Proc Natl Acad Sci U S A* 105(46): 17688-17693.

Liu, C. C. & P. G. Schultz (2010). "Adding new chemistries to the genetic code." *Annu Rev Biochem* 79: 413-444.

Ljungberg, U. K., B. Jansson, U. Niss, R. Nilsson, B. E. Sandberg & B. Nilsson (1993). "The interaction between different domains of staphylococcal protein A and human polyclonal IgG, IgA, IgM and F(ab')2: separation of affinity from specificity." *Mol Immunol* 30(14): 1279-1285.

Ljungquist, C., B. Jansson, T. Moks & M. Uhlen (1989). "Thiol-directed immobilization of recombinant IgG-binding receptors." *Eur J Biochem* 186(3): 557-561.

Lo Conte, L., C. Chothia & J. Janin (1999). "The atomic structure of protein-protein recognition sites." *J Mol Biol* 285(5): 2177-2198.

Lo, K. M., Y. Sudo, J. Chen, Y. Li, Y. Lan, S. M. Kong, L. Chen, Q. An & S. D. Gillies (1998). "High level expression and secretion of Fc-X fusion proteins in mammalian cells." *Protein Eng* 11(6): 495-500.

Lofblom, J., J. Feldwisch, V. Tolmachev, J. Carlsson, S. Stahl & F. Y. Frejd (2010). "Affibody molecules: engineered proteins for therapeutic, diagnostic and biotechnological applications." *FEBS Lett* 584(12): 2670-2680.

Lofdahl, S., B. Guss, M. Uhlen, L. Philipson & M. Lindberg (1983). "Gene for staphylococcal protein A." *Proc Natl Acad Sci U S A* 80(3): 697-701.

Lorca, T., J. C. Labbe, A. Devault, D. Fesquet, U. Strausfeld, J. Nilsson, P. A. Nygren, M. Uhlen, J. C. Cavadore & M. Doree (1992). "Cyclin A-cdc2 kinase does not trigger but delays cyclin degradation in interphase extracts of amphibian eggs." *J Cell Sci* 102 (Pt 1): 55-62.

Lowe, C. R. (2001). "Combinatorial approaches to affinity chromatography." *Curr Opin Chem Biol* 5(3): 248-256.

Lowenadler, B., B. Jansson, S. Paleus, E. Holmgren, B. Nilsson, T. Moks, G. Palm, S. Josephson, L. Philipson & M. Uhlen (1987). "A gene fusion system for generating antibodies against short peptides." *Gene* 58(1): 87-97.

Lowenadler, B., B. Nilsson, L. Abrahmsen, T. Moks, L. Ljungqvist, E. Holmgren, S. Paleus, S. Josephson, L. Philipson & M. Uhlen (1986). "Production of specific antibodies against protein A fusion proteins." *EMBO J* 5(9): 2393-2398.

Magliery, T. J. & L. Regan (2004). "Combinatorial approaches to protein stability and structure." *Eur J Biochem* 271(9): 1595-1608.

Makrides, S. C., P. A. Nygren, B. Andrews, P. J. Ford, K. S. Evans, E. G. Hayman, H. Adari, M. Uhlen & C. A. Toth (1996). "Extended in vivo half-life of human soluble complement receptor type 1 fused to a serum albumin-binding receptor." *J Pharmacol Exp Ther* 277(1): 534-542.

Malakauskas, S. M. & S. L. Mayo (1998). "Design, structure and stability of a hyperthermophilic protein variant." *Nat Struct Biol* 5(6): 470-475.

Martin, R. R., J. G. Crowder & A. White (1967). "Human reactions to staphylococcal antigens. A possible role of leukocyte lysosomal enzymes." *J Immunol* 99(2): 269-275.

Matic, G., T. Bosch & W. Ramlow (2001). "Background and indications for protein A-based extracorporeal immunoadsorption." *Ther Apher* 5(5): 394-403.

Mazor, Y., T. Van Blarcom, S. Carroll & G. Georgiou (2010). "Selection of full-length IgGs by tandem display on filamentous phage particles and Escherichia coli fluorescence-activated cell sorting screening." *FEBS J* 277(10): 2291-2303.

Mazor, Y., T. Van Blarcom, B. L. Iverson & G. Georgiou (2008). "E-clonal antibodies: selection of full-length IgG antibodies using bacterial periplasmic display." *Nat Protoc* 3(11): 1766-1777.

Mazor, Y., T. Van Blarcom, R. Mabry, B. L. Iverson & G. Georgiou (2007). "Isolation of engineered, full-length antibodies from libraries expressed in Escherichia coli." *Nat Biotechnol* 25(5): 563-565.

Moks, T., L. Abrahmsen, B. Nilsson, U. Hellman, J. Sjoquist & M. Uhlen (1986). "Staphylococcal protein A consists of five IgG-binding domains." *Eur J Biochem* 156(3): 637-643.

Movitz, J. (1976). "Formation of extracellular protein A by Staphylococcus aureus." *Eur J Biochem* 68(1): 291-299.

Murby, M., L. Cedergren, J. Nilsson, P. A. Nygren, B. Hammarberg, B. Nilsson, S. O. Enfors & M. Uhlen (1991). "Stabilization of recombinant proteins from proteolytic degradation in Escherichia coli using a dual affinity fusion strategy." *Biotechnol Appl Biochem* 14(3): 336-346.

Murby, M., E. Samuelsson, T. N. Nguyen, L. Mignard, U. Power, H. Binz, M. Uhlen & S. Stahl (1995). "Hydrophobicity engineering to increase solubility and stability of a recombinant protein from respiratory syncytial virus." *Eur J Biochem* 230(1): 38-44.

Murby, M., M. Uhlen & S. Stahl (1996). "Upstream strategies to minimize proteolytic degradation upon recombinant production in Escherichia coli." *Protein Expr Purif* 7(2): 129-136.

Myhre, E. B. & G. Kronvall (1977). "Heterogeneity of nonimmune immunoglobulin Fc reactivity among gram-positive cocci: description of three major types of receptors for human immunoglobulin G." *Infect Immun* 17(3): 475-482.

Myhre, E. B. & G. Kronvall (1980). "Demonstration of a new type of immunoglobulin G receptor in Streptococcus zooepidemicus strains." *Infect Immun* 27(3): 808-816.

Myhre, E. B. & G. Kronvall (1981). "Specific binding of bovine, ovine, caprine and equine IgG subclasses to defined types of immunoglobulin receptors in Gram-positive cocci." *Comp Immunol Microbiol Infect Dis* 4(3-4): 317-328.

Nadaud, P. S., M. Sarkar, B. Wu, C. E. MacPhee, T. J. Magliery & C. P. Jaroniec (2010). "Expression and purification of a recombinant amyloidogenic peptide from transthyretin for solid-state NMR spectroscopy." *Protein Expr Purif* 70(1): 101-108.

Nakamura, M., M. Mie & E. Kobatake (2011). "Construction of a functional IgG-binding luciferase fusion protein for the rapid detection of specific bacterial strains." *Analyst* 136(1): 71-72.

Nakamura, Y., S. Shibasaki, M. Ueda, A. Tanaka, H. Fukuda & A. Kondo (2001). "Development of novel whole-cell immunoadsorbents by yeast surface display of the IgG-binding domain." *Appl Microbiol Biotechnol* 57(4): 500-505.

Newcombe, A. R., C. Cresswell, S. Davies, K. Watson, G. Harris, K. O'Donovan & R. Francis (2005). "Optimised affinity purification of polyclonal antibodies from hyper immunised ovine serum using a synthetic Protein A adsorbent, MAbsorbent A2P." *J Chromatogr B Analyt Technol Biomed Life Sci* 814(2): 209-215.

Nilsson, B. & L. Abrahmsen (1990). "Fusions to staphylococcal protein A." *Methods Enzymol* 185: 144-161.

Nilsson, B., L. Abrahmsen & M. Uhlen (1985). "Immobilization and purification of enzymes with staphylococcal protein A gene fusion vectors." *EMBO J* 4(4): 1075-1080.

Nilsson, B., T. Moks, B. Jansson, L. Abrahmsen, A. Elmblad, E. Holmgren, C. Henrichson, T. A. Jones & M. Uhlen (1987). "A synthetic IgG-binding domain based on staphylococcal protein A." *Protein Eng* 1(2): 107-113.

Nilsson, F. Y. & V. Tolmachev (2007). "Affibody molecules: new protein domains for molecular imaging and targeted tumor therapy." *Curr Opin Drug Discov Devel* 10(2): 167-175.

Nilsson, J., M. Bosnes, F. Larsen, P. A. Nygren, M. Uhlen & J. Lundeberg (1997a). "Heat-mediated activation of affinity-immobilized Taq DNA polymerase." *Biotechniques* 22(4): 744-751.

Nilsson, J., M. Larsson, S. Stahl, P. A. Nygren & M. Uhlen (1996). "Multiple affinity domains for the detection, purification and immobilization of recombinant proteins." *J Mol Recognit* 9(5-6): 585-594.

Nilsson, J., P. Nilsson, Y. Williams, L. Pettersson, M. Uhlen & P. A. Nygren (1994). "Competitive elution of protein A fusion proteins allows specific recovery under mild conditions." *Eur J Biochem* 224(1): 103-108.

Nilsson, J., S. Stahl, J. Lundeberg, M. Uhlen & P. A. Nygren (1997b). "Affinity fusion strategies for detection, purification, and immobilization of recombinant proteins." *Protein Expr Purif* 11(1): 1-16.

Nord, K., E. Gunneriusson, J. Ringdahl, S. Stahl, M. Uhlen & P. A. Nygren (1997). "Binding proteins selected from combinatorial libraries of an alpha-helical bacterial receptor domain." *Nat Biotechnol* 15(8): 772-777.

Nord, K., E. Gunneriusson, M. Uhlen & P. A. Nygren (2000). "Ligands selected from combinatorial libraries of protein A for use in affinity capture of apolipoprotein A-1M and taq DNA polymerase." *J Biotechnol* 80(1): 45-54.

Nord, K., J. Nilsson, B. Nilsson, M. Uhlen & P. A. Nygren (1995). "A combinatorial library of an alpha-helical bacterial receptor domain." *Protein Eng* 8(6): 601-608.

Nord, K., O. Nord, M. Uhlen, B. Kelley, C. Ljungqvist & P. A. Nygren (2001). "Recombinant human factor VIII-specific affinity ligands selected from phage-displayed combinatorial libraries of protein A." *Eur J Biochem* 268(15): 4269-4277.

Nygren, P. A. (2008). "Alternative binding proteins: affibody binding proteins developed from a small three-helix bundle scaffold." *FEBS J* 275(11): 2668-2676.

Nygren, P. A., M. Eliasson, L. Abrahmsen, M. Uhlen & E. Palmcrantz (1988). "Analysis and use of the serum albumin binding domains of streptococcal protein G." *J Mol Recognit* 1(2): 69-74.

Nygren, P. A., C. Ljungquist, H. Tromborg, K. Nustad & M. Uhlen (1990). "Species-dependent binding of serum albumins to the streptococcal receptor protein G." *Eur J Biochem* 193(1): 143-148.

Nygren, P. A. & A. Skerra (2004). "Binding proteins from alternative scaffolds." *J Immunol Methods* 290(1-2): 3-28.

Nygren, P. A. & M. Uhlen (1997). "Scaffolds for engineering novel binding sites in proteins." *Curr Opin Struct Biol* 7(4): 463-469.

O'Neil, K. T., R. H. Hoess, D. P. Raleigh & W. F. DeGrado (1995). "Thermodynamic genetics of the folding of the B1 immunoglobulin-binding domain from streptococcal protein G." *Proteins* 21(1): 11-21.

Oeding, P., A. Grov & B. Myklestad (1964). "Immunochemical Studies on Antigen Preparations from Staphylococcus Aureus. 2. Precipitating and Erythrocyte-Sensitizing Properties of Protein a (Antigen a) and Related Substances." *Acta Pathol Microbiol Scand* 62: 117-127.

Ohlson, S., R. Nilsson, U. Niss, B. M. Kjellberg & C. Freiburghaus (1988). "A novel approach to monoclonal antibody separation using high performance liquid affinity chromatography (HPLAC) with SelectiSpher-10 protein G." *J Immunol Methods* 114(1-2): 175-180.

Ojala, K., J. Koski, W. Ernst, R. Grabherr, I. Jones & C. Oker-Blom (2004). "Improved display of synthetic IgG-binding domains on the baculovirus surface." *Technol Cancer Res Treat* 3(1): 77-84.

Olsson, A., M. Eliasson, B. Guss, B. Nilsson, U. Hellman, M. Lindberg & M. Uhlen (1987). "Structure and evolution of the repetitive gene encoding streptococcal protein G." *Eur J Biochem* 168(2): 319-324.

Orlova, A., M. Magnusson, T. L. Eriksson, M. Nilsson, B. Larsson, I. Hoiden-Guthenberg, C. Widstrom, J. Carlsson, V. Tolmachev, S. Stahl & F. Y. Nilsson (2006). "Tumor imaging using a picomolar affinity HER2 binding affibody molecule." *Cancer Res* 66(8): 4339-4348.

Palmer, B., K. Angus, L. Taylor, J. Warwicker & J. P. Derrick (2008). "Design of stability at extreme alkaline pH in streptococcal protein G." *J Biotechnol* 134(3-4): 222-230.

Parks, T. D., K. K. Leuther, E. D. Howard, S. A. Johnston & W. G. Dougherty (1994). "Release of proteins and peptides from fusion proteins using a recombinant plant virus proteinase." *Anal Biochem* 216(2): 413-417.

Pazehoski, K. O., P. A. Cobine, D. J. Winzor & C. T. Dameron (2011). "Evidence for involvement of the C-terminal domain in the dimerization of the CopY repressor protein from Enterococcus hirae." *Biochem Biophys Res Commun* 406(2): 183-187.

Podlaski, F. J. & A. S. Stern (2000). "Site-specific immobilization of antibodies to protein G-derivatized solid supports." *Methods Mol Biol* 147: 41-48.

Popplewell, A. G., M. G. Gore, M. Scawen & T. Atkinson (1991). "Synthesis and mutagenesis of an IgG-binding protein based upon protein A of Staphylococcus aureus." *Protein Eng* 4(8): 963-970.

Poullin, P., N. Announ, B. Mugnier, S. Guis, J. Roudier & P. Lefevre (2005). "Protein A-immunoadsorption (Prosorba column) in the treatment of rheumatoid arthritis." *Joint Bone Spine* 72(2): 101-103.

Raeder, R., R. A. Otten & M. D. Boyle (1991). "Comparison of albumin receptors expressed on bovine and human group G streptococci." *Infect Immun* 59(2): 609-616.

Ramstrom, M., A. Zuberovic, C. Gronwall, J. Hanrieder, J. Bergquist & S. Hober (2009). "Development of affinity columns for the removal of high-abundance proteins in cerebrospinal fluid." *Biotechnol Appl Biochem* 52(Pt 2): 159-166.

Reiersen, H. & A. R. Rees (1999). "An engineered minidomain containing an elastin turn exhibits a reversible temperature-induced IgG binding." *Biochemistry* 38(45): 14897-14905.

Reiersen, H. & A. R. Rees (2000). "Sodium sulphate reactivates a protein A minidomain with a short elastin beta-turn." *Biochem Biophys Res Commun* 276(3): 899-904.

Reis, K. J., E. M. Ayoub & M. D. Boyle (1984). "Streptococcal Fc receptors. II. Comparison of the reactivity of a receptor from a group C streptococcus with staphylococcal protein A." *J Immunol* 132(6): 3098-3102.

Ren, G., R. Zhang, Z. Liu, J. M. Webster, Z. Miao, S. S. Gambhir, F. A. Syud & Z. Cheng (2009). "A 2-helix small protein labeled with 68Ga for PET imaging of HER2 expression." *J Nucl Med* 50(9): 1492-1499.

Renberg, B., J. Nordin, A. Merca, M. Uhlen, J. Feldwisch, P. A. Nygren & A. E. Karlstrom (2007). "Affibody molecules in protein capture microarrays: evaluation of multidomain ligands and different detection formats." *J Proteome Res* 6(1): 171-179.

Renberg, B., I. Shiroyama, T. Engfeldt, P. K. Nygren & A. E. Karlstrom (2005). "Affibody protein capture microarrays: synthesis and evaluation of random and directed immobilization of affibody molecules." *Anal Biochem* 341(2): 334-343.

Richman, D. D., P. H. Cleveland, M. N. Oxman & K. M. Johnson (1982). "The binding of staphylococcal protein A by the sera of different animal species." *J Immunol* 128(5): 2300-2305.

Rigaut, G., A. Shevchenko, B. Rutz, M. Wilm, M. Mann & B. Seraphin (1999). "A generic protein purification method for protein complex characterization and proteome exploration." *Nat Biotechnol* 17(10): 1030-1032.

Roben, P. W., A. N. Salem & G. J. Silverman (1995). "VH3 family antibodies bind domain D of staphylococcal protein A." *J Immunol* 154(12): 6437-6445.

Ronnmark, J., H. Gronlund, M. Uhlen & P. A. Nygren (2002a). "Human immunoglobulin A (IgA)-specific ligands from combinatorial engineering of protein A." *Eur J Biochem* 269(11): 2647-2655.

Ronnmark, J., M. Hansson, T. Nguyen, M. Uhlen, A. Robert, S. Stahl & P. A. Nygren (2002b). "Construction and characterization of affibody-Fc chimeras produced in Escherichia coli." *J Immunol Methods* 261(1-2): 199-211.

Roque, A. C., M. A. Taipa & C. R. Lowe (2005). "An artificial protein L for the purification of immunoglobulins and fab fragments by affinity chromatography." *J Chromatogr A* 1064(2): 157-167.

Rozak, D. A., P. A. Alexander, Y. He, Y. Chen, J. Orban & P. N. Bryan (2006). "Using offset recombinant polymerase chain reaction to identify functional determinants in a common family of bacterial albumin binding domains." *Biochemistry* 45(10): 3263-3271.

Rozak, D. A. & P. N. Bryan (2005). "Offset recombinant PCR: a simple but effective method for shuffling compact heterologous domains." *Nucleic Acids Res* 33(9): e82.

Sagawa, T., M. Oda, H. Morii, H. Takizawa, H. Kozono & T. Azuma (2005). "Conformational changes in the antibody constant domains upon hapten-binding." *Mol Immunol* 42(1): 9-18.

Samuelsson, E., T. Moks, B. Nilsson & M. Uhlen (1994). "Enhanced in vitro refolding of insulin-like growth factor I using a solubilizing fusion partner." *Biochemistry* 33(14): 4207-4211.

Samuelsson, E. & M. Uhlen (1996). "Chaperone-like effect during in vitro refolding of insulin-like growth factor I using a solubilizing fusion partner." *Ann N Y Acad Sci* 782: 486-494.

Samuelsson, E., H. Wadensten, M. Hartmanis, T. Moks & M. Uhlen (1991). "Facilitated in vitro refolding of human recombinant insulin-like growth factor I using a solubilizing fusion partner." *Biotechnology (N Y)* 9(4): 363-366.

Sato, S., T. L. Religa & A. R. Fersht (2006). "Phi-analysis of the folding of the B domain of protein A using multiple optical probes." *J Mol Biol* 360(4): 850-864.

Sauer-Eriksson, A. E., G. J. Kleywegt, M. Uhlen & T. A. Jones (1995). "Crystal structure of the C2 fragment of streptococcal protein G in complex with the Fc domain of human IgG." *Structure* 3(3): 265-278.

Schneewind, O., A. Fowler & K. F. Faull (1995). "Structure of the cell wall anchor of surface proteins in Staphylococcus aureus." *Science* 268(5207): 103-106.

Scott, M. A., J. M. Davis & K. A. Schwartz (1997). "Staphylococcal protein A binding to canine IgG and IgM." *Vet Immunol Immunopathol* 59(3-4): 205-212.

Shimizu, A., M. Honzawa, S. Ito, T. Miyazaki, H. Matsumoto, H. Nakamura, T. E. Michaelsen & Y. Arata (1983). "H NMR studies of the Fc region of human IgG1 and IgG3 immunoglobulins: assignment of histidine resonances in the CH3 domain and identification of IgG3 protein carrying G3m(st) allotypes." *Mol Immunol* 20(2): 141-148.

Sidhu, S. S. & S. Koide (2007). "Phage display for engineering and analyzing protein interaction interfaces." *Curr Opin Struct Biol* 17(4): 481-487.

Silverman, G. J., C. S. Goodyear & D. L. Siegel (2005). "On the mechanism of staphylococcal protein A immunomodulation." *Transfusion* 45(2): 274-280.

Sisson, T. H. & C. W. Castor (1990). "An improved method for immobilizing IgG antibodies on protein A-agarose." *J Immunol Methods* 127(2): 215-220.

Sjobring, U., L. Bjorck & W. Kastern (1991). "Streptococcal protein G. Gene structure and protein binding properties." *J Biol Chem* 266(1): 399-405.

Sjodahl, J. (1977). "Structural studies on the four repetitive Fc-binding regions in protein A from Staphylococcus aureus." *Eur J Biochem* 78(2): 471-490.

Sjolander, A., R. Andersson, M. Hansson, K. Berzins & P. Perlmann (1995). "Genetic restriction and specificity of the immune response in mice to fusion proteins containing repeated sequences of the Plasmodium falciparum antigen Pf155/RESA." *Immunology* 84(3): 360-366.

Sjolander, A., P. A. Nygren, S. Stahl, K. Berzins, M. Uhlen, P. Perlmann & R. Andersson (1997). "The serum albumin-binding region of streptococcal protein G: a bacterial fusion partner with carrier-related properties." *J Immunol Methods* 201(1): 115-123.

Sjolander, A., S. Stahl & P. Perlmann (1993). "Bacterial Expression Systems Based on a Protein A and Protein G Designed for the Production of Immunogens: Applications to Plasmodium falciparum Malaria Antigens." *ImmunoMethods* 2(1): 79-92.

Sjoquist, J. & G. Stalenheim (1969). "Protein A from Staphylococcus aureus. IX. Complement-fixing activity of protein A-IgG complexes." *J Immunol* 103(3): 467-473.

Smeesters, P. R., D. J. McMillan & K. S. Sriprakash (2010). "The streptococcal M protein: a highly versatile molecule." *Trends Microbiol* 18(6): 275-282.

Stahl, S. & P. A. Nygren (1997). "The use of gene fusions to protein A and protein G in immunology and biotechnology." *Pathol Biol (Paris)* 45(1): 66-76.

Stahl, S., P. A. Nygren & M. Uhlen (1997). "Detection and isolation of recombinant proteins based on binding affinity of reporter: protein A." *Methods Mol Biol* 63: 103-118.

Stahl, S., A. Sjolander, P. A. Nygren, K. Berzins, P. Perlmann & M. Uhlen (1989). "A dual expression system for the generation, analysis and purification of antibodies to a repeated sequence of the Plasmodium falciparum antigen Pf155/RESA." *J Immunol Methods* 124(1): 43-52.

Starovasnik, M. A., A. C. Braisted & J. A. Wells (1997). "Structural mimicry of a native protein by a minimized binding domain." *Proc Natl Acad Sci U S A* 94(19): 10080-10085.

Starovasnik, M. A., M. P. O'Connell, W. J. Fairbrother & R. F. Kelley (1999). "Antibody variable region binding by Staphylococcal protein A: thermodynamic analysis and location of the Fv binding site on E-domain." *Protein Sci* 8(7): 1423-1431.

Starovasnik, M. A., N. J. Skelton, M. P. O'Connell, R. F. Kelley, D. Reilly & W. J. Fairbrother (1996). "Solution structure of the E-domain of staphylococcal protein A." *Biochemistry* 35(48): 15558-15569.

Stone, G. C., U. Sjobring, L. Bjorck, J. Sjoquist, C. V. Barber & F. A. Nardella (1989). "The Fc binding site for streptococcal protein G is in the C gamma 2-C gamma 3 interface region of IgG and is related to the sites that bind staphylococcal protein A and human rheumatoid factors." *J Immunol* 143(2): 565-570.

Stork, R., E. Campigna, B. Robert, D. Muller & R. E. Kontermann (2009). "Biodistribution of a bispecific single-chain diabody and its half-life extended derivatives." *J Biol Chem* 284(38): 25612-25619.

Strandberg, L., S. Hober, M. Uhlen & S. O. Enfors (1990). "Expression and characterization of a tripartite fusion protein consisting of chimeric IgG-binding receptors and beta-galactosidase." *J Biotechnol* 13(1): 83-96.

Svensson, H. G., H. R. Hoogenboom & U. Sjobring (1998). "Protein LA, a novel hybrid protein with unique single-chain Fv antibody- and Fab-binding properties." *Eur J Biochem* 258(2): 890-896.

Tanaka, G., H. Funabashi, M. Mie & E. Kobatake (2006). "Fabrication of an antibody microwell array with self-adhering antibody binding protein." *Anal Biochem* 350(2): 298-303.

Tashiro, M., R. Tejero, D. E. Zimmerman, B. Celda, B. Nilsson & G. T. Montelione (1997). "High-resolution solution NMR structure of the Z domain of staphylococcal protein A." *J Mol Biol* 272(4): 573-590.

Terpe, K. (2003). "Overview of tag protein fusions: from molecular and biochemical fundamentals to commercial systems." *Appl Microbiol Biotechnol* 60(5): 523-533.

Tolmachev, V., H. Wallberg, K. Andersson, A. Wennborg, H. Lundqvist & A. Orlova (2009). "The influence of Bz-DOTA and CHX-A"-DTPA on the biodistribution of ABD-fused anti-HER2 Affibody molecules: implications for (114m)In-mediated targeting therapy." *Eur J Nucl Med Mol Imaging* 36(9): 1460-1468.

Ton-That, H., K. F. Faull & O. Schneewind (1997). "Anchor structure of staphylococcal surface proteins. A branched peptide that links the carboxyl terminus of proteins to the cell wall." *J Biol Chem* 272(35): 22285-22292.

Uhlen, M., G. Forsberg, T. Moks, M. Hartmanis & B. Nilsson (1992). "Fusion proteins in biotechnology." *Curr Opin Biotechnol* 3(4): 363-369.

Uhlen, M., B. Guss, B. Nilsson, S. Gatenbeck, L. Philipson & M. Lindberg (1984). "Complete sequence of the staphylococcal gene encoding protein A. A gene evolved through multiple duplications." *J Biol Chem* 259(3): 1695-1702.

Uhlen, M. & T. Moks (1990). "Gene fusions for purpose of expression: an introduction." *Methods Enzymol* 185: 129-143.

Uhlen, M., B. Nilsson, B. Guss, M. Lindberg, S. Gatenbeck & L. Philipson (1983). "Gene fusion vectors based on the gene for staphylococcal protein A." *Gene* 23(3): 369-378.

Ulmer, T. S., B. E. Ramirez, F. Delaglio & A. Bax (2003). "Evaluation of backbone proton positions and dynamics in a small protein by liquid crystal NMR spectroscopy." *J Am Chem Soc* 125(30): 9179-9191.

Urbas, L., P. Brne, B. Gabor, M. Barut, M. Strlic, T. C. Petric & A. Strancar (2009). "Depletion of high-abundance proteins from human plasma using a combination of an affinity and pseudo-affinity column." *J Chromatogr A* 1216(13): 2689-2694.

Verdoliva, A., F. Pannone, M. Rossi, S. Catello & V. Manfredi (2002). "Affinity purification of polyclonal antibodies using a new all-D synthetic peptide ligand: comparison with protein A and protein G." *J Immunol Methods* 271(1-2): 77-88.

Verwey, W. F. (1940). "A Type-Specific Antigenic Protein Derived from the Staphylococcus." *J Exp Med* 71(5): 635-644.

Wadensten, H., A. Ekebacke, B. Hammarberg, E. Holmgren, C. Kalderen, M. Tally, T. Moks, M. Uhlen, S. Josephson & M. Hartmanis (1991). "Purification and characterization of recombinant human insulin-like growth factor II (IGF-II) expressed as a secreted fusion protein in Escherichia coli." *Biotechnol Appl Biochem* 13(3): 412-421.

Wahlberg, E., C. Lendel, M. Helgstrand, P. Allard, V. Dincbas-Renqvist, A. Hedqvist, H. Berglund, P. A. Nygren & T. Hard (2003). "An affibody in complex with a target protein: structure and coupled folding." *Proc Natl Acad Sci U S A* 100(6): 3185-3190.

Walker, P. A., L. E. Leong, P. W. Ng, S. H. Tan, S. Waller, D. Murphy & A. G. Porter (1994). "Efficient and rapid affinity purification of proteins using recombinant fusion proteases." *Biotechnology (N Y)* 12(6): 601-605.

Wallberg, H., P. K. Lofdahl, K. Tschapalda, M. Uhlen, V. Tolmachev, P. K. Nygren & S. Stahl (2011). "Affinity recovery of eight HER2-binding affibody variants using an anti-idiotypic affibody molecule as capture ligand." *Protein Expr Purif* 76(1): 127-135.

Watanabe, H., H. Matsumaru, A. Ooishi, Y. Feng, T. Odahara, K. Suto & S. Honda (2009). "Optimizing pH response of affinity between protein G and IgG Fc: How electrostatic modulations affect protein-protein interactions." *J Biol Chem* 284(18): 12373-12383.

Waugh, D. S. (2005). "Making the most of affinity tags." *Trends Biotechnol* 23(6): 316-320.

Webster, J. M., R. Zhang, S. S. Gambhir, Z. Cheng & F. A. Syud (2009). "Engineered two-helix small proteins for molecular recognition." *Chembiochem* 10(8): 1293-1296.

Wells, J. A., B. Cunningham, A. Braisted, S. Atwell, W. Delano, M. Ultsch, M. Starovasnik & A. Vos (2002). From big molecules to smaller ones. *Peptide Science – Present and Future.* Y. Shimonishi, Springer Netherlands: 1-5.

Wikstrom, M., T. Drakenberg, S. Forsen, U. Sjobring & L. Bjorck (1994). "Three-dimensional solution structure of an immunoglobulin light chain-binding domain of protein L. Comparison with the IgG-binding domains of protein G." *Biochemistry* 33(47): 14011-14017.

Winter, G., A. D. Griffiths, R. E. Hawkins & H. R. Hoogenboom (1994). "Making antibodies by phage display technology." *Annu Rev Immunol* 12: 433-455.

Wunderlich, M. & F. X. Schmid (2006). "In vitro evolution of a hyperstable Gbeta1 variant." *J Mol Biol* 363(2): 545-557.

Yang, H., J. Cao, L. Q. Li, X. Zhou, Q. L. Chen, W. T. Liao, Z. M. Wen, S. H. Jiang, R. Xu, J. A. Jia, X. Pan, Z. T. Qi & W. Pan (2008). "Evolutional selection of a combinatorial phage library displaying randomly-rearranged various single domains of immunoglobulin (Ig)-binding proteins (IBPs) with four kinds of Ig molecules." *BMC Microbiol* 8: 137.

Zhang, L., K. Jacobsson, J. Vasi, M. Lindberg & L. Frykberg (1998). "A second IgG-binding protein in Staphylococcus aureus. "*Microbiology* 144 (Pt 4): 985-991.

Zheng, D., J. M. Aramini & G. T. Montelione (2004). "Validation of helical tilt angles in the solution NMR structure of the Z domain of Staphylococcal protein A by combined analysis of residual dipolar coupling and NOE data." *Protein Sci* 13(2): 549-554.

Zhou, H. X., R. H. Hoess & W. F. DeGrado (1996). "In vitro evolution of thermodynamically stable turns." *Nat Struct Biol* 3(5): 446-451.

5

Episomal Vectors for Rapid Expression and Purification of Proteins in Mammalian Cells

Giovanni Magistrelli, Pauline Malinge,
Greg Elson and Nicolas Fischer
NovImmune SA, 14 Chemin des Aulx, Plan-les-Ouates,
Switzerland

1. Introduction

Research projects in life sciences aim at studying and better understanding biological systems. Over the past 50 years, tremendous advances in molecular biology and biochemistry have provided essential tools to dissect biological processes down to the molecular level. Most of the time, when studying the structure and function of proteins, obtaining sufficient quantities of the native form of the protein isolated from the relevant cells or tissue is not feasible. The development of recombinant DNA technologies to clone and express genes encoding proteins of interest has revolutionized the design and execution of research projects (Cohen et al., 1973). Indeed, access to purified recombinant proteins enables a wide spectrum of studies, ranging from structural characterization of protein-protein- and protein-nucleic acid interactions to immunization programs to generate antibodies as research tools. The availability of sufficient quantities of purified recombinant proteins is often key to success. Furthermore, recombinant approaches for the production of proteins have profoundly impacted biomedical research and drug development as they have opened the possibility of producing clinical grade proteins as drugs. This has subsequently paved the way for the emergence and fast development of protein biologics that today represent a very successful and quickly expanding class of drugs (Saladin et al., 2009; Chiverton et al., 2010).

A variety of expression systems using prokaryotic and eukaryotic cells as well as in vitro translation systems can, in principle, be envisaged for any protein of interest. However, when the folding and the extent of post-translational modifications of the recombinant protein are critical, the use of a system that best maintains the characteristics of the native protein is preferential. In the context of drug development programs, the biological activity of the protein is of paramount importance. Eukaryotic expression systems, and in particular the use of mammalian host cells, is therefore attractive for the production of human recombinant proteins considered either as therapeutics or therapeutic targets (Andersen et al., 2002). In addition, different forms of a given protein, such as truncations, protein fusions or modifications obtained via site-directed mutagenesis, are often needed in the course of a project and thus require the expression and purification of many protein variants in short periods of time. Therefore, the flexibility and speed of a particular system have also to be taken into consideration. Ideally, an expression system should combine high yield, ease of purification, high product quality and short timelines.

In this chapter, we describe the design and use of multicistronic episomal protein expression vectors combined with improved cell culture methods and single step affinity purification in order to meet these requirements. This approach is rapid (4-6 weeks) and can be used in any laboratory equipped for mammalian cell culture and standard protein purification for the production, in the milligram per liter range, of biologically active recombinant proteins from cell culture supernatants.

2. Approaches for recombinant protein expression in mammalian cells

When studying human or mammalian proteins, expression in mammalian cells not only provides the optimal machinery for proper folding and post-translational modifications, but also facilitates the expression of large and multimeric protein complexes. Several approaches for recombinant protein expression in mammalian cells can be envisaged that dramatically differ in overall yield, workload and timelines (Colosimo et al., 2000). Small-scale transient transfections offer a fast and flexible approach for producing microgram quantities of proteins in a short period of time (days). Different methods have been described for the delivery of plasmid DNA into cells that drives the transient expression of the gene of interest. Up-scaling this approach is feasible in order to produce larger amounts of proteins (milligrams to grams) in a short time. However large-scale transient expression requires significant quantities of both exponentially growing cells and DNA, as well as specialized equipment and is thus not easy to implement in all laboratories (Geisse et al., 2005; Backliwal et al., 2008). Using transient approaches, a new transfection has to be performed for each protein production batch.

The most commonly used strategy for large-scale protein expression (milligram to kilogram scale) is the establishment of stable cell lines, in which the expression plasmid incorporates into the host cell genome (Hacker et al., 2009). The plasmid also includes a marker that allows the selection and clonal amplification of cells that have stably integrated the expression plasmid (Costa et al., 2010). Once the genetic stability of the cell line has been established, it can be expanded, cryopreserved and used for multiple production runs thus maximizing batch to batch consistency of the expressed protein. The main limitation is that stable cell line generation is time-consuming and laborious. It is therefore well suited for the production of proteins at industrial scale or for the production of therapeutic proteins, but not for covering the evolving needs of a research project.

Semi-stable expression offers a compromise between transient transfection and stable cell line generation. In this case, following transfection with a plasmid containing a selectable marker, pools of cells are expanded under selective pressure to obtain large volumes of cells expressing the protein of interest in a relatively short time (weeks). The main advantages and limitations of the methods described above are summarized in Table 1.

Approach	Advantages	Limitations
Transient transfection *Small scale*	Fast (days) Small amounts of DNA and cells required	Reduced yields (micrograms) Increased variability between batches
Transient transfection *Large scale*	Fast (days) Intermediate yields (milligrams to grams)	Large amounts of DNA and cells required Specialized equipment required
Semi stable pools	Relatively fast (weeks) Small amounts of DNA and cells required Intermediate yields (milligrams to grams) Single pool can be used for several production runs	Heterogeneous cell population
Stable cell lines	Homogenous cell population High yields (grams to kilograms) Unlimited number of production runs Increased product consistency	Time consuming (months) Labour intensive cell line screening and characterization

Table 1. Characteristics of different mammalian cell expression systems.

3. Expression vector design

3.1 Episomal vectors

The generation and amplification of semi-stable cell pools is performed under selective pressure, for instance using an antibiotic resistance gene (Lufino et al. 2008; Wong et al. 2009). After transfection, cells that have integrated plasmid DNA into their genome in a location that enables expression of the selectable marker will grow and expand. Depending on the genome integration site, the expression level of the selection marker and the gene of interest can vary significantly. Episomal vectors present the advantage that they can replicate and propagate extrachromosomally in the transfected cells without the need for genomic integration. Episomal vectors contain sequences from DNA viruses, such as bovine papilloma virus 1, BK virus or Epstein-Barr virus. The expression of viral early genes in the host cell such as the Epstein-Barr virus nuclear antigen 1 (EBNA-1) activates the viral origin of replication that is present in the vector, allowing its independent replication. This leads to an efficient retention of multiple copies of the plasmid expressing the gene of interest despite a non-equal partitioning between the dividing cells (Van Craenenbroeck et al., 2000). This high retention rate combined with the selective pressure ensures that expanded cells contain the expression construct. In addition, the high plasmid copy number leads to amplification of the gene of interest and higher protein expression similar to transient transfection experiments (Mazda et al., 1997). Here we focus on the use of the pEAK8 vector that encodes the puromycin resistance gene as a selection marker, the Epstein-Barr virus nuclear antigen 1 (EBNA-1) and the oriP origin of replication (Magistrelli et al., 2010).

3.2 Multicistronic expression vectors

Vectors that can drive multiple gene expression have several advantages. Firstly, they can be used for the production of multimeric protein complexes resulting from the assembly of different polypeptides. Such protein complexes are frequently found in nature and their structural and functional properties can differ from those of their individual subunits. A series of single and dual promoter vectors was generated, based on the pEAK8 episomal vector described above (Figure 1). These vectors incorporated a multicistronic design enabling the co-expression of 2 to 4 independent genes in addition to the antibiotic resistance genes and viral elements of the original episomal vector. The genes encoding the protein of interest can be cloned downstream of the EF1 or SRα promoters that drive strong gene transcription. One or two subsequent internal ribosome entry sites (IRES) drive the translation of the second and third genes (Komar et al., 2005). The gene located after the first IRES is BirA and encodes a biotin ligase that can add a biotin molecule to a protein fused to the biotin acceptor peptide AviTag™ (Tirat et al., 2006). In all the vectors described here, enhanced green fluorescent protein (EGFP) was placed after the last IRES. In a multicistronic transcript, the last cistron is in principle the least translated. Thus, EGFP expression can be used as a reporter to indicate whether the genes of interest are also expressed, although it has to be noted that there is not necessarily a correlation between the expression levels of the protein of interest and EGFP.

In this chapter, we focus on the expression of extracellular proteins or protein complexes. Their secretion into the culture medium is mediated by a leader sequence that can be either the original leader sequence of the protein to be expressed or a generic one. We successfully used the CD33 and Gaussia P. leader sequences for a variety of proteins (Magistrelli et al., 2010). However, significantly different yields can be observed depending on the choice of leader sequence and this parameter should therefore be considered in order to optimize expression levels that are not satisfactory. When biotinylation of the secreted protein is desired, the biotin ligase must also be secreted so that it can add a biotin to the AviTag™ either during the secretion process or in the extracellular milieu. It is therefore mandatory to add a leader sequence to the BirA gene in order to obtain a biotinylated product.

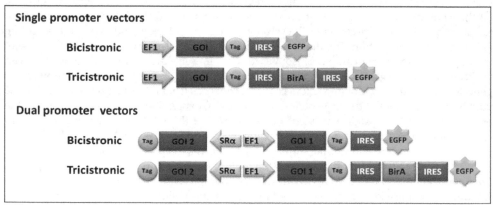

GOI, gene of interest; EF1, EF1 promoter; SRα, SRα promoter; IRES, internal ribosome entry site; BirA, biotin ligase; EGFP, enhanced green fluorescent protein; Tag, peptidic tag (His, AviTag™, HA, Flag or StrepTag).

Fig. 1. Single promoter and dual promoter multicistronic vector design.

4. Generation of semi-stable cell pools

4.1 Transfection and selection

After molecular cloning of the gene - or genes - of interest in one of the vectors described above, the constructs can be verified by DNA sequencing. The plasmids are then transfected into mammalian cells using a liposome-based transfection reagent such as TransIT-LT1 (Mirus, Madison, WI). The transfection step requires only small quantities of DNA and cells, typically 2×10^5 cells and 2 µg of plasmid DNA per well and the transfection is carried out in a 6-well plate. Although different mammalian cell lines can be used, in the examples given below, transformed human embryo kidney monolayer epithelial cells (PEAK cells) were transfected. These cells stably express the EBNA-1 gene, further supporting the episomal replication process, are semi-adherent and can be grown under standard conditions in a cell culture incubator (5% CO_2; 37 °C in DMEM medium supplemented with 10% fetal calf serum). After 24h, cells were placed under selective conditions by adding medium containing 0.5–2 µg/mL puromycin, as cells harbouring the episomal vector are resistant to this antibiotic. 48h after transfection, its efficiency can be evaluated via the brightness of the EGFP signal as well as the proportion of EGFP positive cells in the wells, using epifluorescence microscopy.

4.2 Amplification and production

Cells are maintained in a serum-containing medium, which allows for fast growth, high viability and fast expansion without adaptation to serum-free medium. The selection and amplification process of the pool can easily be monitored by the increase in EGFP signal either using epifluorescence microscopy or flow cytometry (Figure 2).

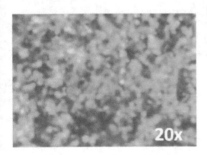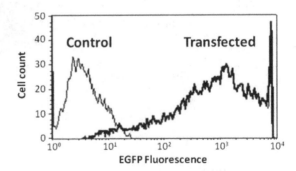

Fig. 2. EGFP expression in transfected pools of PEAK cells after 2 weeks of selection and propagation, monitored by epifluorescence microscopy (left) or flow cytometry (right).

At this stage the expression of the protein of interest can be tested by ELISA or western blot analysis of the supernatant. This early evaluation point at the beginning of the selection process is not absolutely required but provides an indication that the protein can be expressed and secreted in this system. After one week, the cells are transferred to larger vessels and kept under selective pressure in order to expand the transfected cell population. Two to three weeks after transfection, cells can be used to seed Tri-flasks (Nunc) or disposable CELLine bioreactors (Integra) for the production step (Figure 3). Tri-flasks are cell culture vessels that contain three levels for cells to adhere and thus maximize cell density in a limited space. The CELLine is a two compartment bioreactor that can be used in a standard cell culture incubator. The smaller compartment (15 mL) contains the cells and is separated from a larger (one liter) medium-containing compartment by a semi-permeable membrane with a cut-off size of 10 kDa (Bruce et al., 2002; McDonald et al., 2005). This system allows for the diffusion of nutrients, gases and metabolic waste products, while retaining cells and secreted proteins in the smaller compartment. It is also possible to use serum-free medium in the cell compartment and complete medium in the larger compartment. This allows for the secretion of the protein of interest into serum-free medium, which facilitates the purification process by decreasing the amount of contaminants. As the medium and cell compartments can be accessed independently, medium can be replaced or complemented with fresh medium without losing cells or the protein of interest. It overcomes the limitations of standard cell culture containers and offers the advantage that the secreted protein remains concentrated in smaller volume facilitating downstream purification. Using both systems, the culture is maintained for 7–10 days before harvest of the supernatant. As the medium contains serum, the cells maintain high viability and several production runs can be generated using the same cells and containers.

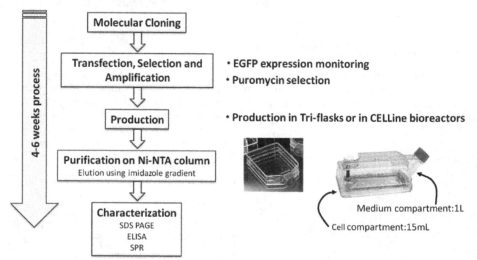

Fig. 3. Schematic representation and timelines of the overall process.

5. Protein purification

In order to streamline the overall process, an important objective is to efficiently purify the secreted recombinant proteins with a single immobilized metal ion affinity chromatography step (IMAC). This affinity purification approach is well established for proteins containing a hexahistidine tag (Block et al., 2009). However, optimization was required to efficiently purify recombinant protein from supernatants containing 10% FCS. After harvest, the cell culture supernatants are clarified by centrifugation and filtered through a 0.22 μm membrane. The supernatant from Tri-flasks or other standard cell culture vessels have to be concentrated 20–40 times using a concentration device such as a SartoFlow 200 (Sartorius) with a membrane having an appropriate cut-off size to retain the protein of interest. As mentioned above, this step is not required using the CELLine bioreactor due to the low volume recovered from the cell compartment. In addition, the concentration step increases the concentration of both the protein of interest and high molecular weight contaminants such as bovine serum albumin or immunoglobulins. In contrast, the supernatant retrieved from the cell compartment of the CELLine bioreactor contains concentrated recombinant protein and reduced levels of contaminants as they cannot cross the 10 kDa membrane separating the two chambers of the reactor. This increased recombinant protein to contaminant ratio greatly enhances the purification efficiency by IMAC. The concentrated supernatant is then supplemented with 100 mM imidazole and loaded onto Ni–NTA affinity chromatography resin (Qiagen). The relatively high concentration of imidazole minimizes binding of contaminants to the resin. After a wash step, proteins are eluted at a flow rate of 2 mL/min using a 30 mL imidazole gradient (20–400 mM imidazole) on an ÄKTA Prime chromatography system (GE Healthcare). The elution gradient further improves the purity of the recombinant protein but can be replaced by a step elution approach if a chromatography system is not available. The eluted fractions can be analyzed by SDS-PAGE or ELISA to determine their recombinant protein content. The fractions of interest are pooled and desalted on PD-10 columns (GE Healthcare) equilibrated with phosphate buffered saline or another appropriate buffer. The desalted proteins can then be quantified using various techniques and their purity analyzed by SDS-PAGE.

During the expansion step of the cell pools, several vials of cells can be cryopreserved in liquid nitrogen. These frozen pools can be thawed at a later stage and rapidly expanded under selective conditions to produce recombinant protein without the need for a new transfection step. This possibility further accelerates the timelines for the generation of additional recombinant protein batches and is a clear advantage over transient transfection approaches (Table 1).

Protein	Yield (mg/L)	Tags
Human IL-17F homodimer	5 - 6	His
Human IL-17A homodimer	6 - 9	His - StrepTag
Human IL-17A/F heterodimer	0.4 - 1	His - StrepTag
Rat IL-17A homodimer	2 - 4	His - StrepTAg
Human IL-6	11 - 14	His - AviTag™
Mouse IL-6	10 - 12	His - AviTag™
Human IL-6 receptor	15 - 16	His - AviTag™
Mouse IL-6 receptor	13 - 14	His - AviTag™
Human IL-6 / IL-6 receptor complex	7 - 9	His
Human CD16b	7 - 8	His - AviTag™
Human CD79A homodimer	4 - 6	His - AviTag™
Human CD79B homodimer	0.5 - 1	His - AviTag™
Human CD79A/B heterodimer	0.8 - 1	His - AviTag™
Human LIF	5 - 6	His
Mouse LIF	4 - 6	His
Mouse FcγRIV	4 - 5	His - AviTag™

Table 2. Proteins successfully expressed using the process described in this chapter.

We have applied the process described above to express and purify 16 mammalian proteins (Table 2). In most cases, several milligrams of highly pure recombinant protein were obtained per liter of cell culture supernatant (Magistrelli et al., 2010). The proteins carried various tags at the N- or C-terminus and could be biotinylated in vivo during the production step via an AviTag™. Most of these proteins were shown to be functionally active in cell-based assays.

Three case studies are reported below, illustrating the expression, purification and characterization of monomeric, homodimeric and heterodimeric recombinant proteins.

6. Case studies

6.1 Monomeric proteins: Recombinant biotinylated human CD16b

Human CD16b, also called FcγRIIIb, is a member of the receptor family for the Fc of immunoglobulins (Takai, 2002). CD16b is a low affinity, GPI-anchored, monomeric receptor expressed on neutrophils and eosinophils. In order to characterize the binding of human IgG to this receptor, a DNA fragment encoding the extracellular portion of hCD16b fused to a C-terminal hexahistidine and an AviTag™ was cloned into the single promoter tricistronic vector that drives the co-expression of BirA and EGFP (Figure 1). After transfection and selection, hCD16b was purified in a single step from the cell culture supernatant as described above. The elution fractions corresponding to the main elution peak of the chromatogram were pooled and desalted into phosphate buffered saline. SDS-PAGE analysis showed that hCD16b was highly pure and had the expected molecular weight of 43 kDa (Figure 4).

In order to verify that hCD16b was biotinylated during expression, an aliquot of each pooled fraction was added to streptavidin coated microplates. After incubation, the plate was washed and the immobilized hCD16b detected using a horseradish peroxydase (HRP)-coupled anti-hexahistidine antibody (Figure 5). The ELISA signals correlated with the intensity of the bands on the SDS-PAGE gel and indicated that hCD16b was efficiently biotinylated, thus facilitating its immobilization on a solid surface.

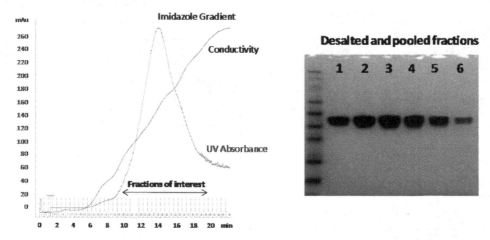

Fig. 4. Purification of hCD16. Chromatogram of the gradient elution step (left). The indicated fractions were collected in several pools, desalted and analyzed by SDS-PAGE (right).

The ability of purified hCD16b to bind IgG was tested by Surface Plasmon Resonance (SPR) using a Biacore 2000 instrument (GE Healthcare). A CM5 biosensor chip was coated with streptavidin followed by injection of hCD16b, resulting in efficient immobilization of the receptor on the chip surface. Kinetic experiments were performed by injecting various concentrations of hIgG1, ranging from 83 nM to 1.3 µM (Figure 5). Dose dependent binding was observed and the affinity of the interaction could be determined (KD=7.18×10^{-5} ± 3.2×10^{-5} M). The affinity of human IgG1 for hCD16b was found to be similar to previously described values (Bruhns et al., 2009).

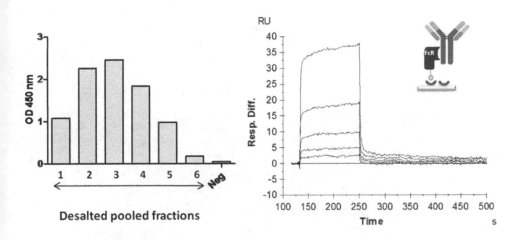

Fig. 5. Characterization of hCD16b by ELISA (left) and by Surface Plasmon Resonance on a Biacore 2000 system (right). Pooled fractions of purified hCD16b were captured on a streptavidin-coated ELISA plate and detected with and anti-hexahistidine antibody (left). The sensorgrams were obtained by injecting hIgG1 on hCD16b immobilized on a streptavidin surface (right). A schematic representation of the IgG-CD16b interaction on the chip surface is represented in the top right corner.

6.2 Homodimeric proteins: Recombinant human IL-17F

Interleukin 17F (IL-17F) is a member of the IL-17 cytokine family and has been shown to play a pro-inflammatory role, particularly in asthma (Kawaguchi et al., 2009). IL-17F is a secreted protein that forms a homodimer linked by a disulfide bond (Hymowitz et al., 2001). This protein was expressed using the single promoter bicistronic variant of the vector (Figure 1). The transfection, selection and purification process described above was applied and the purity of dimeric IL-17F was confirmed by denaturing SDS-PAGE in reducing and non-reducing conditions (Figure 6). As expected the IL-17F monomer and disulphide-linked dimer had an apparent molecular weight of 20 kDa and 40 kDa, respectively.

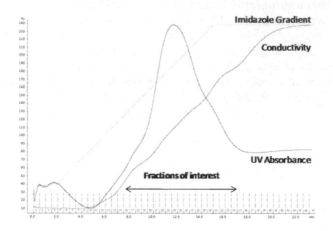

Desalted pooled fractions

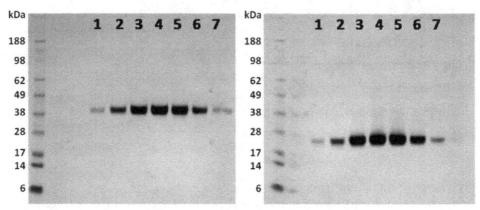

Fig. 6. Purification of human IL-17F homodimer. Chromatogram of the gradient elution step (top). The indicated fractions were collected in several pools, desalted and analyzed by SDS-PAGE in non-reducing conditions (bottom left) and in reducing conditions (bottom right).

The biological activity of purified hIL-17F was assessed in a cell-based assay (Yao et al., 1995). Human fibroblasts secrete IL-6 in response to hIL-17F and this activity can be inhibited by a neutralizing antibody directed against hIL-17F. The neutralizing activity of the antibody was tested in the assay using hIL-17F purified from the supernatant or a commercial source of the cytokine (PeproTech). The results indicated that both cytokines have equivalent biological activities and can be neutralized in a similar fashion by the antibody (Figure 7).

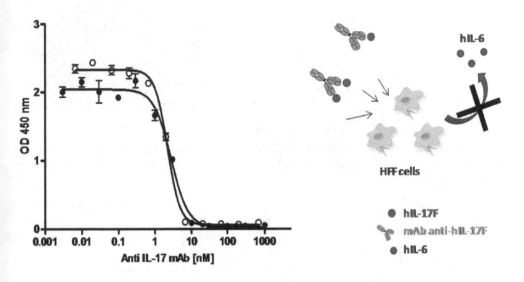

Fig. 7. Neutralization of the biological activity of commercially available hIL-17F (closed circles) and purified hIL-17F (open circles) using an anti-hIL-17F monoclonal antibody (left). Schematic representation of the cell-based assay (right).

6.3 Heterodimeric proteins: Recombinant human CD79A/B

As an example demonstrating successful heterodimeric protein complex expression, the sequences encoding the extracellular domains of human CD79A and CD79B were cloned into a dual promoter, tricistronic vector (Figure 1). The native CD79A/B heterodimer is expressed on B lymphocytes and is the signalling component of the B cell receptor complex (Chu et al., 2001). For recombinant expression, the two proteins were fused to different tags. A hexahistidine tag was introduced at the C-terminus of CD79B for purification by IMAC and an AviTag™ at the C-terminus of CD79A for *in vivo* biotinylation. During the purification step, CD79A/B heterodimer and CD79B homodimer complexes were purified via the hexahistidine tag on CD79B. The purified protein complexes analyzed by SDS-PAGE presented a diffuse pattern due to glycosylation of the proteins (Figure 8). As only CD79A is biotinylated, only the heterodimeric complex can be specifically immobilized on streptavidin coated surfaces.

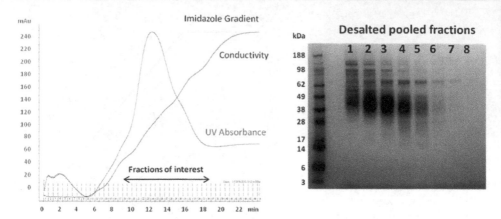

Fig. 8. Purification of human CD79A/B heterodimer. Chromatogram of the gradient elution step (left). The indicated fractions were collected in several pools, desalted and analyzed by SDS-PAGE (right).

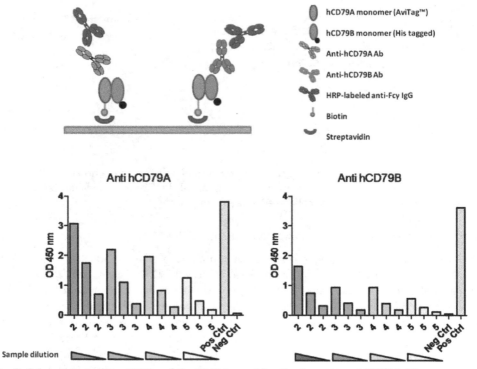

Fig. 9. Schematic representation of the ELISA used for the detection of CD79A/B heterodimer in the pooled elution fractions (top). ELISA signals obtained using anti-CD79A (bottom left) and anti-hCD79B (bottom right) antibodies and dilutions of the pooled fractions (pools 2 to 5). Biotinylated hCD79A and hCD79B homodimers were used as positive and negative controls.

To confirm the presence of the CD79A/B heterodimer in the purified fractions, aliquots were incubated in streptavidin-coated ELISA plates. After washing, commercial anti-CD79A and anti-CD79B antibodies were added to different wells and detected using a HRP-labeled Fcγ specific antibody (Figure 9). Positive signals obtained with both anti-CD79A and anti-CD79B antibodies demonstrated that the heterodimer was efficiently produced and captured via biotin-streptavidin interaction.

7. Conclusions

A number of considerations can influence the choice of system for the expression of recombinant proteins, but the final intended use of the protein is a key determining factor. For most applications, the recombinant protein should closely mimic the structural and functional properties of the native protein. For this reason, mammalian expression that provides the appropriate folding and complex post-translational and secretion machineries represents the system of choice for the study of human proteins, in particular for therapeutic applications (Andersen et al., 2002). High yields, flexibility and speed are also important parameters that are difficult to combine in a single and ideal expression system. Transient expression via plasmid transfection into mammalian cells provides maximal flexibility and speed, at the expense of yield. Indeed, only microgram amounts of protein can be obtained unless performed at large scale, a procedure that has its own technical challenges and is therefore not easily implemented in most laboratories (Hacker et al. 2009). At the other end of the spectrum, the establishment and selection of stable cell lines supporting high expression levels is time consuming but is clearly a system of choice when large amounts of protein are required. In addition, the clonal nature of the cell line increases product homogeneity and batch-to-batch consistency, two highly desirable features for industrial applications. However, neither approach is fully satisfactory when conducting research projects that involve the development of protein-protein interaction assays, structural characterization, immunization or screening procedures. Such activities require milligram amounts of protein and often multiple variants, fusions or tagged version of the same protein have to be generated.

The expression system described in this chapter contributes to bridging the gap between yield and speed by providing several attractive features: integration-free maintenance of the expression vector via autonomous episomal replication; single or dual promoter multicistronic vector design for the co-expression of proteins; secretion of biotin ligase for single site *in vivo* protein biotinylation; co-expression of EGFP for the monitoring of transfection efficiency, selection and amplification of cell pools; cryopreservation of cell pools for additional batch productions; single step affinity purification and use of disposable bioreactors. The latter element, although not strictly required, significantly enhances the overall quality of the process by providing highly concentrated supernatants, containing lower levels of serum derived contaminants, thus improving the performance of the affinity chromatography step. This 4-6 weeks process requires standard cell culture and protein purification equipment and can therefore be implemented in most laboratories. Beyond speed and yield, the possibility to obtain single site biotinylated proteins facilitates the development of protein-protein interaction assays via simple biotin-streptavidin oriented immobilization of one of the interacting partners.

Finally, as illustrated by several examples, the mammalian cell machinery offers the possibility to produce homodimeric and heterodimeric protein complexes in significant quantities.

In our laboratory, the availability of this approach has significantly simplified and streamlined the production of high quality recombinant proteins and supported multiple aspects of our research programs. We therefore believe that it could also benefit other research groups and become more widely used for the expression of recombinant proteins.

8. References

Andersen, D.C.; Krummen, L. (2002). Recombinant protein expression for therapeutic applications, *Current Opinion in Biotechnology*, Vol.13, pp. 117-123.

Backliwal, G.; Hildinger, M.; Chenuet, S.; Wulhfard, J.M. De;Wurm, F.M. (2008). Rational vector design and multi-pathway modulation of HEK 293E cells yield recombinant antibody titers exceeding 1 g/L by transient transfection under serum-free conditions, *Nucleic Acids Research*, Vol.36, e96.

Block, H.; Maertens, B.; Spriestersbach, A.; Brinker, N.; Kubicek, J.; Fabis, R.; Labahn, J.; Schafer, F. (2009). Immobilized-Metal Affinity Chromatography (IMAC): A review, *Methods in Enzymology*, Vol.463, pp. 439-473.

Bruce, M.P.; Boyd, V.; Duch, C.; White, J.R. (2002). Dialysis-based bioreactor systems for the production of monoclonal antibodies – Alternatives to ascites production in mice, *Journal of Immunological Methods*, Vol.264, pp. 59-68.

Bruhns, P.; Iannascoli, B.; England, P.; Mancardi, D.A.; Fernandez, N.; Jorieux, S.; Daëron, M. (2009). Specificty and affinity of human Fcγ receptors and their polymorphic variants for human IgG subclasses, *Blood*, Vol.113, No.16, pp. 3716-3725.

Chiverton, L.M. (2010). Modern challenges in therapeutic protein production, *Expert Review Proteomics*, Vol.7, No.5, pp. 635-637.

Chu, P.; Arber, D. (2001). CD79: a review, Applied Immunochemistry and Molecular Morphology, Vol.9, No.2, pp. 97-106.

Cohen, S.N.; Chang, A.C.Y.; Boyer, H.W.; Helling, R.B. (1973). Construction of biologically functional bacterial plasmids in vitro, *Proceedings of the National Academy of Sciences*, Vol.70, No.11, pp. 3240-3244.

Colosimo, A.; Goncz, K.K.; Holmes, A.R.; Kunzelmann, K.; Novelli, G.; Malone, R.W.; Bennett, M.J.; Gruenert, D.C. (2000). Transfer and expression of foreign genes in mammalian cells, *Biotechniques*, Vol.28, pp. 314-318, 320-322, 324.

Costa, A.R.; Rodrigues, M.E.; Henriques, M.; Azeredo, J. ; Oliveira, R. (2010). Guidelines to cell engineering for monoclonal antibody production, *European Journal of Pharmaceutics and Biopharmaceutics*, Vol.74, pp. 127-138.

Geisse, S.; Henke, M. (2005). Large-scale transient transfection of mammalian cells: a newly emerging attractive option for recombinant protein production, *Journal of Structural and Functional Genomics*, Vol.6, pp. 165-170.

Hacker, D.L.; De Jesus, M.; Wurm, F.M. (2009). 25 years of recombinant proteins from reactor-grown cells- where do we go from here? *Biotechnology Advances*, Vol.27, No.6, pp. 1023-1027.

Hymowitz, S.G.; Filvaroff, E.H.;Yin, J.P.; Lee, J.; Cai, L.; Risser, P.; Maruoka, M.; Mao, W.; Foster, J.; Kelley, R.F.; Pan, G.; Gurney, A.L.; M.de Vos, A.; Starovasnik, M.A. (2001). IL-17s adopt a cysteine knot fold: structure and activity of a novel cytokine, IL-17F, and implications for receptor binding, *European Molecular Biology Organization Journal*, Vol.20, No.19, pp. 5332-5341.

Kawaguchi, M.; Kokubu, F.; Fujita, J.; Huang, S.K.; Hizawa, N. (2009). Role of interleukin-17F in asthma, *Inflammation and Allergy – Drug Targets*, Vol.8, No.5, pp. 383-389.

Komar, A.A.; Hatzoglou, M. (2005). Internal ribosome entry sites in cellular mRNAs: Mystery of their existence, *Journal of Biological Chemistry*, Vol.280, pp. 23425-23428.

Lufino, M.; Edser, P.; Wade-Martins, R. (2008). Advances in High-capacity Extrachromosomal Vector Technology: Episomal Maintenance, Vector Delivery, and Trangene Expression, *Molecular Therapy*, Vol.16, No.9, pp. 1525-1538.

Magistrelli, G.; Malinge, P.; Lissilaa, R.; Fagete, S.; Guilhot, F.; Moine, V.; Buatois, V.; Delneste, Y.; Kellenberger, S.; Gueneau, F.; Ravn, U.; Kosco-Vilbois, M.; Fischer, N. (2010). Rapid, simple and high yield production of recombinant proteins in mammalian cells using a versatile episomal system, *Protein Expression and Purification*, Vol.72, pp. 209-216.

Mazda, O.; Satoh, E.; Yasutomi, K.; Imanishi, J. (1997). Extremely efficient gene transfection into lympho-hematopoietic cell lines by Epstein-Barr virus-based vectors, *Journal of Immunological Methods*, Vol.204, pp. 143-151.

McDonald, K.A.; Lo Ming Hong; Trombly, D.M.; Qing Xie, Jackman, A.P. (2005). Production of human α-1-antitrypsin from transgenic rice cell culture in a membrane bioreactor, *Biotechnological Progress*, Vol.21, pp. 728-734.

Saladin, P.M.; Zhang, B.D.; Reichert, J.M. (2009). Current trends in the clinical development of peptide therapeutics, *The Investigational Drugs Journal*, Vol.12, No.12, pp. 779-784.

Takai, T. (2002). Roles of Fc receptors in autoimmunity, *Nature Review Immunology*, Vol.2, No.8, pp. 580-592.

Tirat, A.; Freuler, F.; Stettler, T.; Mayr, L.M.; Leder, L. (2006). Evaluation of two novel tag-based labelling technologies for site-specific modifications of proteins, *International Journal of Biological Macromolecules*, Vol.39, pp. 66-76.

Van Craenenbroeck, C.; Vanhoenacker, P.; Haegeman, G. (2000). Episomal vectors for gene expression in mammalian cells, *European Journal of Biochemistry*, Vol.267, pp. 5665-5678.

Wong, S.P.; Argyros, O.; Coutelle, C.; Harbottle, R.P. (2009). Strategies for the episomal modification of cells, *Current Opinion in Molecular Therapy*, Vol.11, No.4, pp. 433-441.

Yao, Z.; Painter, S.L.; Fanslow, W.C.; Ulrich, D.; Macduff, B.M.; Spriggs, M.K.; Armitage, R.J. (1995). Human IL-17: a novel cytokine derived from T cells, Journla of Immunology, Vol.155, pp. 5483-5486.

6

The Denaturing and Renaturing are Critical Steps in the Purification of Recombinant Protein in Prokaryotic System

Di Xiang[1], Yan Yu[2] and Wei Han[1]
*[1]Laboratory of Regeneromics, School of Pharmacy,
Shanghai Jiaotong University, Shanghai,
[2]Shanghai Municipality Key Laboratory of Veterinary Biotechnology,
School of Agriculture and Biology, Shanghai Jiaotong University, Shanghai,
China*

1. Introduction

Proteins were first described in 1838 by the Dutch chemist Gerardus Johannes Mulder and named by the Swedish chemist Jöns Jacob Berzelius. [1] The term protein comes from the Greek word proteios, meaning "primary". [2] Protein carries primary functions of the organisms, including both structural and physiological functions. One of the appropriate samples of its importance is the enzyme, which catalyzes almost all biochemical process in vivo. [3] Actually, lots of the most vital biological functions of an organism are fulfilled by proteins, so the isolation and expression of proteins become an important issue for both research and application.

Most proteins consist of 20 different L-α-amino acids, which are linked by peptide bonds, and finally form linear chains. There are 4 levels of structures of a protein. The primary structure of protein lies in the sequence of amino acids, the secondary structure of protein means the basic elements like regularly repeating local structures stabilized by hydrogen bonds, the tertiary structure of protein is its 3-demintional structure, and the quaternary structure exists in some proteins, in which a few single polypeptide chains aggregate together and function as a whole complex. [4]

The sequence of amino acids in a protein is defined by the sequence of its genetic code. [5] As a commonly accepted opinion, the primary structure of a protein determines its higher level structure; however, the biological function of a protein is not decided by the amino acids composition, but most by its 3-dimentional structure. That is also why, not like DNA or RNA, to know the exact sequence of amino acids of a protein cannot ensure the successful production of the biological active one. During expression, isolation or purification processes of a protein, the natural structure of the protein must be preserved or recovered for it to exhibit normal biological activities. In this meaning, to develop a general technology or a method that can be used to purify every protein becomes an impossible task.

The production of recombinant proteins in prokaryotic system is a powerful tool that has been developed in the research and production of functional proteins for many years. To achieve successful production of a recombinant protein, the genetic code which decides the final amino acids sequence in the target protein is firstly sequenced. A vector containing the coding sequence is then cloned or synthesized, and transduced into bacteria, such as Escherichia *coli* (E. *coli*). The protein synthesis system in bacteria then recognizes the specific sequence, and proteins will be produced by the bacteria. Since the post-translational system are different in bacteria than the protein's original organism, the over-expressed recombinant protein are normally not functional, and will aggregate to form inclusion bodies. The inclusion bodies are easily to be isolated, and functional proteins will then be recovered from these inclusion bodies by different denaturing and renaturing techniques. Fortunately, there are some basic principles that can be applied to the expression and purification of protein, and in this chapter, some of them will be discussed. [6]

In general, to achieve successful expression and purification of recombinant protein, the usual steps will be (1) obtain target gene; (2) prepare the expressing vector; (3) clone the target gene into the expressing vector with accuracy ensured by sequencing; (4) transduce host bacteria; (5) induce the expression of target protein; (6) analyze the expressed proteins; and (7) cultivate the engineered strain in large scale and (8) purify the expressed protein.

2. First step: The preparation of engineered bacteria

2.1 The purpose of target protein determines its form

In order to purify a protein, one must first ask themselves a question: what is the purpose of the target protein? If some wants to investigate the biological functions of a protein, it's appropriate to express and purify the protein in its full length or matured form, with natural bioactivity. Or, if the target protein is for antibody production, segment of the protein will serve the purpose well. Since proteins are synthesized as linear polypeptides and undergo post-translational modification and maturation, the denaturing and renaturing must also be considered in the expression and purification of recombinant proteins.

In order to investigate the biological functions of a protein in vitro or in vivo, full length, matured proteins need to be expressed and purified. One good example that can be easily produced and purified in E. *coli* is recombinant human interleukin-1 receptor antagonist (rhIL-1Ra) [7]. rhIL-1Ra can be produced by E. *coli* at large scale in soluble form. The recombinant protein in supernatant of the lyzed bacteria can be purified by 2-steps of ion-exchange chromatography to yield high-output biological functional proteins. However, some other proteins, such as recombinant human Midkine (rhMK), [8] recombinant mouse monokine induced by IFN-γ (rmMig) [9], recombinant human Reg 4 (rhReg-4) [10] or recombinant human Chemerin (rhChemerin), [11] easily form inclusion bodies in E. *coli*, hence, necessary denaturing and renaturing steps must be applied. In some other cases, like recombinant murine CXCL14 (rmCXCL14), [12] full length protein is hard to be purified, and an additional tag will help to get final products with high yield and high purity.

2.2 The preparation of engineered bacteria

2.2.1 Information collection

Once the target protein is decided, the initial step of the expression and purification of a protein is to capture target gene of a protein. This can be simply achieved by acquiring corresponding information of the genetic code of the gene that determines the amino acids sequence of the target protein from some database. The fragment of target genes can be easily synthesized or cloned, and cloned into different commercial available expressing vectors. The packed vector, containing the target gene, the selective markers and the elements needed for expressing in bacteria is then cloned into expressing strains, such as E. *coli* BL21 (DE3). Once the engineered strain is successfully constructed, it will then use its own resources to produce the alien protein.

2.2.2 The vectors and strains are the keys ensuring successful expression and purification

E. *coli* was discovered by Theodor Escherich in 1885. [13] It is a Gram-negative, rod-shaped bacterium that is commonly found in the lower intestine of endotherms. [14] E. *coli* is one of the most adopted bacteria in the expression and purification of recombinant protein, since it can be easily cultivated in laboratory in a relative cheap way, and has been intensively investigated as a widely studied prokaryotic model organism.

The pET system is developed by Novagen Company as a powerful system for the cloning and expression of recombinant proteins in E. *coli*. In this system, target genes are cloned in pET plasmids under control of T7 transcription. The expression is induced by T7 RNA polymerase in the host cell. When fully induced, target protein may consist more than 50% of the total cell protein in a few hours after induction. On the other hand, in the un-induced state, the expression level of target gene is low. pET system and its host bacteria E. *coli* BL21 (DE3) can be used as an optimized system in the expression and purification of recombinant protein. In order to choose a suitable pET system, a good source of knowledge comes from the manual of pET system offered by Novagen. [15]

3. The cultivation of engineered bacteria and the extraction of recombinant proteins

3.1 The cultivation conditions of E. *coli* should be tested and verified

Although the expression of recombinant protein in E. *coli* is normally high and consists large part of total bacteria protein, the expression could also be low, and the cultivation conditions of bacteria must be tested, including different temperature, vortex speed and induction conditions.

Soluble, highly expressed target proteins are priorities in some cases, because the yield of target protein is high, and the biological activity is normally preserved; however, the down-stream purification may be difficult, because the target protein is mixed with all soluble bacteria protein. Another side effect is that target protein may be digested by protease. On the other hand, if the target protein exists in inclusion bodies, it is non-biological active, but can be easily isolated with high purity, and the recombinant protein will not harmed by the presence of protease. After the optimized denaturing and renaturing processes, the downstream process may be much easier than soluble expressed target protein. In each case, the expressing condition must serve the downstream purification steps.

Normally, the conditions that lead to high growth speed and high expression result in aggregation of target proteins, for there are short time for the protein to fold correctly and so the inclusion bodies will be formed. Strong induction condition easily leads to high-expressing of target protein, and also leads to the formation of inclusion bodies. [16] On the contrary, lowering the cultivation temperature, adding some additives in cultivation medium and change medium can reduce the formation of inclusion bodies, and the target protein will present in soluble form. [17]

3.2 The extraction of crude protein from cell lysates

Different methods can be applied to isolate the target proteins from the cell lysates maximally. Normally the first step after the induction of target protein is to lyze the bacteria. This could be achieved by different methods including frozen-thaw, ultrasonication, lysozyme treatment, or the combination of them. Depending on the forms of expressed protein, the target protein can be found in supernatant or precipitate of the cell lysate. This could be identified easily by Western Blotting.

In this step, the target protein should be verified by Western Blotting using certain antibodies, and the proportion of the target protein in total cell lysates should be calculated. If the expression level of the recombinant protein is too low, the downstream process will become very difficult. Optimization needs to be considered to enhance the production of the target protein.

The goal of the extraction of recombinant protein is to facilitate the downstream process. The more concentrated proteins in this step, the easier the later purification will be.

4. The denaturing and renaturing of recombinant protein

4.1 The purification of soluble recombinant protein

If the protein exists in soluble form, it will probably possess natural bioactivity. The protein then needs no denaturing and renaturing treatments. For instance, when packed into pET11a expressing vector, rhIL-1ra is induced in a soluble form in E. *coli*. A good example of how to express and purify soluble rhIL-1Ra is indicated below. [18]

The interleukin-1 receptor antagonist (IL-1Ra) is a protein encoded by the IL1RN gene. IL-1Ra was initially called the IL-1 inhibitor and was discovered in 1984. [24] IL-1Ra binds to the cell surface interleukin-1 receptor (IL-1R) hence blocks IL-1 signal pathway.

The full length of secreted rhIL-1Ra gene (gi: 186385) is 534bp with a CDS of 534bp, whereas the first 75 bps encode a 25-amino acid signal peptide. A DNA fragment corresponding to bp 76 to 534 in the gene IL1RN which encodes 152 amino acid residues in the matured rhIL-1Ra is cloned and packed into pET11a for the protein expression in E. *coli*. After induction, rhIL-1Ra mainly exists in the supernatant of the cell lysates.

The detailed processes of the expression and purification of rhIL-1a are listed below:

•   Inoculate a single colony of transformed bacteria into LB medium containing 100 µg/mL ampicillin at 37 °C overnight with vigorous shaking. Inoculate large-scale of LB medium containing 100 µg/mL ampicillin with small-scale culture mentioned above at 37°C for 2 h with vigorous shaking. 1 mM IPTG was used to induce the strain to produce rhIL-1Ra afterwards. The incubation continued for additional 4 h at 37 °C and the cell pellets were spun down.

- The supernatant after ultrasonication was collected. 0.284 g ammonium sulfate was used for each milliliter of supernatant to precipitate rhIL-1Ra. Precipitated protein was solubilized 10-fold into the loading buffer 1 (1 mM EDTA, 20 mM Tris-HCl, pH 9.0). After centrifugation the supernatant was collected and named as rhIL-1RaST (Fig. 1)

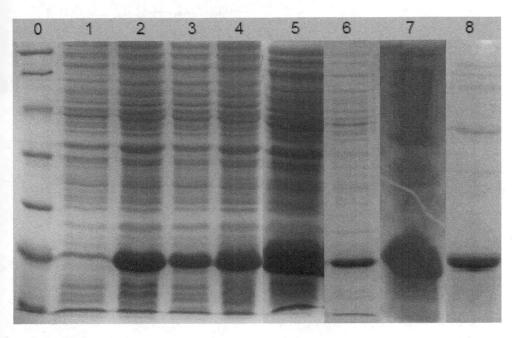

Fig. 1. Picture of Commassie-Blue staining after SDS-Page of induced rhIL-1Ra. (Lane 0: Makers; 1: Engineered E. coli without IPTG inducement; 2: IPTG induced E. *coli*; 3: Inclusion bodies, 4: Supernatant after unltrasonication; 5: Solublized inclusion bodies in 8M Urea; 6: rhIL-1RaIB; 7: Supernatant precipitated by ammonium sulfate; 8: rhIL-1RaST)

- rhIL-1Ra can then be purified by running the solution over DEAE-Sepharose and S-Sepharose, successively. First rhIL-1Ra solution was run over DEAE-Sepharose and flow through solution is collected. rhIL-1Ra purity was determined by Commassie-Blue staining after SDS-Page. The solution was then adjusted to pH 4.5. After centrifugation the supernatant was applied to a cation-exchange S-Sepharose. A 0 - 1 M NaCl gradient (1%/min) in loading buffer 2 (32.26 mM HAc, 17.74 mM NaAc, 1 mM EDTA, pH 4.5) was applied to elute bounded rhIL-1Ra from the column. According to UV absorption peak fractions were collected. The protein purity was determined by Commassie-Blue staining after SDS-Page (Fig. 2, 3, 4)

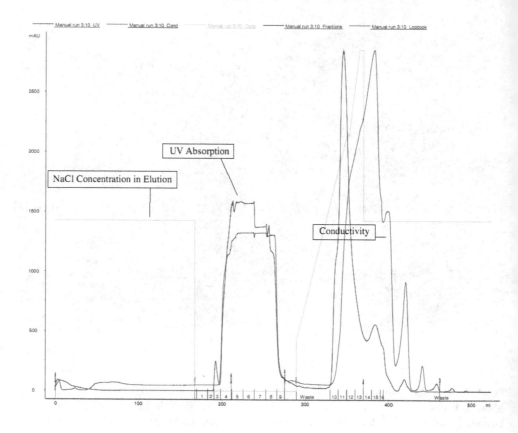

Fig. 2. Purification of rhIL-1RaST by DEAE-Sepharose (Flow through solution marked with 4/5/6/7/8/9 were collected)

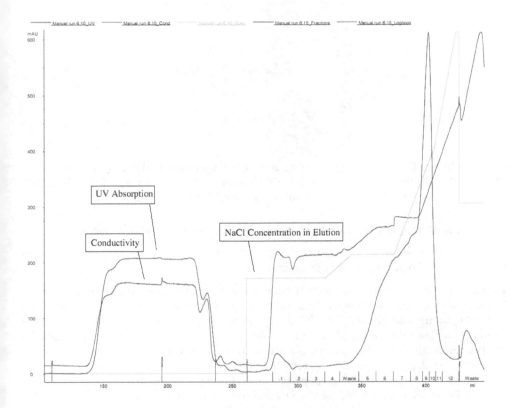

Fig. 3. Purification of rhIL-1RaST by S-Sepharose (According to UV absorption peak, fractions 6, 7, 8, 9 and 10 were collected and determined for its rhIL-1Ra concentration and purity)

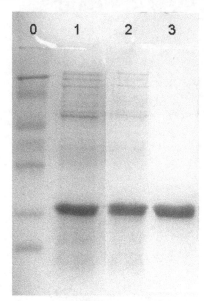

Fig. 4. Picture of Commassie-Blue staining after SDS-Page of purified rhIL-1Ra (Lane 0: Markers; 1: Flow through solution of rhIL-1RaST after DEAE-Sepharose purification; 2: rhIL-1Ra before S-Sepharose purification; 3: rhIL-1Ra after S-Sepharose purification)

4.2 The denaturing and renaturing of recombinant protein in inclusion bodies

If the protein exists in inclusion bodies form, denaturing and renaturung treatments are mandatory for the purification of this protein.

An ideal denaturing and renaturing system must serve the following purpose: high-output, low-loss and normal bio-activity final products. In industrial level, the ideal renaturing technology must also be easily amplified to produce large-scale products. Must commonly used denaturing solution is 6 M guanidine chloride or 8 M urea. There are some renaturing methods including dialysis, dilution, ultra-filtration. Some additives can be applied into the renaturing system to enhance the refolding efficacy such as denaturing agents, PEG, detergents, redox system, L-Arg, chaperone, etc. However, there are no general guidelines that can be applied to all refolding systems and one must test the efficiency of each method to ensure the optimized output. Some commercial available kits can be used to test the optimized renaturing system.

Here are a few samples of protein expressed in inclusion bodies:

Monokine induced by IFN-γ (Mig) is a CXC-chemokine (CXCL9). It plays important roles in regulation of immune activities, and knowledge of the protein in areas of allograft transplants, autoimmune diseases, and cancer therapy is evolving quickly. Mature murine MIG is a protein of 106 amino acids with molecular weight of 14.4 kDa. It contains 4 cysteines and 2 disulfide-bonds. The mature mouse MIG cDNA covers the coding sequences of the gene without the signal peptide coding sequences. After induction, rmMig mainly exists in inclusion bodies of the cell lysates. [19]

Chemerin is a novel chemokine that binds to the G protein-coupled receptor (GPCR) ChemR23. It is secreted as a precursor and executes pro-inflammatory functions when the last six amino acids are removed from its C-terminus by serine proteases. Mature human Chemerin is a protein of 137 amino acids with molecular weight of 16 kDa. The mature human Chemerin cDNA covers the coding sequences of the gene without the signal peptide coding sequences and the last 21 bps coding the peptide need to be removed during maturation. After induction, rhChemerin mainly exists in inclusion bodies of the cell lysates. [20]

Regenerating gene (Reg) IV is a newly discovered member of the regenerating gene family belonging to the calcium (C-type) dependent lectin superfamily and was firstly isolated

Recombinant Protein	rmMig	rhChemerin	rhReg-4	rhMK	rmCXCL-14
Protein Forms	Full length without signal peptide	Full length without signal peptide and last 6 amino acids	Full length without signal peptide	Full length without signal peptide	Full length without signal peptide
Tags	N/A	N/A	N/A	N/A	His-Tag
Engineered Bacteria	E. coli BL21 (DE3)	E. coli BL21 (DE3)	E. coli BL21 (DE3)	E. coli BL21 (DE3)	E. coli BL21 (DE3)
Expressing Vector	pET28a	pET28a	pET30a	pET30a	pET28a
Cultivation Medium and Temerature	LB @ 37 °C, 3 h	LB @ 42 °C, 4 h	LB @ 37 °C, 4 h	LB @ 42 °C, 3 h	LB @ 37 °C, 6 h
Induction	1 mM IPTG	1 mM IPTG	1 mM IPTG	1 mM IPTG	0.1 mM IPTG
Cell Lysate Preparation	Ultraso-nication	Ultraso-nication	Ultraso-nication	Ultraso-nication	Ultrasonication
Forms	Inclusion bodies	Inclusion bodies	Inclusion bodies	Inclusion bodies	Inclusion bodies
Denaturing	8 M urea, 20 mM Na2HPO4, 1 mM EDTA, 50 mM NaCl, pH 9.0.	6 M guanidine chloride, 1 mM EDTA, 50 mMNaCl, 50 mM Tris–HCl; pH 8.0.	6 M guanidine chloride, 1 mM EDTA, 50 mM NaCl, 50 mM Tris–HCl, pH 8.0.	6 M guanidine chloride, 20 mM Na2HPO4, 1 mM EDTA, 50 mM NaCl, 0.1 mM PMSF, pH 8.0.	6 M guanidine hydrochloride, 50 mM imidazole, 0.5 M NaCl, 0.02 M Tris-HCl, 0.1% Tween-20, 1 mM PMSF, pH 7.9.

Renaturing	The protein solution was dropped by pumping into 10-fold volume refolding Buffer (20 mM NaPO4, 1 mM EDTA, pH 6.0) under vigorous agitation. pH of the solution was maintained at 6.0. Then an equal volume dilution Buffer (20 mM NaPO4, 1 mM EDTA, pH 8.5) was added into the folding buffer. The refolded protein solution was set at 4°C for about 24 h.	The protein was refolded by slow dilution of the denaturing buffer into 100 volumes of renaturing buffer (1 mM reduced glutathione (GSH), 0.1 mM oxidized glutathione (GSSG), 0.5 M guanidine chloride, 0.4 M sucrose, 0.1 M Tris–HCl; pH 9.5), and concentrated eightfold using the tangential flow concentrator with a 10 kDa cutoff filter (Minimate TFF System; Pall, USA). The protein solution was further diluted into 10 volumes of dilution buffer (1 mM GSH, 0.1 mM GSSG, 0.1 M Tris; pH 9.5) and the pH was adjusted to 7.5 with 1 M HCl.	The denatured protein was added dro-pwise to a total of 500 ml of the optimized refolding buffer and left to stand for 24 h. The buffer at pH 10.5 consists of 50 mM Tris–HCl, 10 mM KCl, 240 mM NaCl, 2 mM MgCl2, 2 mM CaCl2, 0.5 M gua-nidine–HCl, 400 mM sucrose, 500 mM L-argi-nine–HCl, 1 mM reduced glutathione, and 0.1 mM oxidized glutathione. The refolding buffer was concentrated and changed to the phosphate buffer (20 mM, pH 6.5) using the tangential flow concen-trator with a 10 kDa cutoff filter (Mini-mate TFF System, Pall, USA).	The protein refolding was proceeded by drop-wise dilution into defined protein folding buffer The protein solution was dropped by pumping into 10-fold volume refolding buffer (20 mM Na2HPO4 and 1 mM EDTA, pH 7.4) under vigorous (magnetic stirrer) agitation. pH value of the solution was maintained at 7.4. Then an equal volume dilution buffer (20 mM Na2HPO4 and 1 mM EDTA, pH 8.0) was added into the folding buffer, and final pH of the solution was maintained at 8.0. This protein solution was stored at 4 °C for 24 h.	Three ml Ni-Sepharose (Pharmacia, USA) was loaded onto column and recharged by 5 column volumes (CV) of 50 mM NiSO4 and equilibrated by 3 CV of denatured binding buffer. Cell lysate was loaded onto the purifi-cation column at rate of 0.5 ml/min at 4 °C. Then, the column was washed with 10 CV of denatured wash buffer (60 mM imidazole, 0.5 M NaCl, 0.02 M Tris-Cl pH 7.9, 0.1% Tween 20, and 6 M Guanidine hydro-chloride) before elution by denatured elute buffer (1 M imidazole, 0.5 M NaCl, 0.02 M Tris-Cl pH 7.9, 0.1% Tween-20, and 6 M Guanidine hydrochloride). The intermolecular disulfi-de bonds were reduced by adding dithiothreitol (DTT) into protein solu-tion to final concen-tration of 0.2 M. After overnight incubation at 4 °C, the solution was gradually diluted with refolding buffer (20 mM Tris-Cl pH 7.9, 0.1 M NaCl, 1 mM reducing glutathione, 1 mM EDTA, 10% sucrose, 0.1% Triton-X114) to 20-fold of the initial volume over 1 h with gentle stirring at room temperature and then incubated at room temperature for 1 h. Precipitate was removed by centrifugation at 20,000×g for 30 min.

Table 1.

from a cDNA library of ulcerative colitis (UC) tissues by Hartupee et al. The Human Reg IV is synthesized as a 158 amino acid (aa) precursor with a 22 aa signal sequence and a 136 aa mature chain. The molecular weight (MW) of mature Reg IV protein is about 16 kDa. The mature form of human Reg IV is encoded by the cDNA sequences of the gene from 354 to 743 bp without the signal peptide coding sequences. After induction, rhReg IV mainly exists in inclusion bodies of the cell lysates. [21]

Midkine is a heparin-binding growth factor, which plays important roles in the regulation of cell growth and differentiation. The non-tagged recombinant human midkine (rhMK) is therefore required to facilitate its functional studies of this important growth factor. The mature human midkine protein is a protein contains 122 amino acids with a molecular weight of 17 kDa. The mature form of human midkine is encoded by the full-length sequence from 289 to 654 bp without the signal peptide coding sequences. After induction, rhMidkine mainly exists in inclusion bodies of the cell lysates. [22]

Mouse CXCL14/BRAK is a monocyte-selective chemokine which is expressed in almost all normal tissues. The mouse cxcl14 mRNA encodes 99 amino acids, in which N-terminal 22 amino acids serve as a signal peptide which is cleaved as the protein is secreted. The remaining C-terminal 77 amino acids construct the mature chemokine with a calculated molecular weight of 9.4 kDa and pI 9.9. The mature form of mouse CXCL14 is encoded by the full-length sequence from 445 to 678 bp without the signal peptide coding sequences. After induction, rmCXCL14 mainly exists in inclusion bodies of the cell lysates. [23]

The detailed processes of the denaturing and renaturing of those proteins were already published, and can be brifly introduced below Table 1.:

5. The purification of recombinant protein: Chromatography technologies

Like the denaturing and renaturing, lots of technologies had been developed to purify recombinant. Among which chromatography technologies are the most powerful tools, and can give satisfactory purity and yield of final protein products. According to the proteins pI, Ion exchange chromatography can be adopted. According to the hydrophilic and hydrophobic properties and molecular size, HPLC can be used. pET system carries certain fusion tag and can be easily identified by certain types of Affinity Chromatography. The most important guideline is, the purification of recombinant protein may not be achieved by simply using one technology mentioned above; on the contrary, the combination of them can lead to optimized results.

Like the above section, the detailed processes of the purification of the above-mentioned proteins were already published, and can be briefly introduced in table 2.

Recombinant Protein	rmMig	rhChemerin	rhReg-4	rhMK	rmCXCL-14
Protein Forms	Full length without signal peptide	Full length without signal peptide and last 6 amino acids	Full length without signal peptide	Full length without signal peptide	Full length without signal peptide

Recombinant Protein	rmMig	rhChemerin	rhReg-4	rhMK	rmCXCL-14
Tags	N/A	N/A	N/A	N/A	His-Tag
Purification	The supernatant of refolding was loaded on an S-Sepharose column with a volume of 20 ml. The column was pre-equilibrated with Buffer A (20 mM NaPO4, 1 mM EDTA, pH 7.2). Sample was loaded at speed of 0.5 ml/min and the column was then washed with 2 column volumes of Buffer A. The column bound proteins were eluted using a programmed gradient of Buffer B (20 mM NaPO4, 1 mM EDTA, 1 M NaCl, pH 7.2) at a speed of 3 ml/min.	The supernatant of refolding was loaded onto a Q Sepharose FF column equilibrated with wash buffer 1 (20 mM Tris–HCl, 1 mM EDTA, 25 mM NaCl; pH 7.5) at a speed of 5 ml/min. The flow-through was collected and adjusted with glacial acetic acid to pH 4.5. After centrifugation, the supernatant was loaded onto an S-Sepharose FF column equilibrated with wash buffer 2 (50 mM NaAc–HAc buffer, 1 mM EDTA, 25 mM NaCl; pH 4.5) at a speed of 5 ml/min. The protein was eluted using a 0.3–1.0 M NaCl gradient in wash buffer 2 (1%/min) at a speed of 1 ml/min.	The supernatant of refolding was loaded on an S-Sepharose column in a volume of 20 ml. The column was pre-equilibrated with Buffer A (20 mM phosphate buffer at pH 6.5, 1 mM EDTA). Sample was loaded at speed of 3 ml/min and the column was then washed with 2 column volumes of the Buffer A. The column bound proteins were eluted using a programmed gradient of Buffer B (20 mM phosphate buffer at pH 6.5, 1 mM EDTA, 1 M NaCl) at a speed of 1 ml/min.	The supernatant of refolding was loaded on an S-Sepharose column with a volume of 20 ml. The column was pre-equilibrated with Buffer A (20 mM Na2HPO4 and 1 mM EDTA, pH 8.0). Sample was loaded at a flow rate of 0.5 ml/min and the column was then washed with 2 column volumes of Buffer A. The column-bound proteins were eluted using a programmed gradient of Buffer B (20 mM Na2HPO4, 1 mM EDTA, and 1 mM NaCl, pH 8.0) at a flow rate of 1 ml/min.	The supernatant of refolding was loaded onto an SP Sepharose column with a volume of 20 ml, which had been pre-equilibrated with Buffer A (20 mM Tris-HCl, pH 8.0), with flow rate maintained at 5.0 ml/min. After washing with 5 CV of Buffer A, protein was eluted using programmed gradient Buffer B (20 mM Tris-Cl and 1.0 M NaCl, pH 8.0), which was linearly increased from 20% to 80% with a flow rate of 1.0 ml/ml. The eluted fractions containing the target protein were desalinated using a Sephadex G25 column (height 60 cm and diameter 1.5 cm), which was pre-equilibrated with 2 CV of Buffer A.
Target Protein Purity	99%	99%	97%	98%	97%
Final Yield	5.2%	14%	28%	6.2%	11%

Table 2.

6. Purified recombinant proteins: The quality control and the storage

6.1 The storage of recombinant proteins

Protein samples are easily to lose their bioactivity, so it's important to test appropriate storage conditions and buffers of this protein. After purification, the composition of storage solution and storage condition can both affect the bio-activity of recombinant proteins. For instance, rhIL-1Ra can be safely stored at 4°C for nearly half a year in 10 mM sodium citrate, 140 mM sodium chloride, 0.5 mM EDTA, while rhChemerin, when stored in PBS, pH 7.4, must be frozen in -80°C refrigerator. A good source of the knowledge of the suitable storage of the recombinant protein comes from the manual of commercialized recombinant protein.

6.2 The quality control of recombinant proteins

After protein expression and purification, some quality control items should be tested to serve the later in vivo and in vitro experiment, or other application. These test items include but not limit to:

Concentration (measured by BSA or Bradford methods, etc.)

Purity (measured by silver staining after SDS-Page, HPLC, or SEC-HPLC etc.)

Sequence confirmation (measured by GC-Mass or etc.)

Endotoxin (measured by TAL or LAL test)

Bioactivity (measurements depends. Bioactivity should be expressed as EC50, if applicable).

7. References

[1] Wikipedia: Protein
[2] Judy Pearsall, Patrick Hanks, New Oxford Dictionary of English, Oxford University Press. ISBN-13: 978-0198604419
[3] Smith AL. Oxford dictionary of biochemistry and molecular biology. Oxford [Oxfordshire]: Oxford University Press. ISBN 0-19-854768-4.
[4] Robert K. Murray, Daryl K. Granner, Peter A. Mayes, Victor W. Rodwell, Harper's Biochemistry, McGraw-Hill Publishing Co.
[5] Robert K. Murray, Daryl K. Granner, Peter A. Mayes, Victor W. Rodwell, Harper's Biochemistry, McGraw-Hill Publishing Co.
[6] The Recombinant Protein Handbook: Protein Amplification and Simple Purification, Amersham Pharmacia Biotech Inc.
[7] D. Xiang, Y. Guo, J. Zhang, J. Gao, H. Lu, S. Zhu, M. Wu, Y. Yu, W. Han, Interleukin-1 receptor antagonist attenuates cyclophosphamide-induced mucositis in a murine model, Cancer Chemoth Pharm, DOI: 10.1007/s00280-010-1439-1.
[8] Z. Zhang, L. Du, D. Xiang, S. Zhu, M. Wu, H. Lu, Y. Yu, W. Han, Expression and purification of bioactive high-purity human midkine in Escherichia coli, J Zhejiang Univ Sci B 10 79-86.
[9] H. Lu, M. Yu, Y. Sun, W. Mao, Q. Wang, M. Wu, Han, W. Expression and purification of bioactive high-purity mouse monokine induced by IFN-gamma in Escherichia coli, Protein Express. Purif. 55 (2007) 132-138.

[10] G. Hu, J. Shen, L. Cheng, D. Xiang, Z. Zhang, M. He, H. Lu, S. Zhu, M. Wu, Y. Yu, X. Wang, W. Han, Purification of a bioactive recombinant human Reg IV expressed in Escherichia coli, Protein Expr Purif 69 186-190.

[11] D. Xiang, J. Zhang, Y. Chen, Y. Guo, A. Schalow, Z. Zhang, X. Hu, H. Yu, M. Zhao, S. Zhu, H. Lu, M. Wu, Y. Yu, A. Moldenhauer, W. Han, Expressions and purification of a mature form of recombinant human Chemerin in Escherichia coli, Protein Expr Purif 69 153-158.

[12] Jingjing Li, Jin Gao, Sunita Sah, Uttam Satyal, Ruliang Zhang, Wei Han, Ya Yu, Expression and Purification of Bioactive High-Purity Recombinant Mouse CXCL14 in Escherichia coli, Appl Biochem Biotechnol (2011) 164:1366–1375.

[13] Feng P, Weagant S, Enumeration of Escherichia coli and the Coliform Bacteria". Bacteriological Analytical Manual (8th ed.). FDA/Center for Food Safety & Applied Nutrition. Retrieved 2007-01-25.

[14] Wikipedia: Escherichia coli

[15] pET System Manual, Novagen Inc.

[16] Gribskov M, Burgess RR. Overexpression and purification of the sigma subunit of E.coil RNA polymerase. Gene,1983,26:109

[17] Xie Y, Wetlaufer D B. Control of aggregation in protein refolding: the temperature-leap tactic. Protein Sci,1996,5(3):517-523

[18] D. Xiang, Y. Guo, J. Zhang, J. Gao, H. Lu, S. Zhu, M. Wu, Y. Yu, W. Han, Interleukin-1 receptor antagonist attenuates cyclophosphamide-induced mucositis in a murine model, Cancer Chemoth Pharm, DOI: 10.1007/s00280-010-1439-1

[19] H. Lu, M. Yu, Y. Sun, W. Mao, Q. Wang, M. Wu, Han, W. Expression and purification of bioactive high-purity mouse monokine induced by IFN-gamma in Escherichia coli, Protein Express. Purif. 55 (2007) 132-138

[20] D. Xiang, J. Zhang, Y. Chen, Y. Guo, A. Schalow, Z. Zhang, X. Hu, H. Yu, M. Zhao, S. Zhu, H. Lu, M. Wu, Y. Yu, A. Moldenhauer, W. Han, Expressions and purification of a mature form of recombinant human Chemerin in Escherichia coli, Protein Expr Purif 69 153-158

[21] G. Hu, J. Shen, L. Cheng, D. Xiang, Z. Zhang, M. He, H. Lu, S. Zhu, M. Wu, Y. Yu, X. Wang, W. Han, Purification of a bioactive recombinant human Reg IV expressed in Escherichia coli, Protein Expr Purif 69 186-190

[22] Z. Zhang, L. Du, D. Xiang, S. Zhu, M. Wu, H. Lu, Y. Yu, W. Han, Expression and purification of bioactive high-purity human midkine in Escherichia coli, J Zhejiang Univ Sci B 10 79-86

[23] Jingjing Li & Jin Gao & Sunita Sah & Uttam Satyal & Ruliang Zhang & Wei Han & Ya n Yu, Expression and Purification of Bioactive High-Purity Recombinant Mouse CXCL14 in Escherichia coli, Appl Biochem Biotechnol (2011) 164:1366–1375

[24] Dinarello CA (December 1994). "The interleukin-1 family: 10 years of discovery". FASEB J. 8 (15): 1314–25.

7

Identification and Characterization of Feruloyl Esterases Produced by Probiotic Bacteria

Kin-Kwan Lai, Clara Vu, Ricardo B. Valladares,
Anastasia H. Potts and Claudio F. Gonzalez
University of Florida,
USA

1. Introduction

A variety of phenolic compounds are naturally available, and contain one or more phenolic rings with or without substituents such as hydroxyl or methoxy groups. The term phytophenol, or phytochemical, is also used due to the widespread distribution of these chemicals throughout the plant kingdom (Huang *et al.*, 2007). Phytophenols are secondary metabolites of plants, which are primarily used in defense against ultraviolet radiation and pathogens (Beckman, 2000). These chemicals also participate in the formation of macromolecular structures in plant cell walls, and are naturally present in the form of monophenols or polyphenols with ester linkages. The presence of phenolic ester linkages limits the hydrolytic activity of enzymes such as xylanases, cellulases, and pectinases, by shielding the site of hydrolysis on plant cell walls from these enzymes. Hydrolyzing the ester linkages within the phytophenols releases the phenolic acids and relaxes the structure of the plant cell wall, aiding in the degradation and maximizing the nutritional value of dietary fiber. Phenolic acids such as ferulic acid, caffeic acid, chlorogenic acid, and rosmarinic acid are studied extensively due to their anti-oxidative, anti-inflammatory, and other health related properties which have been demonstrated both *in vitro* and *in vivo* (Srinivasan *et al.*, 2007). Even though phenolic acids can be easily found in dietary fiber, the ester linkages prevent their absorption in the human intestine. It has been demonstrated that only small monophenolic acids, but not esterified phenolic acids, can be absorbed efficiently by the monocarboxylic acid transporter (Konishi *et al.*, 2005). Thus, an enzymatic step is required to convert the esterified phenolic acids into monophenolic acids prior to absorption. In the presence of water, a specific type of enzyme, feruloyl esterases (FAEs), is able to hydrolyze the phenolic compounds into respective alcohols and phenolic acids. Thus, FAEs become one of the target fields of study to improve the bioavailiability and assimilation of phenolic acids in the human diet.

Feruloyl esterases (EC 3.1.1.73) are classified as a subclass of carboxylic acid esterases (EC 3.1.1.1). Alternate names such as ferulic acid esterases, cinnamoyl ester hydrolases, cinnamoyl esterases, and hydroxycinnamoyl esterases are generally used in the literature. They are also termed hemicellulase accessory enzymes because they can act synergistically with xylanases, cellulases, and pectinases to degrade the hemicellulose of plant cell walls. High substrate preference of FAEs is achieved when the carboxylic ester is in the phenolic/aromatic form, such that an aromatic hydrocarbon is attached to the carbon atom of the carbonyl group of the ester. FAEs are important enzymes in the rumen ecosystem due to their ability to increase the absorption of plant-based energy sources in ruminant animals. In recent years, several FAEs from fungi were partially characterized, but little is known about bacterial or plant FAEs. For both medicinal and industrial applications, there is an increasing amount of research focused on FAEs that are capable of releasing monophenols from plant biomass. To the best of today's knowledge, humans do not synthesize FAEs. However, FAE activity is found in total human gut microbiota (Kroon *et al.*, 1997, Gonthier *et al.*, 2006), indicating that FAEs are present in the intestine and may contribute to the release of phytophenols from dietary fiber. Currently, most characterized FAEs have been identified from fungi, and the lack of FAEs identified in other organisms, particularly intestinal bacteria, has limited their application.

Tertiary structure information on FAEs is scarce, while primary and secondary structure is poorly conserved between fungi and bacteria. Identifying and characterizing FAEs in bacteria is an important challenge. This chapter tells the story of the identification, purification, characterization, and crystallization of FAEs from probiotic bacteria. The potential FAEs were identified based on the enzymatic activity displayed by the bacterial strains as well as bioinformatics analysis. The work discussed herein will provide insight for further exploration of FAEs in other species, enhancing the path for medicinal and industrial application of these enzymes.

2. Identification of feruloyl esterases from bacteria

Our understanding of the relevance of the commensal microbiota in relation to the healthy status of the host is rapidly expanding. However, the mechanisms by which these microorganisms interact with the host are still unclear. Important technological advances such as rapid sequencing methods, bioinformatics, and species identification using 16S rDNA are valuable tools to describe the variability and composition of these small ecosystems. One of the most interesting applications of the study of commensal microbiota is the identification of species potentially responsible for, or correlated with, specific host diseases. For example, there are noticeable changes in the composition of the gut microbial ecosystem of diabetes patients compared to healthy individuals (Vaarala *et al.*, 2008). Some studies indicate that there is a predominant presence of probiotic bacteria such as *Lactobacillus johnsonii*, *Lactobacillus reuteri*, and *Bifidobacterium* species in healthy individuals (Roesch *et al.*, 2009).

Lactic acid bacteria (Lactobacilli) are well known bacteria present in the human intestine and used in probiotic supplements. There are a variety of explanations in the literature as to the mechanisms responsible for the probiotic effects of these bacteria. These mechanisms include competitive exclusion of pathogens, secretion of bioactive compounds, immune system alteration, and host metabolism modification. However, there is no general consensus as to the mechanism of probiotics action, and studies of mechanism typically differ depending on the species or strain of bacteria. For example, a feeding study using the biobreeding diabetes-prone (BB-DP) rat model for type 1 diabetes with the intestinal bacteria *L. johnsonii* showed that oral administration of the probiotic bacterium *L. johnsonii* decreases the incidence of type 1 diabetes, possibly by decreasing the intestinal oxidative stress response (Valladares *et al.*, 2010). The decreased oxidative stress at the intestinal level could be a consequence of multiple factors. For example, the rat chow is formulated with many ingredients containing 6% to 8% (weight to weight) of fiber in the form of sugar beet pulp. The sugar beet pulp is an important source of ferulic acid, a phytophenol with anti-oxidative and anti-inflammatory effects (Couteau et al., 1998). It has been demonstrated that low dosage of cinnamic acids (especially ferulic acid) has been related with the stimulation of insulin secretion (Balasubashini et al., 2003, Adisakwattana et al., 2008), prevention of oxidative stress and lipid peroxidation (Balasubashini et al., 2004, Srinivasan et al., 2007), and inhibition of diabetic nephropathy progression (Fujita et al., 2008). The interaction of select bacteria in the host intestines with dietary fiber, and the possible release of phenolic acids, is an interesting process to be analyzed in order to generate a rationale understanding of the problem.

Despite the fact that total human gut microbiota displays FAE activity, specific bacterial species producing FAEs have not been investigated in depth. FAE activity was identified in several lactobacilli, including L. fermentum, L. reuteri, L. leichmanni, and L. farciminis, however the genes encoding the FAEs were not identified (Donaghy et al., 1998). It is hypothesized that probiotic bacteria could enhance the release of bioactive phenolic acids from dietary fiber by producing the necessary FAE activity and aiding in the assimilation of phenolic acids. It is necessary to identify and characterize the FAEs from probiotic bacteria to further investigate this hypothesis.

2.1 Identification of FAE-producing bacterial strains

A screening assay for detection of FAE activity from *Lactobacillus* strains was first described by Donaghy et al. (Donaghy et al., 1998). A model substrate for FAEs, ethyl ferulate, was embedded in de Man Rogosa Sharpe (MRS) plates. The presence of ethyl ferulate created a turbid appearance in the MRS agar due to the semi-soluble ethyl ferulate at 0.1% (weight to volume) final concentration. Ferulate assay (MRS-EF) plates were inoculated with cells obtained from individual overnight MRS cultures. The plates were incubated at 37°C in a gas pack system for a maximum of 3 days. The formation of halo (clear area) around the colonies due to the hydrolysis of ethyl ferulate indicated the presence of FAE activity. This technique was sucessfully applied by Lai et al. (Lai et al., 2009), which identified that *L. johnsonii*, *L. reutri*, and *L. helveticus* are able to produce FAEs (Fig. 1).

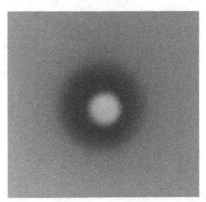

Fig. 1. Halo zone created by *L. johnsonii* colony on MRS-EF screening plate.

2.2 Selection of potential FAEs using bioinformatics analysis

Since FAEs are classified as carboxylic esterases, they display the characteristics of serine esterases (Brenner, 1988, Cygler *et al.*, 1993). These enyzmes have a classically conserved pentapeptide esterase motif with a consensus sequence glycine-X-serine-X-glycine (GlyXSerXGly), where X represents any amino acid. Public databases such as Comprehensive Microbial Resource (CMR) Database (Davidsen *et al.*, 2010) can be used to predict potential FAEs from the genomic sequences of FAE producing strains. Since there are only a handful of bacterial FAEs currently identified, most of the potential FAEs are annotated as hypothetical hydrolases or esterases. Two FAEs, LJ0536 and LJ1228 were previously identified in a probiotic bacterium *L. johnsonii* (Lai *et al.*, 2009). The amino acid sequences of both FAEs can also be used to identify potential FAE homologs using a BLASTP search in NCBI database (Altschul *et al.*, 1997).

2.3 Purification and characterization of FAEs

Apart from fungal FAEs, most of the bacterial FAEs described in the literature are purified directly from the growth media of bacterial strains, without previous knowledge of their coding gene or protein amino acid sequence. In order to obtain a large amount of enzyme to carry out thorough biochemical characterization, it is necessary to express and purify the target FAEs as recombinant proteins.

2.3.1 Cloning and purification of FAEs

After selection of potential FAEs from genomic analysis of FAE-producing strains, the genes of interest can be cloned into an expression vector. pET vectors are one of the most common expression vectors used for the cloning and expression of recombinant proteins in *Escherichia coli*. The pET System is driven by the T7 promoter, so target genes are regulated by the strong bacteriophage T7 transcription and translation signals. In this system, gene expression is effectively induced by the presence of isopropyl β-D-1-thiogalactopyranoside (IPTG). The host cells, such as *E. coli* BL21 DE3, provide T7 RNA polymerase during protein expression. The proteins are expressed with specific tags such as Histidine Tag or S epitope tag, depending on the vector used.

Using pET15 vector as an example, the expression of His_6-tagged proteins is carried out in *E. coli* BL21 using IPTG (1 mM) to induce gene transcription on the recombinant vector p15TV-L. The cells are collected by centrifugation at 8000 RPM (JLA8.1000 rotor, Beckman Coulter) for 25 min. The collected cell mass is resuspended in 25 mL binding buffer (5 mM imidazole, 500 mM NaCl, 20 mM 4-(2-hydroxyethyl)-1-piperazineethanesulfonic acid (HEPES) pH 7.5) and then disrupted by French press (20000 psi). The cell free extract is collected by centrifugation at 4°C, 17500 RPM (JA25.50 rotor, Beckman Coulter) for 25 min. The soluble His_6-tagged proteins are purified by affinity chromatography. All solutions pass through the nickel-nitriloacetic acid (Ni-NTA) column by gravity flow. The Ni-NTA column is first washed with 30 mL of ultra-pure water to wash out unbound nickel ions. It is then pre-equilibrated with 30 mL binding buffer. The cell free extract is then applied to the Ni-NTA column. During this step, the His_6-tagged proteins bind to nickel ions that are immobilized by NTA. The resin is washed with 30 mL of binding buffer to wash out any unbound proteins. 200 mL of wash buffer (20 mM imidazole, 500 mM NaCl, 20 mM HEPES pH 7.5), which contains a higher concentration of imidazole, is used to remove unspecific proteins that are bound to the resin. Imidazole is a competitive molecule that displaces the nickel ions bound to His_6-tagged protein. The His_6-tagged proteins are eluted using 20 mL elution buffer (250 mM imidazole, 500 mM NaCl, 20 mM HEPES pH 7.5). The purified proteins are dialyzed at 4°C for 16 hours. The dialysis buffer is composed of 50 mM HEPES buffer pH 7.5, 500 mM sodium chloride (NaCl), and 1 mM dithiothreitol (DTT). After dialysis, the samples are flash frozen and preserved at -80°C in 200 µL aliquots until needed. The His_6-tag can be removed by treatment with tobacco etch virus (TEV) protease (60 ug TEV protease per 1 mg of target protein) at 4°C for 16 hours. The sample is applied to the nickel affinity chromatography column to eliminate the released His_6-tag. Collected proteins are dialyzed at 4°C against dialysis buffer for 16 hours. The purified proteins without His_6-tag are flash-frozen and preserved in small aliquots at -80°C until needed.

A rapid method to evaluate the FAE activity can be used immediately after purification. 3 µl of the purified proteins (3-5 µl equivalent to 0.1 µg total protein) are dropped on the surface of the MRS-EF screening plate. The formation of halo zones indicates the presence of FAE activity. This system was used to purify the recombinant proteins LJ0536 and LJ1228 (Lai et al., 2009). By using the same strategy, a hypothetical protein LREU1684 was purified from L. reuteri and identified as a FAE.

2.3.2 Enzymatic substrate profile analysis

The change in pH that occurs during ester hydrolysis can be used to screen for substrate preference of FAEs. A pH indicator such as 4-nitrophenol (Janes et al., 1998) can be used to detect the change of pH during a reaction by monitoring the absorbance with a spectrophotometer. From the information in absorbance change, an estimate of enzymatic activity on different ester substrates can be determined. Enzymatic substrate profiles are determined at 25°C using an ester library composed of a variety of aliphatic and phenolic ester substrates (Liu et al., 2001). The purified enzymes are first thawed from -80°C and re-dialyzed against 5 mM N,N-bis(2-hydroxyethyl)-2-aminoethanesulfonic acid (BES) buffer pH 7.2. The reactions are carried out in 96 well plates. Each enzymatic reaction contains 1 mM ester substrate, 0.44 mM 4-nitrophenol (proton acceptor), 4.39 mM BES pH 7.2, 7.1% (volume to volume) acetonitrile, and 30 - 35 µg · mL^{-1} of enzyme in a total volume of 105 µL reaction mixture. The 96 well plates are incubated at 25°C using Synergy™ HT Multi-Detection Microplate Reader (Biotek). The reactions are continuously monitored for 30 min at 404 nm. FAEs such as LJ0536 and LJ1228 display high activity towards phenolic esters (Lai et al., 2009).

This techique is used only to demonstrate enzyme substrate preferences since it allows the use of several substrates in parallel. It utilizes specific conditions to detect the release of hydrogen ion (proton) during hydrolysis. The buffer (BES buffer) and the pH indicator (4-nitrophenol) to be used in this type of assay must have similar affinity (BES buffer pK_a = 7.09; 4-nitrophenol pK_a = 7.15) for the release of protons. In this way, the ratio of protonated buffer and the protonated indicator remains constant. The pH shifts by the proton release during the enzymatic reaction and it is detected as a change in the yellow color of the indicator present in the mixture. Thus, this technique is not flexible enough to adjust to the best conditions (pH, type of buffers, ions, etc.) that many enzymes require to function at their maximum activity. The specific enzyme activity determined using this method does not reflect the true specific enzyme activity. In addition, the stability of several enzymes could be affected by exhaustive dialysis in the reaction buffer. The dialysis was performed using 120 to 150 times the volume of the original enzyme suspension. Consequently, the technique is valid only to demonstrate substrate preferences. An alternate method involving the commerical equipment HydroPlate® instead of the tranditional 4-nitrophenol pH indicator can be used to monitor the pH during ester hydrolysis. The HydroPlate® is a 96 well plate containing immobilized pH sensor layers in each well (PreSens). The sensor contains a stable reference dye and one sensitive to oxygen. Thus, the measured fluorescence used to determine pH are internally referenced for precision across the plate.

2.3.3 Biochemical properties of FAEs

As enzyme reactions are saturable, the biochemical parameters such as K_m (Michaelis constant: amount of substrate required to reach half of V_{max}), V_{max} (maximum rate of reaction or maximum enzyme specific activity, $\mu mol \cdot mg^{-1} \cdot min^{-1}$), K_{cat} (catalytic rate constant, s^{-1}), and K_{cat} / K_m (catalytic efficiency, $M^{-1} \cdot s^{-1}$) can be determined by measuring the initial rate of the reaction over a range of substrate concentrations. The optimal conditions such as pH and temperature should be determined with model substrates before attempting to determine these biochemical parameters. Naphthyl esters and 4-nitrophenyl esters are common model substrates used to determine the biochemical parameters of esterases using saturation kinetics. Hydrolysis of naphthyl esters or 4-nitrophenyl esters generates napthol or 4-nitrophenol respectively, resulting in a specific absorbance at 412 nm. In similar ferulic acid esterase screenings, phenolic esters such as ethyl ferulate and chlorogenic acid should be included to compare the differences in biochemical parameters between aliphatic and phenolic esters. The hydrolysis of these phenolic esters can be monitored at 324 nm (Lai et al., 2009). FAEs have higher catalytic efficiency and affinity towards phenolic esters. However, phenolic esters have higher values of absorbance and are less stable compared to the model substrates such as 4-nitrophenyl esters. The concentration of phenolic esters used to obtain the initial rate of the reaction is usually limited to 0.1 mM due to the upper limit of plate reader absorbance. Using absorbance to obtain enzyme kinetic parameters from phenolic ester hydrolysis can be difficult. An alternative method such as high performance liquid chromatography could be used to estimate the enzymatic activity by monitoring the release of products during phenolic ester hydrolysis (Mastihuba et al., 2002).

3. Structural analysis of FAEs

FAEs belong to a structural group described as α / β fold hydrolases (Ollis et al., 1992). The secondary structure of this group is composed of a minimum of eight β-strands in the center core surrounded by α-helices. The term α / β barrel is also used to describe the structure. The β-strands in the central core and α-helices are mostly parallel. The α-helices and β-strands tend to alternate along the chain of the polypeptide. There are only few structures of FAEs deposited in public databases (PDB) (Berman et al., 2000). All the structures (apo-enzymes or co-crystallized with a substrate) deposited in the PDB belong to two enzymes purified from only two species, a fungus *Aspergillus niger* and a bacterium *Butyrivibrio proteoclasticus*. Five additional structures related to the probiotic bacterium *L. johnsonii* FAE LJ0536 were recently released in the PDB. The recently available structures of bacterial FAEs will allow researchers to identify new FAEs based on structural alignments and conserved structural features.

3.1 GlyXSerXGly is the classical esterase motif

In general, carboxylic acid esterases have one classical esterase motif composed of a consensus sequence GlyXSerXGly. Analysis of the LJ0536 amino acid sequence showed that LJ0536 has two GlyXSerXGly motifs, an exception to the general one motif rule for carboxylesterases. Mutiple sequence alignments indicated that LJ0536 and its homologs, including LREU1684, all have two GlyXSerXGly motifs (Fig. 2). The reason for the presence of two GlyXSerXGly motifs in enzymes has not been addressed clearly in the literature. A recent structural study on LJ0536 (Lai *et al.*, 2011) shows that LJ0536 displays a typical α / β hydrolase fold, which is composed of twelve β-strands and nine α–helices. Only one GlyXSerXGly motif (Gly104-X-Ser106-X-Gly108) of LJ0536 is catalytically active, while the other (Gly66-X-Ser68-X-Gly70) maintains the folding of the protein by hydrogen bonds. The newly identified FAE LREU1684 shares 47% amino acid sequence identity with LJ0536.

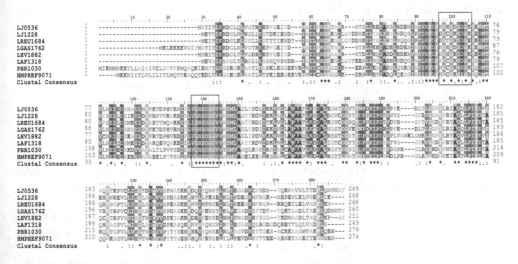

Fig. 2. Multiple sequence alignment of LJ0536 with its respresentative homologs. LJ0536: *L. johnsonii* N6.2, cinnamoyl esterase, GI# 289594369; LJ1228: *L. johnsonii* N6.2, cinnamoyl esterase, GI# 289594371; LREU1684: *L. reuteri* DSM 20016, alpha / beta fold family hydrolase-like protein, GI# 148544890; LGAS1762: *L. gasseri* ATCC 33323, alpha / beta fold family hydrolase, GI# 116630316; LHV1882: *L. helveticus* DPC 4571, alpha / beta fold family hydrolase, GI# 161508065; LAF1318: *L. fermentum* IFO 3956, hypothetical protein, GI# 184155794; PBR1030: *Prevotella bryantii* B14, hydrolase of alpha-beta family, GI# 299776930; HMPREF9071: *Capnocytophaga sp.* oral taxon 338 str. F0234, hydrolase of alpha-beta family protein, GI# 325692879. Two GlyXSerXGly motifs are located and boxed in the sequences. Amino acids are color coded.

3.2 Identification of critical amino acids involved in substrate hydrolysis

Ferulic acid esterase features such as the catalytic triad and the oxyanion hole are usually maintained by several amino acid residues that are highly conserved among homologs. Other critical amino acids involved in substrate recognition and binding are also conserved among closely related homologs, but not necessarily with less related homologs. A technique called alanine scanning, or site-directed mutagenesis, is helpful to determine the conserved amino acids critical for catalysis in proteins with unknown structure. The target amino acids selected for modification are replaced by alanine. Alanine is chosen because the inert alanine methyl functional group generally does not interact with other residues or alter the overall protein structure. To introduce the alanine mutation, 39-nucleotide long complementary primers containing the desired amino acid replacement are used to introduce individual mutations. The protein variants are then constructed by Polymerase Chain Reaction using Finnymes Phusion™ high fidelity DNA polymerase. This approach was used to identify the critical amino acids of LJ0536 (Lai et al., 2011).

The enzymatic activities of alanine variants are impared when the mutated amino acids are critical to function of the proteins. However, the results obtained from alanine scanning may not be useful in distinguishing the specific function of the amino acids, such as the involvement of amino acids in the formation of catalytic triad, oxyanion hole, tertiary structure of the protein, or substrate recognition and binding. The amino acids involved in substrate recognition and binding can be determined by measuring the enzymatic activity of the protein variants with different substrate types. For example, mutation of the amino acids that are only necessary for phenolic ester binding would not impair the enzymatic activity when aliphatic esters are used as substrates (Lai et al., 2011). Ultimately, the tertiary structure of the proteins are still necessary to conclude the findings from alanine scanning.

3.2.1 Catalytic triad of FAEs is composed of serine, histidine, and aspartic acid

Two basic steps are involved during ester hydrolysis: acylation and deacylation (Ding et al., 1994). During acylation, the hydroxyl oxygen of the catalytic serine carries out a nucleophilic attack on the carbonyl carbon of the ester substrate. After the attack, a general base (the histidine of the catalytic triad) deprotonates the catalytic serine and the first tetrahedral intermediate is formed. The hydrogen bonding of the third member of the triad, aspartic acid, plays a critical role in the stabilization of the protonated histidine. The oxyanion of the resulting tetrahedral intermediate is positioned towards the oxyanion hole. The oxyanion hole is created by hydrogen bonding between the substrate carbonyl oxygen anion and the backbone of two nitrogen atoms from other residues of the catalytic pocket. The general base, histidine, transfers the proton to the leaving group. The deprotonation of histidine leads to the protonation of an ester oxygen to release the first product (for example: methanol with methyl ferulate as substrate). As a consequence, the tetrahedral intermediate collapses and the characteristic acylenzyme intermediate is formed. Thus, the residual half of the substrate remains attached to the catalytic serine.

The second step of the reaction, deacylation, takes place in the presence of water. A molecule of water performs a nucleophilic attack on the carbonyl carbon of the remaining substrate in the acylenzyme intermediate. The general base (histidine) immediately deprotonates a molecule of water, leading to the formation of a second tetrahedral intermediate. The catalysis follows a similar pattern described for the acylation. The second tetrahedral intermediate is stabilized by the formation of the oxyanion hole. The proton of the general base moves to the nucleophilic serine. Consequently, the ester oxygen is protonated and the tetrahedral intermediate collapses. The protonation of ester oxygen releases the final product (for example: ferulic acid with methyl ferulate as substrate), and reconstitutes the native serine residue and the original state of the enzyme.

The catalytic center of esterases always consists of a triad composed of a nucleophile (serine or cysteine), a fully conserved histidine, and an acidic residue (aspartic acid). In order for the catalytic triad residues to carry out their roles during hydrolysis as described above, the histidine must be located next to the catalytic serine, while the aspartic acid must be located next to the histidine. The catalytic triad of LJ0536 is composed of serine, histidine, and

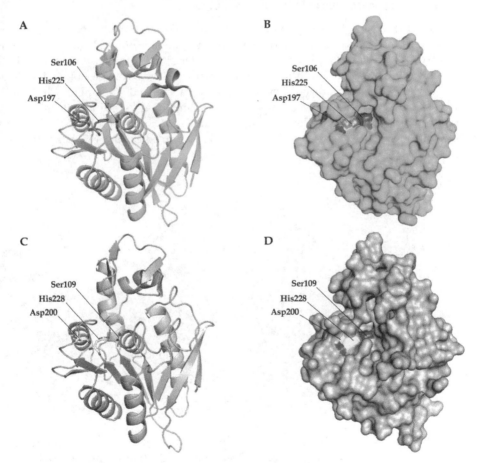

Fig. 3. Three dimensional structures of LJ0536 and LREU1684. (A) Ribbon and (B) surface representation of LJ0536. The catalytic triad of LJ0536 is colored orange. (C) Ribbon and (D) surface representation of LREU1684. The catalytic triad of LREU1684 is colored yellow.

aspartic acid (Ser106, His225, Asp197). Due to the high amino acid sequence identity between LJ0536 and LREU1684, it is expected that both enzymes could have similar tertiary structures. Hypothetical tertiary structure of LREU1684 is predicted using SWISS-MODEL (Arnold *et al.*, 2006). SWISS MODEL is a structure homology-modeling server, which allows users to predict the structure of a protein with a simple input of the peptide sequence. The modeling is generated based on the existing protein structures. The results indicate that the folding of LREU1684 is highly similar to LJ0536 (PDB: 3PF8). The catalytic triad of LREU1684 is arranged in an identical orientation as in LJ0536. It is composed of Ser109, His228, and Asp200 (Fig. 3). The catalytic serine residue (Ser109) is located on top of the sharp turn of an α-helix (nucleophilic elbow). The catalytic triad arrangement of both LJ0536 and LREU1684 follows the general rule of ester hydrolysis.

3.2.2 Classical oxyanion hole aids in substrate binding

Co-crystallization assays of the LJ0536 catalytic serine deficient mutant Ser106Ala (LJ0536-S106A) with various ligands identifies the classical oxyanion hole of LJ0536 (Lai *et al.*, 2011). LJ0536-S106A was co-crystallized with ethyl ferulate (PDB: 3QM1), ferulic acid (PDB: 3PFC), and caffeic acid (PDB: 3S2Z). All these structures show that the oxyanion hole of LJ0536 is formed by the backbone nitrogen atoms of phenylalanine and glutamine (Phe34 and Gln107). The oxyanion hole is an important structural feature, which stabilizes the tetrahedral intermediates during hydrolysis. Structural superimposition of LREU1684 and LJ0536 shows that the oxyanion hole of LREU1684 is formed by the backbone nitrogen atoms of Phe34 and Gln110 (Fig. 4).

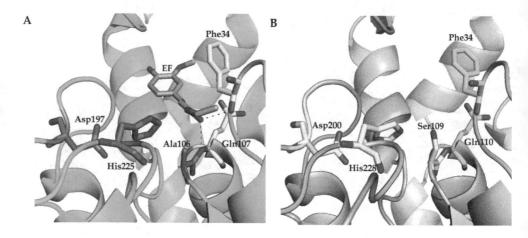

Fig. 4. Binding cavities of LJ0536-S106A co-crystallized with ethyl ferulate and LREU1684. (A) The oxyanion hole of LJ0536 is formed by Phe34 and Gln107 (palecyan). The catalytic triad residues are colored orange. Ethyl ferulate (EF) is colored pink. Dashed lines represent hydrogen bonds. (B) The oxyanion hole of LREU1684 is formed by Phe34 and Gln110 (cyans). The catalytic triad residues are colored yellow.

3.2.3 Specific inserted domain contributes to substrate binding

The study of the LJ0536 structure indicated that a specific α / β inserted domain is critical for substrate binding (Lai *et al.*, 2011). The inserted domain of LJ0536 is formed by a sequence of 54 amino acids from proline to glutamine (Pro131 to Qln184), and is located on top of the binding cavity. The two protruding hairpins from the inserted domain decorate the entrance and form the roof of the catalytic compartment. The phenolic ring of the ester substrate binds in the deepest part of the binding cavity, towards the inserted domain. In addition, three amino acid residues of the inserted domain, Asp138, Tyr169, and Gln145, contribute to the specific phenolic ester binding. The 4-hydroxyl group (ethyl ferulate, ferulic acid, and caffeic acid) and 3-hydroxyl group (caffeic acid) of the phenolic ring of the substrates are hydrogen bonded to Asp138 and Tyr169, respectively. Gln145 creates a bridge-like structure on top of the binding cavity, serving as a physical clamp holding the substrate inside the binding cavity. It also orients a water molecule in the binding cavity, which is important for activating the catalytic serine residue during hydrolysis.

Similar to LJ0536, LREU1684 has an α / β inserted domain formed by a sequence of 53 amino acids from Pro134 to Qln185 (Fig. 5). Asp141, Gln148, and Tyr172 of LREU1684 correspond to Asp138, Gln145, and Tyr169 of LJ0536, respectively. They adopted the same orientations as the residues in LJ0536 (Fig. 6). Thus, it is highly possible that Asp141, Gln148, and Tyr172 of LREU1684 also adopt the functional roles of Asp138, Gln145, and Tyr169 of LJ0536. LREU1684 and LJ0536 have both high amino acid sequence identity and high structural conservation.

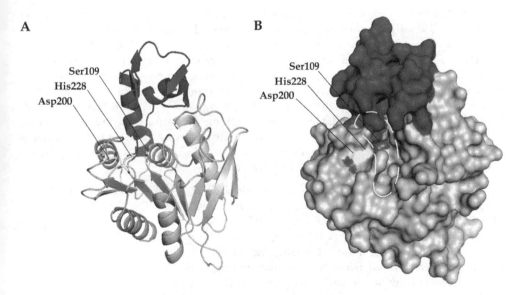

Fig. 5. (A) Ribbon and (B) surface representation of the LREU1684 α / β inserted domain. It is composed of two short β-hairpins and three α–helices. The domain is colored dark blue. The catalytic triad residues are colored yellow. The binding cavity is circled with dashed lines.

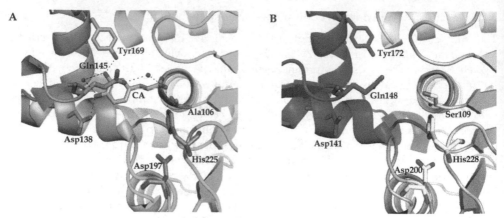

Fig. 6. Substrate binding mechanism of LJ0536 and LREU1684. (A) Binding cavity of LJ0536-S106A co-crystallized with caffeic acid. The phenolic ring of the ester is stabilized in the binding cavity by Asp138, Gln145, and Tyr169. The inserted domain is colored dark green. The catalytic triad residues are colored orange. Caffeic acid (CA) is colored grey. Dashed lines represent hydrogen bonds. (B) Binding cavity of LREU1684. Critical amino acid residues for phenolic ester binding are identified as Asp141, Gln148, and Tyr172. The inserted domain is colored dark blue. The catalytic triad residues are colored yellow.

3.3 Folding of LJ0536 is conserved among homologs

Since LREU1684 is predicted to have tertiary structure and binding mechanism that are similar to LJ0536, it is hypothesized that the other LJ0536 homologs should also contain the structural features of LJ0536. To test this hypothesis, the models of LJ0536 homologs are predicted using SWISS MODEL. The quality of the modeling is estimated by the E-value, QMEAN Z-Score, and QMEANscore4 (Benkert *et al.*, 2011). The E-value is a parameter that describes the number of hits that you expect to find a protein by chance when searching a database. The lower the E-value, the more structurally significant the hit is. The Q-MEAN Z-Score measures the absolute quality of a model. A strongly negative value indicates a model of low quality. The QMEANscore4 represents the probability that the input protein matches the predicted model. The value ranges between 0 and 1. The results obtained using an automatic template search are summarized in Table 1.

All predictions provided good quality models except for the modeling of EVE, a hypothetical protein from *Eubacterium ventriosum* ATCC 27560. EVE has an E-Value of 1.40E-28, a QMEANscore4 of 0.477, and a QMEAN Z-Score of -4.276. BFI-1, a cinnamoyl ester hydrolase from *Butyrivibrio fibrisolvens* E14, has the best quality of model with an E-Value of 1.61E-91, a QMEANscore4 of 0.82, and a QMEAN Z-Score of 0.425. Among all 12 homologs, 10 were predicted to have similar folding to Est1E (Goldstone *et al.*, 2010), a feruloyl esterase from *Butyrivibrio proteoclasticus* (PDB: 2wtmC and 2wtnA). The structures of LJ0536 and Est1E are highly similar as previously studied (Lai *et al.*, 2011). The predictions were validated by including the sequences of LJ0536 in the analysis. The homologs, LBA-1 and BFI-2, do not have a similar Est1E folding. LBA-1 is annotated as α / β superfamily hydrolase in *L. acidophilus* NCFM. It was predicted to be similar to lipase in *Burkholderia cepacia* (PDB: 1YS1). BFI-2 is annotated as a cinnamoyl ester hydrolase in *B. fibrisolvens* E14. It was predicted to be similar to acetyl xylan esterase in *Bacillus pumilus* (PDB: 3FVR).

Protein	PDB match	Sequence Identity [%]	E-value	QMEAN Z-Score	QMEAN score4
LJO-1	2wtmC (1.60Å)	30.9	6.00E-42	-1.223	0.696
LJO-2	2wtmC (1.60Å)	31.3	4.70E-43	-1.844	0.651
LRE	2wtmC (1.60Å)	32.9	1.20E-43	-1.599	0.669
LBA-1	1ys1X (1.10Å)	24.6	2.40E-08	-2.414	0.556
LBA-2	2wtmC (1.60Å)	29.7	4.20E-41	-1.490	0.677
EVE	2wtmC (1.60Å)	25.6	1.40E-28	-4.276	0.477
TDE	2wtmC (1.60Å)	24.4	8.50E-38	-1.362	0.686
BFI-1	2wtnA (2.10Å)	64.6	1.64E-91	0.425	0.820
BFI-2	3fvrC (2.50Å)	20.1	1.70E-32	-3.380	0.539
LPL	2wtmC (1.60Å)	29.7	3.60E-42	-0.949	0.717
LGA	2wtmC (1.60Å)	30.9	1.20E-40	-1.046	0.710
LHV	2wtmC (1.60Å)	29.7	4.80E-42	-1.558	0.672
LAF	2wtmC (1.60Å)	32.0	3.60E-42	-1.819	0.653

Table 1. LJ0536 homologs model automatic prediction using SWISS MODEL. *L. johnsonii* N6.2 cinnamoyl esterase LJ0536 (LJO-1), GI# 289594369. *L. johnsonii* N6.2 cinnamoyl esterase LJ1228 (LJO-2), GI# 289594371. *L. gasseri* ATCC 33323 alpha/beta fold family hydrolase LGAS1762 (LGA), GI# 116630316. *L. acidophilus* NCFM alpha/beta superfamily hydrolase LBA1350 (LBA-1), GI# 58337623. *L. acidophilus* NCFM, alpha/beta superfamily hydrolase LBA1842 (LBA-2), GI# 58338090. *L. helveticus* DPC 4571 alpha / beta fold family hydrolase LHV1882 (LHV), GI#161508065. *L. plantarum* WCSF1 putative esterase LP2953 (LPL), GI# 28379396. *L. fermentum* IFO 3956 hypothetical protein LAF1318 (LAF), GI# 184155794. *L. reuteri* DSM 20016 alpha/beta fold family hydrolase-like protein LREU1684 (LRE), GI# 148544890. *Butyrivibrio fibrisolvens* E14 cinnamoyl ester hydrolase CinI (BFI-1), GI# 1622732. *B. fibrisolvens* E14 cinnamoyl ester hydrolase CinII (BFI-2), GI# 1765979. *Treponema denticola* ATCC 35405 cinnamoyl ester hydrolase TDE0358 (TDE), GI# 41815924. *Eubacterium ventriosum* ATCC 27560 hypothetical protein EUBVEN_01801 (EVE), GI# 154484090. Numbers in parentheses indicate X-ray resolution.

In order to prove that the folding of LJ0536 is conserved in LBA-1 and BFI-2, a second prediction was preformed using Est1E or LJ0536 as the template structure. When Est1E was used as the template, the E-value of LBA-1 improved from 2.40E-08 to 2.70E-32. QMEAN Z-Score and QMEANscre4 decreased from -2.414 to -3.495 and from 0.556 to 0.527, respectively. When the prediction was done using LJ0536 as a template, the E-value improved to 1.2E-32, the QMEAN Z-Score decreased to -2.533, and the QMEANscre4 improved to 0.598. A similar scenario was observed when the protein BFI-2 was analyzed. The results indicated that the folding of LJ0536 is conserved in LBA-1 and BFI-2. The overall structure of LJ0536 is conserved among all homologs studied.

4. Conclusion

FAE application is one of the major fields of study for improving the bioavailability of phenolic acids in food components (phytophenols). After FAE activity on phytophenols in the intestinal tract, released phenolic acids become bioavailable and are absorbed in the intestines and can provide beneficial effects to the host. The identification and crystallization of the first intestinal probiotic bacterium FAE, LJ0536 identified from *L. johnsonii*, provides the fundamental knowledge (protein sequence and structural features) required to further identify FAEs from other species. Using a hypothetical protein LREU1684 as an example, this chapter provides a basic approach on how to identify, purify, and characterize FAEs, predict the model structure, and compare the model with known FAE structures. Further FAE crystallization is required to prove that the structure of LJ0536 is conserved among all homologs.

5. References

Adisakwattana, S., Moonsan, P. and Yibchok-Anun, S. (2008). Insulin-releasing properties of a series of cinnamic acid derivatives in vitro and in vivo. *J Agric Food Chem* 56, 7838-7844.

Altschul, S.F., Madden, T.L., Schäffer, A.A., Zhang, J., Zhang, Z., Miller, W. and Lipman, D.J. (1997). Gapped BLAST and PSI-BLAST: a new generation of protein database search programs. *Nucleic Acids Res* 25, 3389-3402.

Arnold, K., Bordoli, L., Kopp, J. and Schwede, T. (2006). The SWISS-MODEL workspace: a web-based environment for protein structure homology modelling. *Bioinformatics* 22, 195-201.

Balasubashini, M., Rukkumani, R. and Menon, V.P. (2003). Protective effects of ferulic acid on hyperlipidemic diabetic rats. *Acta Diabetol* 40, 118-122.

Balasubashini, M., Rukkumani, R., Viswanathan, P. and Menon, V. (2004). Ferulic acid alleviates lipid peroxidation in diabetic rats. *Phytother Res* 18, 310-314.

Beckman, C.H. (2000). Phenolic-storing cells: keys to programmed cell death and periderm formation in wilt disease resistance and in general defence responses in plants? *Physiol Mol Plant Path* 57, 101-110.

Berman, H.M., Westbrook, J., Feng, Z., Gilliland, G., Bhat, T.N., Weissig, H., *et al.* (2000) The Protein Data Bank. *Nucleic Acids Res*, pp. 235-242.

Brenner, S. (1988). The molecular evolution of genes and proteins: a tale of two serines. *Nature* 334, 528-530.

Couteau, D. and Mathaly, P. (1998). Fixed-bed purification of ferulic acid from sugar-beet pulp using activated carbon: Optimization studies. *Bioresource Technol* 64, 17-25.

Cygler, M., Schrag, J., Sussman, J., Harel, M., Silman, I., Gentry, M. and Doctor, B. (1993). Relationship between sequence conservation and three-dimensional structure in a large family of esterases, lipases, and related proteins. *Protein Sci* 2, 366-382.

Davidsen, T., Beck, E., Ganapathy, A., Montgomery, R., Zafar, N., Yang, Q., *et al.* (2010). The comprehensive microbial resource. *Nucleic Acids Res* 38, D340-345.

Donaghy, J., Kelly, P.F. and McKay, A.M. (1998). Detection of ferulic acid esterase production by Bacillus spp. and lactobacilli. *Appl Microbiol Biotechnol* 50, 257-260.

Fujita, A., Sasaki, H., Doi, A., Okamoto, K., Matsuno, S., Furuta, H., *et al.* (2008). Ferulic acid prevents pathological and functional abnormalities of the kidney in Otsuka Long-Evans Tokushima Fatty diabetic rats. *Diabetes Res Clin Pract* 79, 11-17.

Gonthier, M., Remesy, C., Scalbert, A., Cheynier, V., Souquet, J., Poutanen, K. and Aura, A. (2006). Microbial metabolism of caffeic acid and its esters chlorogenic and caftaric acids by human faecal microbiota in vitro. *Biomed Pharmacother* 60, 536-540.

Huang, Z., Wang, B., Eaves, D.H., Shikany, J.M. and Pace, R.D. (2007). Phenolic compound profile of selected vegetables frequently consumed by African Americans in the southeast United States. *Food Chem* 103, 1395-1402.

Janes, L., Löwendahl, C. and Kazlauskas, R. (1998). Quantitative Screening of Hydrolase Libraries Using pH Indicators: Identifying Active and Enantioselective Hydrolases. *Chem.-Eur. J.* 4, 2324-2331.

Konishi, Y. and Kobayashi, S. (2005). Transepithelial transport of rosmarinic acid in intestinal Caco-2 cell monolayers. *Biosci Biotechnol Biochem* 69, 583-591.

Kroon, P.A., Faulds, C.B., Ryden, P., Robertson, J.A. and Williamson, G. (1997). Release of Covalently Bound Ferulic Acid from Fiber in the Human Colon. *J. Agric. Food Chem.* 45, 661-667.

Lai, K., Lorca, G. and Gonzalez, C. (2009). Biochemical properties of two cinnamoyl esterases purified from a *Lactobacillus johnsonii* strain isolated from stool samples of diabetes-resistant rats. *Appl Environ Microbiol* 75, 5018-5024.

Lai, K., Stogios, P., Vu, C., Xu, X., Cui, H., Molloy, S., *et al.* (2011). An Inserted α/β Subdomain Shapes the Catalytic Pocket of *Lactobacillus johnsonii* Cinnamoyl Esterase. *PLoS One.*

Liu, A.M.F., Somers, N.A., Kazlauskas, R.J., Brush, T.S., Zocher, F., Enzelberger, M.M., *et al.* (2001). Mapping the substrate selectivity of new hydrolases using colorimetric screening: lipases from *Bacillus thermocatenulatus* and *Ophiostoma piliferum*, esterases from *Pseudomonas fluorescens* and *Streptomyces diastatochromogenes*. *Tetrahedron: Asymmetry* 12, 545-556.

Mastihuba, V., Kremnický, L., Mastihubová, M., Willett, J. and Côté, G. (2002). A spectrophotometric assay for feruloyl esterases. *Anal Biochem* 309, 96-101.

Ollis, D.L., Cheah, E., Cygler, M., Dijkstra, B., Frolow, F., Franken, S.M., *et al.* (1992). The alpha/beta hydrolase fold. *Protein Eng* 5, 197-211.

Roesch, L., Lorca, G., Casella, G., Giongo, A., Naranjo, A., Pionzio, A., *et al.* (2009). Culture-independent identification of gut bacteria correlated with the onset of diabetes in a rat model. *ISME J* 3, 536-548.

Srinivasan, M., Sudheer, A. and Menon, V. (2007). Ferulic Acid: therapeutic potential through its antioxidant property. *J Clin Biochem Nutr* 40, 92-100.

Vaarala, O., Atkinson, M. and Neu, J. (2008). The "perfect storm" for type 1 diabetes: the complex interplay between intestinal microbiota, gut permeability, and mucosal immunity. *Diabetes* 57, 2555-2562.

Valladares, R., Sankar, D., Li, N., Williams, E., Lai, K., Abdelgeliel, A., *et al.* (2010). *Lactobacillus johnsonii* N6.2 mitigates the development of type 1 diabetes in BB-DP rats. *PLoS One* 5, e10507.

8

Phosphoproteomics

Francesco Lonardoni and Alessandra Maria Bossi

Verona University,
Italy

1. Introduction

Phosphorylation is the most widespread and studied Post Translational Modification (PTM) in proteins [Collins et al. 2007]. It is involved in almost all cell functions: metabolism, osmoregulation, transcription, translation, cell cycle progression, cytoskeletal rearrangement, cell movement, apoptosis, differentiation, regulation of the signal transduction pathways, intercellular communication during the development and functioning of the nervous system [Graves & Krebs, 1999; Hunter, 2000; Sickmann & Mayer, 2001].

The phosphorylation/dephosphorylation process is regulated by the switch kinases/phosphatases. Kinases add a phosphate group to a receptive side chain of an aminoacid; phosphatases catalyze instead the hydrolysis of a phosphoester bond [Raggiaschi et al., 2005; Thingholm et al., 2009a]. The effect of the addition or subtraction of a phosphate group is the modification of enzymatic activity, protein-protein interaction and cellular localization.

Phosphorylation is not an unique process: often a single protein can display more than a single site suitable for the process, often catalyzed by different kinases. For example, glycogen synthase contains at least 9 phosphorylation sites, and its modulation is performed by at least 5 protein kinases acting on different sites of the protein [Nelson & Cox, 2004].

A misregulation of the phosphorylation processes can cause severe damage to the cells, leading to diseases like cancer, diabetes or neurodegeneration [Clevenger, 2004; Zhu et al., 2002].

The most common kind of phosphorylation in eukaryotes is O-phosphorylation, on serine, threonine and tyrosine with a ratio of 1800/200/1 [Grønborg et al., 2002; Kersten et al., 2006]. Other sites of phosphorylation can be histidine, lysine, arginine, glutamic acid, aspartic acid and cysteine [Sickmann & Mayer, 2001], even though are less studied due to the lability of the chemical bond and the subsequent necessity to use very special techniques to analyse them.

It is esteemed that 2-4% of eukaryotic genes are associated with kinases and phosphatases (there are about 500 kinase and 100 phosphatase genes in the human genome) [Manning et al., 2002; Twyman, 2004; Venter et al., 2001]. Around 100,000 phosphorylation sites may exist in the human proteome, the majority of which are presently unknown [H. Zhang et al., 2002]. The importance of studying the phosphorylation was marked by the success of the cancer drug *Gleevec*, the first to inhibit a specific kinase, which gave definitely an impulse to the research on kinases and their substrates as potential drug targets [Manning et al., 2002].

The comprehensive analysis of the protein phosphorylation should include: identification of phosphorylated proteins and of their sites of phosphorylation, how these phosphorylations modify the biological activity of the protein and kinases and phosphatases involved in the process.

2. A delicate analysis

Working in phosphoproteomics presents a series of hurdles in the analytical strategy.

The first issue concerns the reversible nature of the phosphorylation. The study of the phosphoproteins necessitates their isolation from a cell extract or a sub-cellular compartment. Subsequently to the cell lysis, however, many enzymes like phosphatases become active, determining the degradation of the proteins and detachment of phosphate groups from their sites. Also kinases can express their action, confusing the picture of which phosphate groups are biologically relevant [Raggiaschi et al., 2005]. Working at low temperature helps significantly in slowing down these processes, but it's not enough. In order to stop the action of these enzymes it is essential to add to the cell extracts a specific mix of inhibitors of proteases and phosphatases [Hemmings, 1996; Reinders & Sickmann, 2005; Schmidt et al., 2007; Thingholm et al., 2008a], while to inhibit kinases EDTA, EGTA or kinase inhibitors are added. It is also important to choose an inhibition mix that doesn't interfere with the downstream analytical methods, like the phospho-specific enrichment methods.

Another issue relates with phosphoproteins characterization, mostly performed through Mass Spectrometry (MS) methods after enzymatic digesting. The detection of phosphopeptides (FPs from now on) with MS is hampered by the presence of the non phosphorylated partners; moreover, the efficiency of ionization is higher for the latter ones, also generally more present in the sample (this fact is referred to as "low stoichiometry" of phosphorylation).

It follows that enrichment methods are necessary to extract the phosphoproteins or the FPs from the sample. There are various methods to choose from, depending to the kind of sample and the aims of the study [Kalume et al., 2003; Mann et al., 2002].

3. Detection of phosphoproteins

The detection of phosphoproteins in a sample still relies on optimized "classical" methods.

3.1 Isotopic labeling of phosphoproteins

One of the oldest methods used to study phosphoproteins is the metabolic labeling with ^{32}P and ^{33}P. It consists in nourishing living cells or organisms with substances labeled with these radioactive isotopes which are incorporated in the synthesised proteins. Following lysis of the cells, the protein population is isolated through 1-DE or 2-DE, visualized on the gels with autoradiography or acquired digitally with Phosphorimager systems. This method is still largely employed, because of its simplicity and reliability when somebody works with alive systems in vitro or in vivo [Eymann et al., 2007; Su et al., 2007].

A comparison between the performances of ^{32}P and ^{33}P in labeling the proteins was made [Guy et al., 1994], indicating that ^{33}P gives more neat image and higher resolution, even though after a longer exposition time, respect [32]P.

Apart from the safety and environmental implications in using radioactive isotopes, this method has other drawbacks. First of all it is not compatible with some downstream methods, like MS. Furthermore it can only be applied on viable cells, since the radioactive isotopes have to be taken from the media and metabolized: it cannot be therefore applied on post-mortem tissues or biopsies.

In *in vivo* studies, cells are incubated with ^{32}P, however the presence of ATP reservoirs inside the cells can interfere with the labeling, reducing the efficiency of the method [Steen et al., 2002]. Furthermore, ^{32}P is toxic for the cells, and over time can cause damages [Hu & Heikka, 2000; Hu et al., 2001; Yeargin & Haas, 1995].

In *in vitro* studies, proteins are incubated with specific kinases in the presence of [γ-^{32}P]-ATP and, under specific conditions, the radioactive atom is incorporated into aminoacidic residues. Due to the unnatural presence of kinase respect the target protein, however, it is frequent the phosphorylation of a different target instead of the natural one (promiscuity) [Graham et al., 2007].

3.2 Western blotting employing phosphospecific antibodies

Western blotting is a quite old technique. It is based on the selective binding of an antibody to a protein, transferred from a 1D or 2D gel to a nitrocellulose or polyvinylidene difluoride (PVDF) membrane support, and the subsequent revealing of the antibody marked spot with some visual method [Magi et al., 1999; Towbin et al., 1979]. The key role is played by the antibody, that should be specific for the protein epitope of interest: in this case epitopes are phosphoserine, phosphothreonine and phosphotyrosine. The selectivity and affinity characteristics of the antibody are of major importance, to perform specific recognition and limit false positives. While excellent anti-phosphotyrosine antibodies have been developed (e.g. (PY)20, (PY)100 and 4G10 hybridoma clones), better antibodies are still needed for phosphoserine and phosphothreonine. Antibodies generated against pSer and pThr, in fact, very often necessitate of a consensus sequence flanking the phosphoaminoacid; this might be due to the lower immunogenicity of pSer/pThr compared to pTyr [Schmidt et al., 2007].

Grønborg et al. performed a test for specificity and reliability of anti-phosphoserine and anti-phosphothreonine antibodies [Grønborg et al., 2002]. They made a large scale differential analysis of phosphorylated proteins, succeeding in identifying phosphorylation sites and FPs not identified with dedicated prediction software. The combination of high resolution 2-DE techniques and the Enhanced ChemiLuminescence (ECL) system give improved sensitivity to the method, i.e. intensification of around 1000 times of the light emitted from a spot [Buonocore et al., 1999; Kaufmann et al., 2001].

Although Western blotting is an efficient technique to reveal even small amount of protein (10^{-10} mol), its use in quantitative analysis is limited by the variability of the amount of proteins that can be transferred to the membrane.

3.3 Direct staining of phosphoproteins

The easiest way to detect phosphoproteins in a sample is the direct staining with a phosphospecific dye after a SDS-PAGE gel. Many attempts have been done from the 1970's [Steinberg et al., 2003], but all these methods face problems in terms of sensibility and of non specificity, e.g. the inability to discriminate between phosphorylated and non phosphorylated proteins or to detect phosphotyrosine. Quite recently a novel fluorescent dye has been introduced: Pro-Q Diamond [Schulenberg et al., 2003]. This dye selectively binds to phosphoproteins requiring a very simple experimental protocol. The sensitivity of the stain depends on the number of phosphorylated residues of the proteins: the detection limit was 16 ng for pepsin (1 phosphorylated residue) and 2 ng for casein (8 phosphorylated residues) [Schulenberg et al., 2003]. Thus, the method is very useful for a preliminary screening, but still not sufficient for a comprehensive analysis of the phosphoproteome.

The dye is also compatible with MS and with Multiplex Proteomics (MP), i.e. detecting simultaneously phosphoproteins and total proteins (e.g. these latter stained with SYPRO Ruby dye) on the same 2D gel. The combination of the two staining methods permits to distinguish low represented but highly phosphorylated proteins from highly represented but poorly phosphorylated ones, comparing the results from the two different colorations.

3.4 Detection of phosphoproteins employing protein phosphatases

The presence or absence of a phosphate group on a protein is enough to change its pI and subsequently its position in the 1st dimension of a 2D gel. That's why, by employing a phosphatase enzyme, it is possible to modify the position of a protein spot on a map and thus determine its nature comparing the maps before and after the treatment. Softwares have been also developed that can predict the pI shift due to the addition/removal of a phosphate group, that can be of 1-2 pH units [Kumar et al., 2004]. As an example, Yamagata et al. exploited the specific enzymatic activity of k-phosphatase (kPPase) on phosphoserine, phosphothreonine, phosphotyrosine and phosphohistidine residues to identify novel phosphoproteins in cultured rat fibroblasts [Yamagata et al., 2002].

The methods employing phosphatases, however, are not suitable for quantitative analysis, mainly because of the complexity of the analysis and the variable efficacy of the enzymatic action.

4. Selective enrichment of phosphoproteins and FPs

The identification of the PTM sites on a protein is generally performed by using MS approaches. On most occasions, the only enrichment of the sample in phosphoproteins followed by protease-specific digestion and MS analysis is not sufficient to identify the sites of phosphorylation present (due to the general low stoichiometry of the phosphorylation), thus a second enrichment step at the FP level is often also required.

Some commercial kits for phosphoprotein and FP enrichment are available, offering ease of use and reproducibility; nevertheless it has been clearly demonstrated that the different methods available differ in the specificity of isolation and in the set of phosphoproteins and FPs isolated [Bodenmiller et al., 2007], strongly suggesting that no single method is sufficient for a comprehensive phosphoproteome analysis. Strategies for phospho-specific enrichment are shown in fig.1.

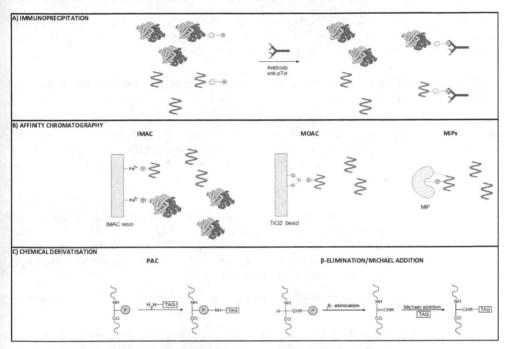

Fig. 1. Selective enrichment of phosphoproteins and phosphopeptides.
A) Immunoprecipitation: phosphoproteins or phosphopeptides are selectively precipitated through the use of appropriate anti-phospho antibodies; at the moment only the use of anti-pTyr antibodies has proven to be robust. B) Affinity chromatography: a resin with immobilized chelated metal or TiO_2 can selectively bind the phosphoric group of peptides and also proteins in IMAC (par. 4.2). A combined approach is SIMAC (IMAC + TiO_2). An alternative technique could be the use of Molecularly Imprinted Polymers (MIPs, par.4.10). C) Chemical derivatization: the phosphoric group reacts with the aminogroup of a tag in PAC or can be subdued to β-elimination and linking of a suitable tag through Michael addition (par. 4.11).

4.1 Phosphoprotein enrichment by Immunoprecipitation

Phospho-specific antibodies can be used to selectively immunoprecipitate phosphorylated proteins depending on the specificity of the antibody. As for Western blotting (see above) anti-phosphotyrosine antibodies are the most reliably and widely used in order to enrich tyrosine-phosphorylated proteins from complex mixtures.

After immunoprecipitation (IP) the phosphotyrosine enriched sample can be analyzed with different analytical methods such as 1-DE and 2-DE [Blagoev et al., 2004; Stannard et al., 2003].

Also in this case, variations of protein phosphorylation levels are very difficult to characterize unless in combination with a particular protein labeling (Stable Isotope Labeling with Aminoacids in Cell culture: SILAC, see par.6.1) with stable isotopes (^{13}C or

[15]N) is used [Ong et al., 2002]. This strategy allowed a quantitative and temporal investigation of tyrosine phosphorylation events of proteins involved in signaling pathways after stimulation with Epidermal Growth Factor (EGF) [Blagoev et al., 2004].

4.2 Phosphopeptide and phosphoprotein enrichment using Immobilized Metal Affinity Chromatography (IMAC)

IMAC exploits the material formed through chelation of a di-, tri- or tetravalent metal by nitrilotriacetic acid (NTA), iminodiacetic acid (IDA) or Tris(carboxymethyl)ethylenediamine (TED) immobilized on a solid support, like porous silica beads [Porath et al., 1975]. The most commonly used resins make use of Fe^{3+}, Ga^{3+} and Al^{3+}, even though Zn^{2+} and Zr^{4+} are used as well [Feng et al., 2007].

This method is routinary employed in FPs enrichment prior to MS analysis, nevertheless it has some limits; the most evident is its undesired ability to bind acidic peptides. This limitation has been largely surpassed by the acidification of the media (to protonate the carboxylic groups) [Posewitz & Tempst, 1999] and esterification of the carboxylic moieties with methanolic HCl before the enrichment step, even if this method introduces complexity due to the variable yield of methylation [Ficarro et al., 2002].

A second limitation is the net bias of the method towards monophosphorylated peptides [Ficarro et al., 2002].

The method is particularly effective when used in combination with an enrichment step at the protein level. This operation can be carried out with methods like IMAC itself [Collins et al., 2005], exploiting however a more suitable solid support, like Sepharose beads. Moreover, secondary interactions have to be reduced, e.g. with the use of denaturing conditions (6M urea). The detachment of the potentially many phosphate groups from the column imposes instead the use of strong eluting buffer as 0.1-0.2 M EDTA [Collins et al., 2005].

Other phosphoprotein enrichment methods are phosphotyrosine IP [Ficarro et al., 2003], Strong Anion eXchange chromatography (SAX) [Trinidad et al., 2006], Strong Cation eXchange chromatography (SCX) [Nühse et al., 2003] and SDS-PAGE [Villen et al., 2007].

4.3 Metal Oxide Affinity Chromatography (MOAC)

Metal oxide chromatography (MOAC) employs mainly Ti, Zr or Al oxides, in the form of solid or coated beads, as chromatographic media to sequestrate FPs. Many different crystalline forms and nanostructured materials have been devised [Leitner, 2010].

The first report about the potentialities of these materials regarded the use of TiO_2-based columns to sequestrate phosphate ions from the water [Connor & McQuillan, 1999; Ikeguchi & Nakamura, 2000].

In 2004, Pinkse et al. published a paper on the use of this material to bind FPs. They devised a 2D-LC-MS online strategy with TiO_2 beads (Titansphere) as first dimension and RP as second one [Pinkse et al., 2004]. Acidic conditions (pH 2.9) promoted the adhesion of FPs to the first column, leaving the non-phosphorylated ones to flow through and to be analysed with nano-LC-RP- MS/MS. Basic conditions (pH 9.0) eluted FPs in a second step. The method was tested on a 153kDa homo-dimeric cGMP-dependent kinase, promoting the discovery a total of 8 phosphosites, 2 of which were novel.

Larsen's group then devised an off-line strategy to bind FPs to a TiO_2 material. The use of additives as 2,5- dihydroxybenzoic acid (DHB), phthalic acid or glycolic acid largely reduced the aspecific attachment of acidic peptides [Jensen & Larsen, 2007; Larsen et al., 2005; Thingholm et al., 2006]. This technique has been named HAMMOC, for hydroxy acid modified metal oxide chromatography.

A significative advantage of TiO_2 is its tolerance towards most buffers and salts used in biochemistry and cell biology laboratories [Jensen & Larsen, 2007]. This is one of the reasons which made TiO_2 so common in large scale phosphoproteomics studies [Dengjel et al., 2007; Molina et al., 2007; Olsen et al., 2006; Olsen et al., 2007; Thingholm et al., 2008a].

Not only titania has been employed as metal oxide for FPs enrichment. Natural substitutes can be oxides belonging to the same group. For example, ZrO_2 microtips have been recently introduced as mean for FPs enrichment. This oxide shows a preference towards singly phosphorylated FPs, while TiO_2 preferentially retains multiply phosphorylated ones [Kweon & Hakansson, 2006]. However, subsequent large-scale studies demonstrated also a lower selectivity versus acidic peptides [Sugiyama et al., 2008], suggesting the necessity of further improvements.

4.4 Sequential elution from IMAC (SIMAC)

The identification of multiply phosphorylated peptides has proven to be a hard task, mostly because of their difficult ionization and subsequent low signal compared to singly- and not-FPs, so to be not selected for the subsequent fragmentation in the mass spectrometer.

To address this issue, in 2007 Martin Larsen's group presented an analytical strategy for sequential elution of mono- and multiphosphorylated peptides. The sequential elution from IMAC (SIMAC) exploits the complementary characteristics of IMAC and TiO_2 in enriching the sample respectively in multiply and singly phosphorylated peptides [Thingholm et al., 2008b].

In particular, the elution from IMAC in acidic conditions (pH 1.0) enriches the sample in mono-FPs, while the basic conditions (pH 11.3) elute subsequently the multi-FPs [Thingholm et al., 2008a]. A further enrichment with TiO_2 chromatography is needed only in the acidic fraction, because the basic one is enough depleted of non-FPs.

The separation of singly and multiply FPs permits then their analysis with pdMS[3] (phosphorylated directed fragmentation) in optimized settings for each group [Raggiaschi et al., 2005; Thingholm et al., 2008b].

The method was tested on a 120µg whole-cell extract from human mesenchimal stem cells (hMSCs) and the results compared with those from TiO_2 enrichment alone. A total of 716 phosphosites was identified with SIMAC, while 350 with TiO_2. Moreover the number of multiply phosphorylated peptides was significantly increased [Thingholm et al., 2008b].

Recently, a new intriguing method for multiply phosphorylated peptides enrichment based on polyarginine-coated diamond nanoparticles was presented, however it has still to be tested on large scale samples and where a low amount of starting material (micrograms) is available [Chang et al., 2008].

4.5 Magnetic beads

Starting from the TiO_2-based chemistry with the idea to find an easier way to perform the extraction of FPs, Chen and Chen [Chen & Chen, 2005] thought to conjugate the properties of magnetic materials with those of TiO_2, coating Fe_3O_4 nanoparticles with TiO_2 through a silanic bridge. The nanobeads are mixed with a tryptic digest, vortexed and captured with a magnet. The captured FPs are subsequently analysed through a laser desorption/ionization from the inorganic particles and a run in MS. The method was named SALDI-MS, from Surface Assisted Laser Desorption/Ionization Mass Spectrometry [Schürenberg et al., 1999; Sunner et al., 1995].

There were subsequent improvements of the method, using for example $Fe_3O_4@C@SnO_2$ core-shell microspheres (the symbol @ means "coated by"), with which scientists were able to detect 77 phosphorylation sites in rat liver cells [Qi et al., 2009].

4.6 Calcium Phosphate Precipitation (CPP)

In 1994, Reynolds et al. presented a strategy for FPs enrichment through Ca^{2+} ions and 50% ethanol precipitation. The precipitated peptides from a tryptic digest of casein mostly contained multiple phosphoserines [Reynolds et al., 1994].

Zhang et al. tested the strategy by using calcium chloride ($CaCl_2$), ammonia solution ($NH_3.H_2O$) and disodium phosphate (Na_2HPO_4) on a rice embryo preparation [X. Zhang et al., 2007]. The dissolved and desalted pellet was then enriched through IMAC. In total, 227 non-redundant phosphorylation sites were identified, of which 213 on serine residues and 14 on threonine.

Through phosphate precipitation directly coupled to LC-MS/MS, Qiangwei Xia et al. identified 466 unique phosphorylation sites (379 on serine and 87 on threonine) in post-mortem Alzheimer disease brain tissue, 70% of which were not reported in Phospho.ELM database [Xia et al., 2008].

In both studies no tyrosine phosphorylated peptides were identified: it is not clear if this is due to the low abundance of these or to the poor selectivity of the method. Anyway the method offers the advantage of being rather simple, column-free and straightforward.

4.7 Ion exchange chromatography (IC)

The simple enrichment of FPs with IMAC, TiO_2 or SIMAC has generally proven to be not enough productive when a deep knowledge of the phosphoproteome is required [Rigbolt et al., 2011]. A fractionation step is also needed. Chromatography techniques based on charge interaction, hydrophilic interaction or a combination of both have been employed for this purpose [Zarei et al., 2011].

Anion exchange chromatography exploits the generally higher affinity of FPs for the positively charged stationary phase due to the intrinsic high negative charge carried by the phosphate group. Strong anion chromatography (SAX) has been employed both as fractionating technique before a FPs enrichment step (e.g. with IMAC or TiO_2) [Nühse et al., 2003; K. Zhang, 2006] and also as sole fractionating technique before LC-MS/MS.

Method	Target	Sample type and amount	Strategy	Results	Reference	Comments
IMAC (Immobilized Metal Affinity Chromatography) IDA or NTA on a polymer matrix or magnetic beads with chelated Fe^{3+}, Ga^{3+}, Al^{3+} or other metal cations. Binding through electrostatic interaction between phosphategroups and metal cations.	protein, peptide	H1 stem cells proteins, 10 mg *D. mela-nogaster* lysate Kc167, 10 mg	SCX-IMAC-RPLC-MS^2 (ETD-OT) SCX-IMAC-RPLC-MS^2 (LTQ-OT)	10844 phosphosites at1% FDR 13720 phosphosites at 0.63% FDR	Swaney et al., 2009 Zhai et al., 2008	Proven to be effective in large scale analysis. Limited capacity and specificity, directed towards multiply phosphorylated peptides, affinity for acidic peptides.
MOAC (Metal Oxide Affinity Chromatography) TiO_2, ZrO_2 or Nb_2O_5 as particles or layered on a Fe_3O_4 magnetic core. Bidentate binding between metal and phosphoric group oxigens.	peptide	HeLa lysate (amount not given) K562 lysate, 400 µg	SCX-TiO_2-RP-MS^2 &MS^3 (LIT-FT-ICR) SCX-TiO_2/Nb_2O_5-RP-MS^2 (MALDI-TOF)	6600 phosphosites, 2244 proteins at <1%FPR 622/642/834 phospho peptides (Ti/Nb/all*) at 4%FPR	Olsen et al., 2006 Ficarro et al., 2008	Robust and chemically inert material, high capacity and fast absorption. Slow flow desorption, most effective for singly phosphorylated peptides.
SCX (Strong Cation Exchange) Ion Exchange Chromatography. Phosphopeptides less retained by stationary phase due to their higher negative charge	peptide	HeLa cell lysate, 300 µg HEK 293 cell lysate, 1 mg	SCX-RP-LC-MS^2&MS^3 (IT) SCX-RP-LC-MS^2 (ETD,LIT)	2002 phosphosites >5000 unique phospho peptides 1%FDR	Beausoleil et al., 2004 Mohammed & Heck, 2011	Coelution with other acidic peptides. Useful as pre-enrichment technique

Method	Target	Sample type and amount	Strategy	Results	Reference	Comments
SAX (Strong Anion Exchange) Ion Exchange Chromatography. Phosphopeptides more retained by stationary phase due to their higher negative charge	peptide	Human liver protein digest, 100 μg *A.thaliana* membrane proteins digest, 500 μg	SAX-RPLC-MS^2&MS^3 (LTQ) SAX-IMAC-RPLC-MS^2 (Q-TOF)	274 phosphosites at 0.96% FDR 299 phospho peptides	Guanghui et al., 2008 Nühse et al., 2003	Useful for peptide fractionation Useful as pre-enrichment technique
HILIC (Hydrophilic Interaction Chromatography) Phosphopeptides more retained by stationary phase due to their higher polarity	peptide	HeLa cell lysate, 1 mg *S.cerevisiae* total protein, 6 mg	HILIC-RP-MS & MS^2(IT) IMAC-HILIC-RP-HPLC-MS^2 (LTQ)	1004 phosphos ites 8764 unique phospho peptides, 2278 proteins	McNulty & Annan, 2008 Albuquerq ue et al., 2008	Coelution with other acidic peptides. Useful as pre-enrichment technique
Chemical derivatisation Phosphoric group directly bound to a tag through phosphoramidate chemistry, diazo chemistry or oxidation-reduction and condensation. Alternatively β-elimination /Michael addition can be performed.	peptide	*D.melanog aster* Kc167 tryptic digest, 1.5 mg Jurkat T cells lysate (amount not reported)	PAC-RP-LC-MS^2 PAC-RP-LC-MS^2 (LTQ)	535 phosphos ites 79 tyrosine phospho proteins	Bodenmill er et al., 2007a Tao et al., 2005	Modulation of peptide properties for MS purposes. Reaction yield, side reactions, large amount of sample required.

Method	Target	Sample type and amount	Strategy	Results	Reference	Comments
Immuno-precipitation Affinity of the phosphosite to the antibody	protein, peptide	Jurkat T cells lysate (amount not reported)	IP(pY-100 antibody)-RP-LC-MS^2 (LCQ-IT)	194 phosphotyrosine sites	Rush et al., 2005	Mostly directed to p-Tyr phosphosites
		HME Cells lysate, 4mg	IP (pY-100 antibody)-RP-LC-MS^2 (LTQ-OT)	481 phosphotyrosine sites at FDR 1%	Heibeck et al., 2009	

*superimposition without redundancies of Ti and Zr detected phosphosites.

Table 1. Comparation of enrichment and fractionation methods for phosphopeptides.
Abbreviations: CID, Collision-Induced Dissociation; DHB, Dihydroxybenzoic acid; ECD, Electron Capture Dissociation; ETD, Electron Transfer Dissociation; FDR, False Discovery Rate; FPR, False positive rate; FT-ICR, Fourier Transform Ion Cyclotron Resonance; b-HPA, b-hydroxypropanoic acid; LIT, Linear Ion Trap; OT, Orbitrap; Q-TOF, Quadrupole-Time-Of-Flight; RP, Reversed Phase; SCX, Strong Cation Exchange; LTQ, Linear Trap Quadrupole; LCQ, Liquid Chromatography Quadrupole; IP, Immunoaffinity Purification.

Nühse et al. for example identified more than 300 phosphorylation sites in the plasma membrane fractions of *Arabidopsis thaliana* using a SAX + IMAC approach [Nühse et al., 2004], while Han and co-workers identified 274 phosphorylation sites in human liver tissue extract without the enrichment step [Han et al., 2008].

A drawback of this method has been reported by Thingholm et al. [Thingholm et al., 2009a], who noted a strong attachment of FPs to the SAX resin, from which they are difficultly recovered.

Its counterpart, strong cation chromatography (SCX), has been largely employed as prefractioning technique for proteins and peptides. The pioneering work on FPs was carried out by Beausoleil et al. in 2004 [Beausoleil et al., 2004]. Tryptic peptides were acidified at pH 2.7, where most of the peptides carry a +2 charge due to C-terminal lysine or arginine and the N-terminal aminogroup. FPs, instead, carry a reduced charge due to the phosphate group, e.g. +1 in monoFPs. The reduced affinity of the resin should therefore leaving FPs to flow more easily through the column. This idea was confirmed by the results, which brought to the identification of 2000 phosphosites in 8mg of nuclear extract of HeLa cells lysate.

Trinidad and co-workers evaluated the efficiency of SCX as prefractionation method prior to IMAC to the efficiency of IMAC and SCX alone, finding a three-fold increased FPs identification in the combined approach respect both methods [Trinidad et al., 2006].

The strength of SCX as prefractioning system was further confirmed by Olsen et al., who identified 6600 phosphosites in 2244 proteins in EGF-stimulated HeLa cells through SCX + TiO_2 [Olsen et al., 2006]. In more recent reports 23000 and 36000 phosphorylation sites have been identified respectively with SCX followed by IMAC and TiO_2 respectively [Huttlin et al., 2010; Rigbolt et al., 2011]. This remarkable increase in number of detected phosphosites is mainly due to instrumental and software improvements [Zarei et al., 2011].

It remains to be assessed if the fractionation/enrichment approach can be suitable in samples where a small amount of starting material is available.

4.8 Hydrophilic interaction chromatography

Hydrophilic interaction chromatography (HILIC) is a separation technique for biomolecules [Alpert, 1990]. The method relies on the interaction of the analytes, e.g. peptides, with a neutral stationary phase through hydrogen bonding. They are eluted from the column with a decreasingly organic mobile phase according to their polarities (hydrophilicity).

McNulty and Annan introduced this method as prefractionation step before IMAC [McNulty & Annan, 2008]. They tested the FPs enrichment capacity of HILIC compared to SCX and IMAC alone, as well as the fractionation ability of HILIC before and after IMAC treatment. The analysis of 1 mg of tryptic digest of HeLa cells evidenced the prefractionation vocation of the method. The use of IMAC prior to HILIC gave in fact a high level of nonspecific binding due to not FPs, while the reversed approach improved the selectivity for FPs to more than 99%.

ERLIC (Electrostatic Repulsion - Hydrophilic Interaction Chromatography), or Ion Pair normal phase, is a newly developed chromatographic method able to enrich and fractionate FPs in a single step [Alpert, 2008].

At pH 2.0 the C-termini and the carboxylic side chains of Asp and Glu are neutral, thus the peptides are generally positively charged and are not retained by a positively charged stationary phase. The presence of a phosphate group, however, reduces the net charge of FPs; this is not enough to overcome the overall repulsion for the stationary phase due to the basic residues, however an organic mobile phase, e.g. acetonitrile 70%, promotes their hydrophilic interaction with the column. The attraction for the column is enhanced in multiply phosphorylated peptides.

Zarei et al. evidenced in fact a higher efficiency of ERLIC compared to SCX and HILIC in fractionating multiply phosphorylated peptides [Zarei et al., 2011], more retained by the stationary phase.

In a recent article [Chen et al., 2011] Xi Chen et al. compared the FPs profiles obtained from a HeLa cell lysate by using 4 HPLC methods after a phosphor-enrichment with IMAC. Even though with any of the four methods (SCX, HILIC, ERLIC with volatile and not volatile solvent) 4-5000 peptides could be identified, the combination of all the methods gave a total number of 9069 unique FPs, with a considerable amount of non-overlapping unique FPs. The four methods are thus complementary to get a full coverage of the phosphoproteome.

4.9 Hydroxyapathite (HAP)

Firstly introduced by Tiselius and co-workers in 1956 [Tiselius et al., 1956], hydroxyapathitc (HAP) chromatography has been very popular until the 1980s, when new matrices have been introduced. HAP is a crystalline form of calcium phosphate with formula $Ca_{10}(PO_4)_6(OH)_2$, which binds FPs in virtue of both anionic and cathionic interactions. In particular Ca^{2+} binds phosphoric groups, and more strongly than it does with other acidic groups. The binding to the matrix is thus proportional to the number of phosphogroups on the peptides and can be exploited for a sequential elution of them according to their degree of phosphorylation. Mamone et al. [Mamone et al., 2010] exploited this material to analyse a tryptic digest of standard proteins. The advantages of HAP are the low cost of the material and the possibility to elute stepwise the peptides according to the degree of phosphorylation. Yet the system has to be tested on more complex biological samples.

4.10 Molecularly Imprinted Polymers (MIPs)

Recently [Emgenbroich et al., 2008; Helling et al., 2011], Borje Sellergren's group has proposed a method for the depletion of phosphotyrosine containing peptides based on Molecularly Imprinted Polymers (MIPs). FPs imprinting was performed by an epitope approach [Nishino et al., 2006; Rachkov & Minoura, 2001; Titirici et al., 2003], i.e. using a part of the analyte of interest (in this case the N- and C- protected phosphotyrosine) as a template to prepare a MIP able to fish FPs. The Solid Phase Extraction analysis showed a 18-fold preferential retention for phosphorylated angiotensin II respect the non phosphorylated one, preference not shown in the not imprinted polymer. Moreover, the material showed a preference for a triply phosphorylated peptide over a monophosphorylated one, showing opposite behaviour comparing to TiO_2 material. Of course a more extensive analysis and improvements are needed in order to tackle more complex samples.

4.11 Chemical derivatisation strategies

The chemical derivatisation strategies exploit the typical properties of the phosphate groups in the peptides, like the lability of the phosphoesteric bond and the subsequent easy substitution of the phosphate with a nucleophilic tag.

A different widely used approach is the methyl esterification of the carboxylic groups, notoriously competitive with the phosphate groups due to their negative charge (fig.2).

4.11.1 Methyl esterification of carboxyl-groups

In IMAC strategy carboxylic groups influence the elution of FPs, due to their charge attracted by the positively charged metal of the resin. For this reason their esterification could prove useful. Ficarro et al. first applied this strategy for the analysis of the cell extract from *S.Cerevisiae* [Ficarro et al., 2002] through IMAC approach: 383 sites of phosphorylation were detected with this method. Some characteristics of this approach have already been depicted in par. 4.2 and will be no further analysed.

Methyl esterification can be also exploited as a mean of isotopic labeling (with CH_3OH and CD_3OH for two different cellular states) for quantification purposes [Ficarro et al., 2003].

4.11.2 Biotin tagging by β-elimination and Michael addition

Many chemical derivatisation strategies have been devised to displace the phosphoryl group and binding a tag to the "naked" peptide. One of these methods [Jaffe et al., 1998] relies on the β-elimination of the phosphate from phosphoserine and phosphothreonine producing dehydroanaline and β-aminobutyric acid respectively. These products can be directly detected using tandem MS or derivatised, for example by Michael addition of a reactive thiol and subsequent binding of a tag. One common tag is biotin, notoriously having great affinity for avidin, liable of immobilization to an affinity column [Oda et al., 2001].

This technique shows the limits of not being applicable to phosphotyrosine containing peptides and of the occurrence of side reactions like tagging of non phosphorylated serine.

Zhou et al. [Zhou et al., 2001] proposed another method of derivatisation applicable to all residues, even if it involves many steps and it has not been tested on complex samples yet.

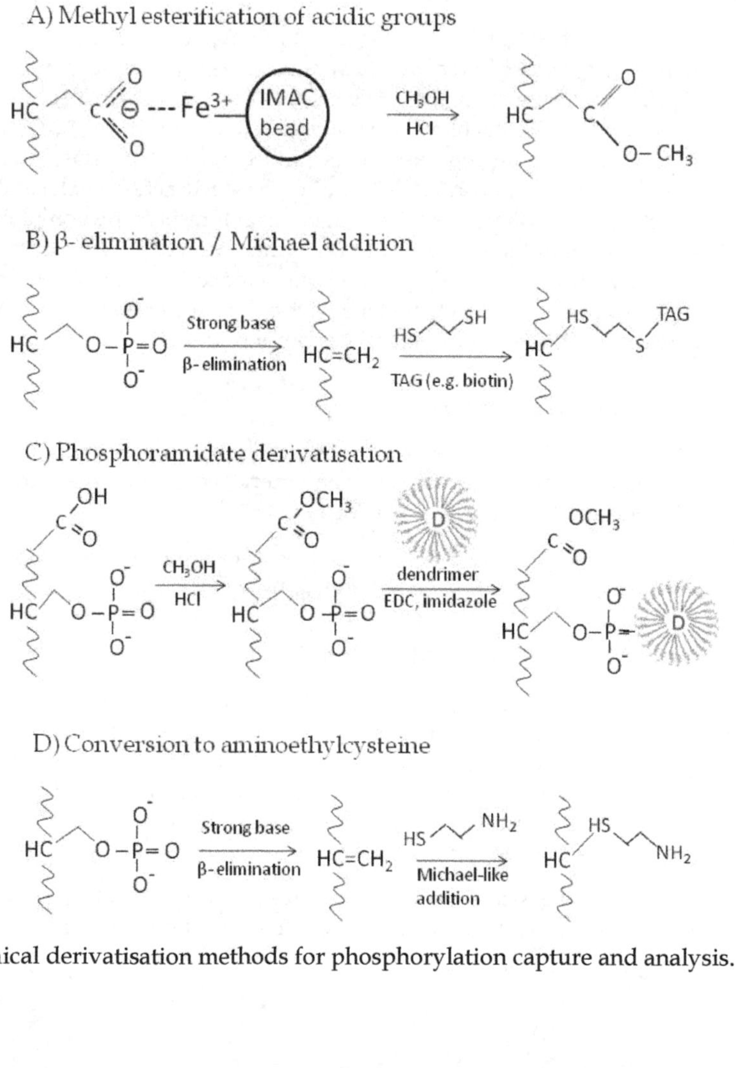

Fig. 2. Chemical derivatisation methods for phosphorylation capture and analysis.

A. IMAC suffers of aspecific binding of acidic peptides to the resin. Esterification of carboxylic groups with Methanol/HCl is a strategy to overcome this problem.

B. In presence of a strong base the β-elimination at phosphoserine and phosphothreonine produces dehydroanaline and β-aminobutyric acid. These can be derivatised with a thiol through a Michael-like addition and a tag can be added. If the tag is biotin, its affinity with avidin can be exploited.

C. FPs can be derivatised to phosphoramidates after esterification of carboxylic groups. If a dendrimer is employed the derivatised peptides can be separated through size-selective methods.

D. β-elimination and Michael-like addition of cysteamine converts phosphoserine and phosphothreonine in lysine analogues that specific enzymes leave in the C-terminus.

4.11.3 Phosphoramidate conversion

In 2005 Aebersold's group proposed a derivatisation procedure based on the carboxyl protection through methyl esterification followed by conjugation to a soluble polymer with phosphoramidate chemistry (PAC) [Tao et al., 2005].

The mixture of peptides is first converted to the corresponding methyl esters. In this step two cellular states can also be differentially labeled for quantitative analysis. Subsequently, the methylated peptides are combined and put to react with EDC, imidazole and a polyamine dendrimer. Phosphopeptides are converted in the corresponding phosphoramidates, easily separated from the non FPs through size selective methods. FPs are recovered with a brief acid hydrolysis and sent to the MS analysis.

When coupled with pervanadate stimulation and an initial antiphosphotyrosine protein precipitation step, this method allowed the identification and quantification of all known plus previously unknown phosphorylation sites in 97 tyrosine proteins in Jurkat T cells.

A modification of this method was proposed [Bodenmiller et al., 2007], exploiting the reaction of the phosphate groups with cystamine and a reducing agent instead of the dendrimer. The –SH group of cystamine reacts with maleimide-activated glass beads, immobilizing the FPs on a solid phase. This method allowed the identification of 229 FPs in the cytosolic proteome of *Drosophila melanogaster* Kc167 cells without any pre-enrichment step.

4.11.4 Conversion to aminoethylcysteine

Even after a good preconcentration step it is difficult the exact assignation of a phosphorylation site, due to the lability of the phosphate group, often lost during the backbone fragmentation in a MS collision, and to the intrinsic low abundance of phosphorylated peptides.

To address this problem, Knight et al. [Knight et al., 2003] devised a derivatisation method based on β-elimination and Michael addition of cysteamine to convert phosphoserine and phosphothreonine in aminoethylcysteine (Aec) and β-methylaminoethylcysteine respectively. Due to the resemblance of Aec to lysine, the use of proteases that recognize this aminoacid (e.g. Lys-C and lysyl endopeptidase) cleaves proteins leaving it in the C-terminus. The system works also with β-methylaminoethylcysteine and permits to identify the exact site of phosphorylation.

The limit of the method is the racemization in the addition step, converting only 50% of the phosphoaminoacids in the appropriate enzyme substrate. In the case of multiply phosphorylated peptides this fact greatly increases the complexity of the peptidic mix arising from the protein.

4.12 Comparison of enrichment methods

From this survey of enrichment methods emerges that no single technique is able to tackle the entire phosphoproteome. Some methods work in the direction of enriching only some species, like the antibodies for phosphotyrosine or the combination β-elimination/ Michael addition selective for phosphoserine and phosphothreonine. Other methods, like calcium precipitation, SAX, SCX and HILIC, work better as preseparation techniques to reduce sample complexity before more specific enrichment methods like IMAC, MOAC, SIMAC and PAC.

Every enrichment technique presents advantages and disadvantages, but also different specificities. Usually, MOAC is more specific for monophosphorylated peptides, due to the strong affinity for the multiply phosphorylated ones, not enough eluted. On the contrary, IMAC is more specific for multiply phosphorylated peptides, but has a low capacity and selectivity when used with highly complex samples. The combined approaches, like SIMAC, seem to be promising but revelate the necessity of a pre-enrichment step [Han et al., 2008; Thingholm et al., 2008a].

A systematic comparison of methods was made by Bodenmiller and co-workers (fig.3). They examined the reproducibility, specificity and efficiency of IMAC, PAC and two protocols for TiO_2 chromatography: $pTiO_2$ (phthalic acid in the loading buffer to quench nonspecific binding) and $dhbTiO_2$ (2,5 dihydrobenzoic acid, quencher too) [Bodenmiller et al., 2007]. Each method was tested through the injection of 1.5 mg of tryptic digest from cytosolic fractions from *Drosophila melanogaster* cells. The authors found a very good reproducibility of all the methods, making them suitable for quantitative analysis. Moreover, none of the methods was able to reveal the entire phosphoproteome, but they show partial overlapping results between each other.

In general a simple and straightforward strategy is desired, with few preparation steps and little sample handling in order to avoid loss of FPs. It is of course critical also the amount of starting material and the expertise of the people performing the extractions. For this reason detailed protocols are needed [Goto & Inagaki, 2007; Thingholm et al., 2006; Thingholm et al., 2009b; Turk et al., 2006].

The graph shows the efficiency and selectivity of IMAC, PAC and TiO_2, applied on a tryptic digest of a cytosolic protein extract of *D.melanogaster* cells [Bodenmiller et al., 2007]. In the starting material no FPs were detected, while the best selectivity in terms of P vs. not-P sites was IMAC.

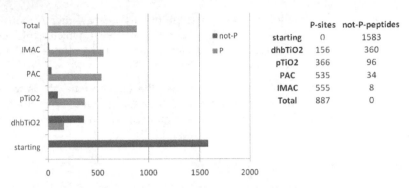

	P-sites	not-P-peptides
starting	0	1583
dhbTiO2	156	360
pTiO2	366	96
PAC	535	34
IMAC	555	8
Total	887	0

Fig. 3. Comparison of phosphorylation enrichment methods.

5. MS-based strategies for phosphoproteome analysis

Mass spectrometry has become the preferential method for peptide and protein identification following the separation steps, also in the PTM analysis [Bennett et al., 2002; Domon & Aebersold, 2006; Loyet et al., 2005].

The first step of a typical MS analysis consists in the cleavage of a single protein or a mixture by using a dedicated enzyme, usually trypsin, which preferentially cleaves the peptide bonds after arginin or lysine. Moreover, the tryptic fragments' weight is 700-3500 Da, a size suitable for MS analysis. The peptides are thus separated by nanoLC and vaporized/ionized through an ESI source. Their mass is evaluated and a second fragmentation, generating MS/MS spectra, permits to evaluate also their aminoacidic sequence. This is possible because of the higher lability of the bonds between aminoacids. Depending on the position of the cleavage along the peptide chain, the MS/MS fragments are classified in a,b,c (starting from the N-terminal) or x,y,z (starting from the C-terminal) according to Roepstorff and Fohlman [Biemann, 1988; Roepstorff & Fohlman, 1984] (fig.4). Only the highest abundance peptides are submitted to MS/MS. This creates a hurdle in phosphoproteomics, because of the lower abundance and difficult ionization of FPs compared to the co-present not-FPs, thus introducing the need of an enrichment step, as explained in section 4.

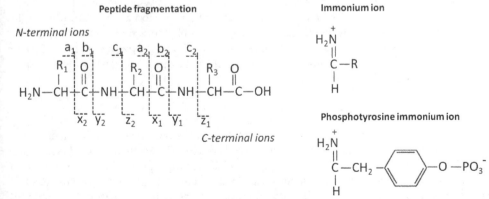

Fig. 4. Common nomenclature of peptide fragment ions.

The information about the peptide sequence is submitted to database-digging softwares as MASCOT [Perkins et al., 1999] or SEQUEST [Ducret et al., 1998], which explore protein databases to find a sequence match with previously annotated proteins and rank the correlations through a probability score.

The peptide fragmentation in MS/MS mostly breaks the inter-residue bonds to generate fragment series. CID generates preferentially y and b ions, while mostly z and c ions are originated by ECD and ETD. Phosphotyrosine immonium ion is diagnostic of tyrosine phosphorylation.

5.1 Collision Induced Dissociation (CID)

The most established method to induce a secondary fragmentation in peptides is the collision induced dissociation (CID). Basically, the peptide ion collides with an inert gas (He or Ar) which transfers its kinetic energy, subsequently redistributed between the atoms bringing to the breaking of the bonds. When a phosphoserine or phosphothreonine is present in the peptide sequence, the phosphoesteric bond is by far the most labile, thus a neutral loss of phosphoric acid H_3PO_4 (98 Da) takes place, originating respectively dehydroalanine and dehydroaminobutiric acid. Given that most part of the energy is employed to break the phosphoesteric bond, far less energy is available for the subsequent fragmentation of the peptide chain [Larsen et al., 2005]. This drains information when the identification of the phosphate group position is needed: only the bare presence or absence of a phosphate is assessed.

To overcome this issue several strategies have been applied. The first one is the introduction of a tertiary fragmentation, specifically directed towards peptides where a phosphate loss is detected. This strategy is named pdMS^3 (phosphorylation directed MS3) [Reinders & Sickmann, 2005]. The information due to the alternative fragmentations of the precursor ion are in this case lost, but they can be kept through another approach, named Multi-Stage Activation (MSA) [Steen et al., 2001]. In this case the ion trap, filled with the selected ion coming from neutral loss, is filled again with the original peptide and both are fragmented at the same time, originating a superimposed MS^2 / MS^3 spectra more information-rich.

Partial neutral loss happens also on phosphotyrosine residues, which leave a HPO_3 group (80Da) originating a characteristic phosphotyrosine immonium ion at m/z 216 (fig.4). The phosphoesteric bond is however in this case more stable, thus not compromising the information collection. Steen et al., for example, used the diagnostic fragment at m/z 216 for the selective detection of phosphotyrosine-containing peptides in chicken ovalbumin and murine MAP-kinase 2 [Steen et al., 2001].

5.2 Electron Capture Dissociation (ECD)

The limit of the peptide backbone poor fragmentation in the presence of a phosphate group was overcome in 1998 with a new fragmentation strategy. Electron Capture Dissociation (ECD) is a method developed by Zubarev and colleagues to improve the fragmentation of multiply charged protein and peptide ions [Zubarev et al., 1998]. These ions capture easily a thermal electron (<0.2 eV), which induces a non ergodic fragmentation, i.e. without vibrational energy redistribution like in CID. The result is a fragmentation mostly at S-S and

N-Cα backbone bonds, leaving intact the PTM bonds. The generated ions are c and z type (fig.4) [Kleinnijenhuis et al., 2007, Stensballe et al., 2000]. The method has some drawbacks, like a bigger affinity for disulphide bonds and a difficult fragmentation of N-terminal proline, which has two bonds to break. Moreover it can be carried out only with expensive FT-ICR instruments (up to 1$ million) to generate the static magnetic field for the electrons, which reduces its wide scale diffusion.

5.3 Electron Transfer Dissociation (ETD)

The efforts to find an ECD-like method without the need of expensive instruments brought to the advent of ETD (Electron Transfer Dissociation). In this approach the electron is transferred to multiply charged peptides (charge >2+) through a radical anion with low electron affinity, like anthracene or azobenzene [Schroeder et al., 2005; Syka et al., 2004]. The method can be implemented on linear quadrupole ITs, with the natural drawback of a reduced resolution and accuracy [Syka et al., 2004]. Molina et al. carried out a large scale analysis of human embryonic kidney 293T cells, identifying 1435 phosphosites, 80% of which were novel. Moreover, they identified 60% more FPs with ETD compared to CID, mainly due to the 40% more fragment ions [Molina et al., 2007]. It has to be remarked the little overlap between the two fragmentation techniques, that was exploited to develop an integrated approach. Since ETD works better with high charge peptides, Lys-C was thought to give better results than trypsin, cleaving the peptides only at C-terminal lysine. Surprisingly the results didn't match the expectations, probably due to the high number of missed cleavages in the tryptic lysate [Molina et al., 2007]. Another way to generate highly charged peptides was attempted by Larsen et al, who added 0.1% m-nitrobenzyl alcohol (m-NBA) to the LC-MS solvent [Kjeldsen et al., 2007]. This approach increased the predominant charge from 2+ to 3+, improving the ETD results. The approach is currently being tested on more complex samples. Another fact to remark is the evolution of the software, born for the CID approach, in the direction of meeting the features of spectra generated by new enzymes and fragmentations [Kim et al., 2010, http://www.matrixscience.com]

	Fragmentation agent	Generated ions	Instruments	Pros	Cons
CID	Inert gas (He, Ar)	y, b	ESI-MS	Better fragmentation of low charge peptides (2+)	Fragmentation mostly at the phosphogroup
ECD	Thermal electron	z, c	FT-ICR	Fragmentation only along the peptide bond	Need of expensive FT-ICR
ETD	Low electron affinity anion (e.g.anthracene)	z, c	IT, Q-TOF	Fragmentation only along the peptide bond	Less sensitive than CID

Table 2. Comparison of fragmentation methods.

All the methods show good results with a class of peptides, suggesting that an integrated approach CID/ECD or CID/ETD could be more effective [Molina et al., 2007].

6. Quantitative approaches for phosphoproteome analysis

In order to take a dynamic picture of the phosphorylation events in a particular pathway, it is desirable monitoring which sites are phosphorylated and to which extent following a stimulus. To achieve this goal some quantification methods are available and can be classified on the basis of the analysis step in which the quantitative information is generated: a differential isotopic label can be introduced in the cell culture, e.g. with labeled aminoacids (SILAC), in the protein mixture (ICAT), in the enzymatic digestion (^{18}O labeled water), or in the peptide mixture (iTRAQ), otherwise, in label-free experiments, the quantitative information is extracted at the MS level (fig.7). A thorough review about quantitation strategies has been published by Bantscheff et al. [Bantscheff et al., 2007].

		LEVEL			
		CELL	PROTEINS	PEPTIDES	MS
METHOD	SILAC	X			
	^{18}O		X		
	ICAT		X	X	
	iTRAQ			X	
	LABEL-free				X

Fig. 7. Strategies for quantitative analysis of protein phosphorylation.

An isotopic label can be introduced in different moments of the analysis or not at all, in label-free experiments.

6.1 Metabolic labeling

Metabolic labeling was first described in 1999 [Oda et al., 1999]. In 2002 Mann and coworkers introduced the term Stable Isotope Labeling by Aminoacids in Cell culture (SILAC) and used it for quantitative analysis of protein phosphorylation in 2003 [Ibarrola et al., 2003]. The typical experiment consists of growing a cell population in a medium containing an essential aminoacid labeled with a stable isotope (^{15}N or ^{13}C), and growing in parallel another cell population in a medium on non-labeled aminoacid. Usually labeled arginine and lysine are used, in order to ensure that every peptide from one culture contains a label after tryptic digestion. After several doublings, the cells are harvested from both cultures, and the protein extracts mixed together. After proteolysis, peptides can be analysed by MS and differences in the abundance of a peptide in the two cell extracts are shown through the different heights of two mass shifted peaks.

Recently, with this method Olsen et al. [Bodenmiller et al., 2007] reported the most comprehensive analysis of the effects of EGF stimulation on phosphoproteome dynamics in HeLa cells. This strategy has allowed the drawing of some detailed maps of time-resolved signaling pathways [Blagoev et al., 2004; Bose et al., 2006; Goss et al., 2006; Olsen et al., 2010]. The major limitation of SILAC stays in the cost of labeled aminoacids.

6.2 Protein and peptide labeling

Post-biosynthetic labeling of proteins and peptides is performed by chemical or enzymatic derivatization in vitro.

Enzymatic labeling exploits the incorporation of ^{18}O atoms from marked water during protein digestion. Trypsin and Glu-C introduce two heavy oxygen atoms, resulting in a 4 Da mass shift, generally sufficient for the differentiation of isotopomers. This method has been applied for quantitative proteomic purposes [Dengjel et al., 2007], but complete labeling is difficult to obtain.

Chemical modification can be carried out at protein or peptide level introducing a tag on a chemically reactive side chain of an aminoacid [Ong et al., 2005], in practice only cysteine and lysine are used for this purpose. A group of labeling reagents targets the N-terminus and the ε-aminogroup in the lysine side chain. They mostly exploit the N-hydroxysuccinimide (NHS) chemistry or other active esters and acid anhydrides, like in the Isotope-Coded Protein Label (ICPL) [Schmidt et al., 2005], isotope Tags for Relative and Absolute Quantification (iTRAQ) [Ross et al., 2004], Tandem Mass Tags (TMT) [Thompson et al., 2003] and acetic/succinic anhydride [Che & Fricker, 2002; Glocker et al., 1994; X.Zhang et al., 2002].

iTRAQ is a commercially available reagent, allowing to follow the evolution of biological systems over multiple time points. It was used, for example, to quantify 222 tyrosine phosphorylation sites across seven time points following EGF stimulation [Wolf-Yadlin et al., 2007].

Carboxylic groups of side chains of aspartic and glutamic acid as well as of the C-termini of peptidic chains can be isotopically labeled by esterification using deuterated alcohols, for example d0 and d3 methanolic HCl [Goodlett et al., 2001; Syka et al., 2004]. This reaction is particularly interesting, because the methylation is also a step used in the IMAC enrichment method to reduce aspecific binding of acidic peptides to the resin (par. 4.2).

General drawbacks of the chemical derivatization methods are the production of not desired side products, that negatively influence the quantification results and the cost of some of the mentioned reagents.

6.3 Absolute quantification using internal standards

The use of isotope-labeled internal standards in the field of proteomics is known with the name AQUA: Absolute QUAntification of proteins [Gerber et al.,2003].

The simplest protocol requires adding a known amount of a stable isotope-labeled peptide to the protein digest and in comparing the signal of it in the mass spectra respect the other peak areas [Pan et al., 2005].

There are some drawbacks with this approach. First of all the high dynamic range of concentrations of peptides makes difficult to find an appropriate concentration of standard for every analyte; second, it's likely to find an isobaric peptide to our standard in the peptide mixture, therefore limiting its specificity. These problems, however, have been addressed with the approach called Multiple Reaction Monitoring (MRM) [Kirkpatrick et al., 2005], in which the triple quadrupole MS monitors both peptide and its fragments mass during the experiment. The combination of retention time, peptide mass and fragment mass practically eliminates the ambiguities, extending the dynamic range to 4-5 orders of magnitude [Bondarenko et al., 2002]. The real value of the quantification through the AQUA approach is naturally biased by the manipulation of sample before adding the standard: the amount of protein determined may therefore not reflect its actual expression level in the cell.

6.4 Label-free quantification

There are two approaches for label-free quantification of proteins. The first one relies on the measure of the area of a MS peak of a peptide related to a protein: the increase of this area means also an increased amount of the protein. This approach is called eXtracted Ion Chromatogram (XIC), because a single ion peak area is extracted from a plot of signal intensities against time in the chromatogram [Bondarenko et al., 2002; Wang et al., 2006]. Signal intensities of the same peptide in different experiments is then compared to extract quantitative information, for example the stoichiometry of phosphorylation [Steen et al., 2005].

The other approach measures the amount of peptides generated from a protein: the more is the amount of a protein the more are the tandem-MS generated peptides. Relative quantification is thus achieved by comparing the number of spectra generated from a protein in different experiments. It is necessary a normalization, for example depending from the protein mass, creating therefore Protein Abundance Indexes (PAIs) [Rappsilber et al., 2002]. The relationship between number of peptides observed and protein amount had been found to be logarithmic (emPAI) [Ishihama et al., 2005; Lu et al., 2007].

7. Non-MS approaches to elucidate cellular signaling networks

7.1 Antibody-based approaches

In order to monitor previously identified phosphorylation sites, arrays employing phosphospecific antibodies have been used to investigate dozens of phosphorylation sites simultaneously [Sheenan et al., 2005; Belluco et al., 2005]. The general hurdle of these techniques is the limited availability of dedicated antibodies, however further improvements could extend the use of microarray technology in phosphoprotein studies [Schmelzle & White, 2006].

Methods were developed to monitor the phosphorylation status of tyrosine [Gembitsky et al., 2004] and the kinetics of phosphorylation [Khan et al., 2006] in proteins in a multiplex format.

In order to evaluate the phosphorylation dynamics on a cellular scale, flow citometry approaches have been also devised to monitor up to 11 phosphorylation events in parallel [Irish et al., 2004; Krutzik et al., 2005; Sachs et al., 2005]. Again, the main limit of this approach is the availability of suitable fluorescent-labeled antibodies.

7.2 Interaction of phosphoproteins and phosphorylated sites

The phosphorylation-related events include also protein-protein interactions in the cell signaling network. To investigate these phenomena, Jones et al. [Jones et al., 2006] devised a protein array to study the binary interactions between 61 fluorescent-labeled, tyrosine phosphorylated peptides from EGFR receptors with approximately 150 SH2 and PTB domains. By measuring the fluorescence at different titration points they determined the K_D values for every peptide-receptor couple.

Another approach was followed by Yaoi et al. [Yaoi et al., 2006], that immobilized SH2 domains on microspheres to extract interacting proteins and phosphoproteins from a complex mixture of different cell lines.

Both approaches revealed new insights in the cellular signaling networks.

7.3 Kinase screening on peptide and protein arrays

Peptide microarrays consist of synthetic peptide sequences deposited onto glass slides or attached to a derivatised surface, usually in triplicate, with peptides having substitutions in the phosphorylation sites as controls. The *in vitro* phosphorylation reaction is performed in the presence of radiolabeled ATP, the array exposed to a film and the image captured. The method assumes that phosphorylation of peptides should be in most of the cases similar to that of the same sequence in the intact protein, due to the fact that many phosphorylation sites are in accessible and flexible regions of the protein structure [Nühse et al., 2004].

Collins et al. used this approach for phosphorylation investigation of synaptic proteins, finding 28 unique phosphorylation sites [Collins et al., 2005].

The *in vitro* phosphorylation can naturally be different from the *in vivo* action, but the screening can select and give priority to some phosphorylation sites for further investigation.

The same approach can be used for immobilized proteins or protein domains.

Ptacek et al. [Ptacek et al., 2005] immobilized yeast proteins on high density (4400 proteins in duplicate) arrays on glass slides. They screened 87 kinases, finding that each kinase recognized up to 256 substrates, with a media of 47 substrates per kinase. These data allowed the construction of a global kinase-substrate interaction network. There is of course a concern about non specific phosphorylation, but also the perspective of a high throughput analysis for mapping phosphorylation networks.

8. Bioinformatics

The knowledge discovery process in proteomics has been greatly boosted in the last years by the introduction of new bioinformatic tools.

Widely developed phosphoproteomics databases are for example PhosphoSite [Hornbeck et al., 2004], containing around 100000 non-redundant phosphorylation sites (as well as other modifications, given that the cell signaling is not exclusively phosphocentric), and Phosida [Gnad et al., 2007], containing temporal phosphorylation data from cell stimulation in time-course experiments.

These databases permit not only the data mining, but also the interpretation of the data in the context of biological regulation, diseases, tissues, subcellular localization, protein domains, sequences, motifs, etc. [http://www.phosphosite.org]

9. Conclusion

Phosphoproteomics is a rapidly growing field, owing this evolution to the importance of the protein phosphorylation in many biological processes and its alteration in many diseases.

The analysis is usually performed with MS-based methods, supported by enrichment steps at the protein or peptide level. The improvement of MS has been enormous, with increase in resolution, mass accuracy, larger dynamic range and more sensitivity and speed, driving the progress in this field. Of course it must be mentioned also the evolution in bioinformatics, with the developing of adequate software for literature mining, prediction algorithms, post-analysis annotation and so on. The aim of phosphoproteomics is not only to find phosphorylation sites, but also monitor the dynamic of phosphorylation following stimuli to characterize completely signaling networks.

Nowadays the phosphoproteome of highly complex samples has been tackled [Nühse et al., 2003; Olsen et al., 2006; Trinidad et al., 2006; Villen et al., 2007], but further development of the methods, including of the bioinformatic tools to integrate complex databases, is needed for a more thorough knowledge of the mechanisms of phosphorylation networks.

10. References

Albuquerque, C.P.; Smolka, M.B.; Payne, S.H.; Bafna, V.; Eng, J. & Huilin Zhou (2008) A Multidimensional Chromatography Technology for In-depth Phosphoproteome Analysis. *Molecular and Cellular Proteomics*, Vol.7, No.7, (Jul 2008), pp.1389–1396

Alpert, A.J. (1990) Hydrophilic-interaction chromatography for the separation of peptides, nucleic acids and other polar compounds. *Journal of Chromatography*, Vol.499, (Jan 1990), pp.177–196

Alpert, A.J. (2008) Electrostatic repulsion hydrophilic interaction chromatography for isocratic separation of charged solutes and selective isolation of phosphopeptides. *Analytical Chemistry*, Vol.80, No.1 (Jan 2008), pp.62-76, ISSN: 0003-2700

Bantscheff, M.; Schirle, M.; Sweetman, G.; Rick, J. & Kuster B. (2007) Quantitative mass spectrometry in proteomics: a critical review. *Analytical Bioanalytical Chemistry*, Vol.389, No.4 (Oct 2007), pp.1017–1031.

Beausoleil, S.A.; Jedrychowski, M.; Schwartz, D.; Elias, J.E.; Villen, J.; Li, J.; Cohn, M.A.; Cantley, L.C. & Gygi, S. P. (2004) Large-scale characterization of HeLa cell nuclear phosphoproteins. *Proceedings National Academy Science USA.*, Vol.101, No.33 (Aug 2004), pp.12130-12135, ISSN: 0027-8424

Belluco, C .; Mammano, E.; Petricoin, E.; Prevedello, L.; Calvert, V.; Liotta, L.; Nitti, D. & Lise, M. (2005) Kinase substrate protein microarray analysis of human colon cancer and hepatic metastasis. *Clinica Chimica Acta*, Vol.357, No.2 (24 Jul 2005), pp.180-183

Bennett, K.L.; Stensballe, A.; Podtelejnikov, A.V.; Moniatte, M. & Jensen, O.N. (2002) Phosphopeptide detection and sequencing by matrix-assisted laser desorption/ionization quadrupole time-of-flight tandem mass spectrometry. *Journal Mass Spectrometry*, Vol.37, No.2, (Feb 2002), pp.179–190

Biemann, K. (1988) Contributions of mass spectrometry to peptide and protein structure. *Biomedical Environmental Mass Spectrometry*, Vol.16, No.1-12 (Oct 1988), pp.99–111

Blagoev, B., Ong, S. E., Kratchmarova, I., Mann, M., (2004) Temporal analysis of phosphotyrosine-dependent signaling networks by quantitative proteomics. *Nature Biotechnology*, Vol.22, No.9 (Sept 2004), pp.1139–1145

Blagoev, B.; Ong, S.E.; Kratchmarova, I. & Mann, M. (2004) Temporal analysis of phosphotyrosine-dependent signaling networks by quantitative proteomics. *Nature Biotechnology*, Vol.22, No.9 (Sep 2004), pp.1139–1145, ISSN: 1087-0156

Bodenmiller, B.; Mueller, L.N.; Mueller, M.; Domon, B. & Aebersold, R. (2007) Reproducible isolation of distinct, overlapping segments of the phosphoproteome. *Nature Methods*, Vol.4, No.3 (March 2007), pp.231–237, ISSN: 1548-7091

Bodenmiller, B.; Mueller, L.N.; Pedrioli, P.G.; Pflieger, D.; Jünger, M.A.; Eng, J.K.; Aebersold, R. & Tao, W.A. (2007) An integrated chemical, mass spectrometric and computational strategy for (quantitative) phosphoproteomics: application to Drosophila melanogaster Kc167 cells. *Molecular Biosystems*, Vol.3, No.4 (Apr 2007), pp.275–286

Bondarenko, P.V.; Chelius, D. & Shaler, T.A. (2002) Identification and relative quantitation of protein mixtures by enzymatic digestion followed by capillary reversed-phase liquid chromatography-tandem mass spectrometry. *Analytical Chemistry*, Vol.74, No.18 (15 Sept 2002), pp.4741–4749

Bose, R.; Molina, H.; Patterson, A.S.; Bitok, J.K.; Periaswamy, B.; Bader, J.S.; Pandey, A. & Cole, P.A. (2006) Phosphoproteomic analysis of Her2/neu signaling and inhibition. *Proceedings National Academy Science USA*, Vol.103, No.26 (27 Jun 2006), pp.9773–9778

Buonocore, G.; Liberatori, S.; Bini, L.; Mishra, O.P.; Delivoria-Papadopoulos, M.; Pallini, V. & Bracci, R. (1999) Hypoxic response of synaptosomal proteins in term guinea pig fetuses. *Journal of Neurochemistry*, Vol.73, No.5 (Nov 1999) 2139–2148, ISSN: 0022-3042

Carr, S.A.; Huddleston, M.J.; Annan, R.S. (1996) Selective detection and sequencing of phosphopeptides at the femtomole level by mass spectrometry. *Analytical Biochemistry*, Vol.239, No.2 (1 Aug 1996), pp.180–192

Chang, C.K., Wu, C.C., Wang, Y.S. & Chang, H.C. (2008) Selective extraction and enrichment of multiphosphorylated peptides using polyarginine-coated diamond nanoparticles. *Analytical Chemistry*, Vol.80, No.10 (Apr 2008), pp.3791–3797

Che, F.Y. & Fricker, L.D. (2002) Quantitation of neuropeptides in Cpe(fat)/Cpe(fat) mice using differential isotopic tags and mass spectrometry. *Analytical Chemistry*, Vol.74, No.13 (1 Jul 2002), pp.3190–3198

Chen, C.T. & Chen, Y.C. (2005) Fe3O 4 / TiO2 Core /Shell Nanoparticles as Affinity Probes for the Analysis of Phosphopeptides Using TiO2 Surface-Assisted Laser Desorption/ Ionization Mass Spectrometry. *Analysis*, Vol.77, No.18, (Sep 2005), pp.5912-5919.

Chen, X.; Wu, D.; Zhao, Y.; Wong, B.H.C.; Guo, L. (2011) Increasing phosphoproteome coverage and identification of phosphorylation motifs through combination of different HPLC fractionation methods. *Journal of Chromatography B*, Vol.879, No.1 (Jan 2011), pp.25-34, ISSN: 1570-0232

Chi, A.; Huttenhower, C.; Geer, L.Y.; Coon, J.J.; Syka, J.E.; Bai, D.L.; Shabanowitz, J.; Burke, D.J.; Troyanskaya, O.G. & Hunt, D.F. (2007) Analysis of phosphorylation sites on proteins from Saccharomyces cerevisiae by electron transfer dissociation (ETD) mass spectrometry. *Proceedings National Academy Science USA*, Vol.104, No.7 (13 Feb 2007), pp.2193-2198

Clevenger, C. V., (2004) Roles and Regulation of Stat Family Transcription Factors in Human Breast Cancer. *American Journal of Pathology*, Vol.165, No.5 (Nov 2004), pp.1449-1460

Collins, M.O.; Yu, L.; Husi, H.; Blackstock, W.P.; Choudhary, J.S. & Grant, S.G. (2005) Robust enrichment of phosphorylated species in complex mixtures by sequential protein and peptide metal-affinity chromatography and analysis by tandem mass spectrometry. *Sci. STKE*, Vol.2005, No.298 (Aug 2005), pp.I6.

Collins, M.O.; Yu, L.; Coba, M.P.; Husi, H.; Campuzano, I.; Blackstock, W.P.; Choudhary, J.S. & Grant, S.G. (2005) Proteomic analysis of in vivo phosphorylated synaptic proteins. *Journal of Biological Chemistry*, Vol.280, No.7 (18 Feb 2005), pp.5972-5982Collins, M.O.; Lu Yu & Choudhary J.S. (2007). Analysis of protein phosphorylation on a proteome-scale. *Proteomics*, Vol.7, No.16, (Aug 2007), pp.2751-2768

Connor, P. A. & McQuillan, A. J. (1999) Phosphate Adsorption onto TiO2 from Aqueous Solutions: An in Situ Internal Reflection Infrared Spectroscopic Study. *Langmuir*, Vol.15, No.8 (Mar 1999), pp.2916-2921

Dengjel, J.; Akimov, V.; Olsen, J.V.; Bunkenborg, J.; Mann, M.; Blagoev, B. & Andersen, J.S. (2007) Quantitative proteomic assessment of very early cellular signaling events, *Nature Biotechnology*, Vol.25, No.5 (May 2007), pp.566-568, ISSN: 1087-0156

Domon, B. & Aebersold, R. (2006) Mass spectrometry and protein analysis. *Science*,Vol. 312, No.5771 (14 Apr 2006), pp.212-217

Ducret, A.; Van Oostveen, I.; Eng, J.K.; Yates, J.R.III & Aebersold, R. (1998) High throughput protein characterization by automated reverse-phase chromatography/ electrospray tandem mass spectrometry. *Protein Science*, Vol.7, No.3, (Mar 1998), pp.706-719

Emgenbroich, M.; Borrelli, C.; Shinde, S.; Lazraq, I.; Vilela, F.; Hall, A.J.; Oxelbark, J.; De Lorenzi, E.; Courtois, J.; Simanova, A.; Verhage, J.; Irgum, K.; Karim, K. & Sellergren, B. (2008) A phosphotyrosine-imprinted polymer receptor for the recognition of tyrosine phosphorylated peptides. *Chemistry - A European Journal*, Vol.14, No.31, (Oct 2008), pp.9516-29

Eymann, C.; Becher, D.; Bernhardt, J.; Gronau, K.; Klutzny, A. & Hecker, M. (2007) Dynamics of protein phosphorylation on Ser/Thr/Tyr in Bacillus subtilis. *Proteomics*, Vol.7, No.19 (Aug 2007), pp.3509-3526

Feng, S.; Ye, M.; Zhou, H.; Jiang, X.; Jiang, X; Zou, H. & Gong, B. (2007) Immobilized zirconium ion affinity chromatography for specific enrichment of phosphopeptides in phosphoproteome analysis. *Molecular and Cellular Proteomics*, Vol.6, No.9 (Jun 2007), pp.1656-1665

Ficarro, S. B.; McCleland, M.L.; Stukenberg, P.T.; Burke, D.J.; Ross, M.M.; Shabanowitz, J.; Hunt, D.F. & White, F.M. (2002) Phosphoproteome analysis by mass spectrometry and its application to Saccharomyces cerevisiae. *Nature Biotechnology*, Vol.20, No.3 (Mar 2002), pp.301-305, ISSN: 1087-0156

Ficarro, S.; Chertihin, O.; Westbrook, V.A.; White, F.; Jayes, F.; Kalab, P.; Marto, J.A.; Shabanowitz, J.; Herr, J.C.; Hunt, D.F. & Visconti, P.E. (2003) Phosphoproteome analysis of capacitated human sperm - Evidence of tyrosine phosphorylation of a kinase-anchoring protein 3 and valosin-containing protein/p97 during capacitation. *Journal of Biological Chemistry*, Vol.278, No.13 (Mar 2003), pp.11579–11589, ISSN: 0021-9258

Fischer, E. H. & Krebs E. G. (1955) Conversion of phosphorylase b to phosphorylase a in muscle extracts. *Journal of Biological Chemistry*, Vol.216, No.1 (Sep 1955), pp.121–132

Gembitsky, D.S.; Lawlor, K.; Jacovina, A.; Yaneva, M. & Tempst, P. (2004) A prototype antibody microarray platform to monitor changes in protein tyrosine phosphorylation. *Molecular and Cellular Proteomics*, Vol.3, No.11 (Nov 2004), pp.1102-1118

Gerber, S.A.; Rush, J.; Stemman, O.; Kirschner, M.W. & Gygi, S.P. (2003) Absolute quantification of proteins and phosphoproteins from cell lysates by tandem MS. *Proceedings National Academy of Science USA*, Vol.100, No.12 (10 Jun 2003), pp.6940–6945

Glocker, M.O.; Borchers, C.; Fiedler, W.; Suckau, D. & Przybylski, M. (1994) Molecular characterization of surface topology in protein tertiary structures by amino-acylation and mass spectrometric peptide mapping. *Bioconjugate Chemistry*, Vol.5, No.6 (Nov-Dec 1994), pp.583–590

Gnad, F.; Ren, S.; Cox, J.; Olsen, J.V.; Macek, B.; Oroshi, M. & Mann, M. (2007) PHOSIDA (phosphorylation site database): management, structural and evolutionary investigation, and prediction of phosphosites. *Genome Biology*, Vol.8, No.11, (Nov 2007), pp.R250.1-R250.13

Goodlett, D.R.; Keller, A.; Watts, J.D.; Newitt, R.; Yi, E.C.; Purvine, S.; Eng, J.K.; von Haller, P.; Aebersold, R. & Kolker, E. (2001) Differential stable isotope labeling of peptides for quantitation and de novo sequence derivation. *Rapid Communication Mass Spectrometry*, Vol.15, No.14, (Jun 2001), pp.1214–1221

Goss, V.L.; Lee, K.A.; Moritz, A.; Nardone, J.; Spek, E.J.; MacNeill, J.; Rush, J.; Comb, M.J.; Polakiewicz, R.D. (2006) A common phosphotyrosine signature for the Bcr-Abl kinase. *Blood*, Vol.107, No.12 (15 Jun 206), pp.4888–4897

Goto, H. & Inagaki, M. (2007) Production of a site- and phosphorylation state-specific antibody. *Nature Protocols*, Vol.2, No.10, (Oct 2007), pp.2574– 2581

Graham, M.E.; Anggono, V.; Bache, N.; Larsen, M.R.; Craft, G.E. & Robinson, P.J. (2007) The in vivo phosphorylation sites of rat brain dynamin I , *Journal of Biological Chemistry*, Vol.282, No.20 (May 2007), pp.14695–14707, ISSN: 0021-9258

Graves, J. D. & Krebs, E. G. (1999) Protein Phosphorylation and Signal Transduction. *Pharmacology & Therapeutics* Vol.82, No.2-3, (May-Jun 1999), pp-111–121

Grønborg, M. ; Kristiansen, T.Z., Stensballe, A., Andersen, J.S., Ohara, O., Mann, M., Jensen, O.N. & Pandey, A. (2002) A mass spectrometry-based proteomic approach for identification of serine/threonine-phosphorylated proteins by enrichment with phospho-specific antibodies: identification of a novel protein, Frigg, as a protein kinase A substrate. *Molecular and Cellular Proteomics*, Vol.1, No.7 (Jul 2002), pp.517–527

Guy, G. R.; Philip, R. & Tan, Y. H. (1994) Analysis of cellular phosphoproteins by two-dimensional gel electrophoresis: Applications for cell signaling in normal and cancer cells. *Electrophoresis*, Vol.15, No.1 (Apr 2005), pp.417–440

Han, G., Ye, M., Zhou, H., Jiang, X. Feng, S.; Jiang, X.G.; Tian, R.J.; Wan, D.F. ; Zou, H.F. & Gu, J.R. (2008) Large-scale phosphoproteome analysis of human liver tissue by enrichment and fractionation of phosphopeptides with strong anion exchange chromatography. *Proteomics*, Vol.8, No.7 (Apr 2008), pp.1346–1361, ISSN: 1615-9853

Han, G.; Ye, M.; Zhou, H.; Jiang, X.; Feng, S.; Jiang, X.; Tian, R.; Wan, D.; Zou, H. & Gu, J. (2008) Large-scale phosphoproteome analysis of human liver tissue by enrichment and fractionation of phosphopeptides with strong anion exchange chromatography. *Proteomics*, Vol.8, No.7 (Apr 2008), pp. 1346–1361

Han, G.; Ye, M.; Zhou, H.; Jiang, Xinning; Feng, S.; Jiang, Xiaogang; Tian, R.; Wan, D.; Zou, H. & Gu, J. (2008) Large-scale phosphoproteome analysis of human liver tissue by enrichment and fractionation of phosphopeptides with strong anion exchange chromatography. *Proteomics*, Vol.8, No.7 (Apr 2008), pp.1346–1361

Heibeck, T.H.; Shi-Jian Ding; Opresko, L.K.; Rui Zhao; Schepmoes, A.A.; Feng Yang; Tolmachev, A.V.; Monroe, M.E.; Camp II, D.G.; Smith, R.D.; Wiley, H. S. & Wei-Jun Qian (2009) An Extensive Survey of Tyrosine Phosphorylation Revealing New Sites in Human Mammary Epithelial Cells. *Journal of Proteome Research*, Vol.8, No.8 (Jun 2009), pp.3852–3861

Helling, S.; Shinde, S.; Brosseron, F.; Schnabel, A.; Müller, T.; Meyer, H.E.; Marcus, K. & Sellergren, B. (2011) Ultratrace Enrichment of Tyrosine Phosphorylated Peptides on an Imprinted Polymer. *Analytical Chemistry*, Vol.83 (Feb 2011), pp.1862–1865

Hemmings Jr., H.C. (1996), *Regulatory Protein Modification: Technique and Protocols*, Humana Press, Totowa, New Jersey, USA

Hornbeck, P.V.; Chabra, I.; Kornhauser, J.M.; Skrzypek, E. & Zhang, B. (2004) PhosphoSite: A bioinformatics resource dedicated to physiological protein phosphorylation. *Proteomics*, Vol.4, No.6 (Jun 2004), pp.1551–1561

Hu, V.W. & Heikka, D.S. (2000) Radiolabeling revisited: metabolic labeling with (35)S-methionine inhibits cell cycle progression, proliferation, and survival. *FASEB J.*, Vol.14, No.3 (Mar 2000), pp.448–454

Hu, V.W.; Heikka, D.S.; Dieffenbach, P.B. & Ha, L. (2001) Metabolic radiolabeling: experimental tool or Trojan horse? S-35-Methionine induces DNA fragmentation and p53-dependent ROS production. *FASEB J.*, Vol.15, No.9 (Jul 2001), pp.1562–1568, ISSN: 0892-6638

Hunter, T. (2000). Signaling--2000 and beyond. *Cell*, Vol. 100, No.1, (Jan 2000), pp.113–127

Huttlin, E.L.; Jedrychowski, M.P.; Elias, J.E.; Goswami, T.; Rad, R.; Beausoleil, S.A.; Villen, J.; Haas, W.; Sowa, M.E. & Gygi, S.P. (2010) A Tissue-Specific Atlas of Mouse Protein Phosphorylation and Expression. *Cell*, Vol.143, No.7 (Dec 2010), pp.1174–1189, ISSN: 0092-8674

Ibarrola, N.; Kalume, D.E.; Gronborg, M.; Iwahori, A. & Pandey, A. (2003) A proteomic approach for quantitation of phosphorylation using stable isotope labeling in cell culture. *Analytical Chemistry*, Vol.75, No.22 (15 Nov 2003), pp.6043–6049

Ikeguchi, Y.& Nakamura, H. (2000) Selective enrichment of phospholipids by titania. *Analytical Sciences*, Vol.16, No.5 (May 2000), pp.541-543, ISSN: 0910-6340

Irish, J.M.; Howland, R.; Krutzik, P.O.; Perez, O.D.; Bruserud, O.; Gjertsen, B.T. & Nolan, G.P. (2004) Single cell profiling of potentiated phospho-protein networks in cancer cells. *Cell*, Vol.118, No.2 (23 Jul 2004), pp.217-228

Ishihama,Y.; Oda, Y.; Tabata, T.; Sato, T.; Nagasu, T.; Rappsilber, J. & Mann, M. (2005) Exponentially modified protein abundance index (emPAI) for estimation of absolute protein amount in proteomics by the number of sequenced peptides per protein. *Molecular and Cellular Proteomics*, Vol.4, No.9 (Sept 2005), pp.1265-1272

Jaffe, H.; Veeranna & Pant, H.C. (1998) Characterization of serine and threonine phosphorylation sites in beta-elimination/ethanethiol addition-modified proteins by electrospray tandem mass spectrometry and database searching. *Biochemistry*, Vol.37, No.46 (17 Nov 1998), pp.16211-16224

Jensen, S.S. & Larsen, M.R. (2007) Evaluation of the impact of some experimental procedures on different phosphopeptide enrichment techniques. *Rapid Communication in Mass Spectrometry*, Vol.21, No.22, (Oct 2007), pp.3635-3645, ISSN: 0951-4198

Jones, R.B.; Gordus, A.; Krall, J.A. & Mac Beath, G. (2006) A quantitative protein interaction network for the ErbB receptors using protein microarrays. *Nature*, Vol.439, No.7073 (12 Jan 2006), pp.168-174

Kalume, D. E.; Molina, H. & Pandey, A. (2003) Tackling the phosphoproteome: tools and strategies. *Current Opinion in Chemical Biology*, Vol.7, No.1 (Feb 2003), pp.64-69

Kaufmann, H.; Bailey, J.E. & Fussenegger, M. (2001) Use of antibodies for detection of phosphorylated proteins separated by two-dimensional gel electrophoresis, *Proteomics*, Vol.1, No.2 (Feb 2001), pp.194-199, ISSN: 1615-9853

Kersten, B.; Agrawal, G. K.; Iwahashi, H. & Rakwal, R. (2006) Plant phosphoproteomics: A long road ahead. *Proteomics*, Vol.6, No.20 (Oct 2006), pp.5517-5528

Khan, I.H.; Mendoza, S.; Rhyne, P.; Ziman, M.; Tuscano, J.; Eisinger, D.; Kung, H.J. & Luciw, P.A. (2006) Multiplex analysis of intracellular signaling pathways in lymphoid cells by microbead suspension arrays. *Molecular and Cellular Proteomics*, Vol.5, No.4 (Apr 2006), pp.758-768

Kim, S.; Mischerikow, N.; Bandeira, N.; Navarro, N.J.D.; Wich, L.; Mohammed, S.; Heck, A.J.R. & Pevzner, P.A. (2010) The generating function of CID, ETD, and CID/ETD pairs of tandem mass spectra: applications to database search. *Molecular and Cellular Proteomics*, Vol.9, No.12 (Dec 2010), pp.2840-2852

Kirkpatrick, D.S. ; Gerber, S.A. & Gygi, S.P. (2005) The absolute quantification strategy: a general procedure for the quantification of proteins and post-translational modifications. Methods, Vol.35, No.3 (Mar 2005), pp. 265-273

Kjeldsen, F.; Giessing, A.M.; Ingrell, C.R. & Jensen, O.N. (2007) Peptide sequencing and characterization of post-translational modifications by enhanced ion-charging and liquid chromatography electron-transfer dissociation tandem mass spectrometry. *Analytical Chemistry*, Vol.79, No.24 (15 Dec 2007), pp.9243-9252

Kleinnijenhuis, A.J.; Kjeldsen, F.; Kallipolitis, B.; Haselmann, K.F. & Jensen, O.N. (2007) Analysis of histidine phosphorylation using tandem MS and ion-electron reactions. Analytical Chemistry, Vol.79, No.19 (1 Oct 2007), pp.7450-7456

Knight, Z.A.; Schilling, B.; Row, R.H.; Kenski, D.M.; Gibson, B.W. & Shokat, K.M. (2003) Phosphospecific proteolysis for mapping sites of protein phosphorylation. *Nature Biotechnology*, Vol.21, No.9 (Sept 2003), pp.1047-1054

Krutzik, P.O.; Hale, M.B. & Nolan, G.P. (2005) Characterization of the murine immunological signaling network with phosphospecific flow cytometry. *Journal of Immunology*, Vol.175, No.4 (15 Aug 2005), pp.2366-2373

Kumar, Y.; Khachane, A.; Belwal, M.; Das, S.; Somsundaram, K. & Tatu, U. (2004) ProteoMod: A new tool to quantitate protein post-translational modifications. *Proteomics*, Vol.4, No.6 (Jun 2004), 1672-1683, ISSN: 1615-9853

Kweon, H.K. & Hakansson, K. (2006) Selective zirconium dioxide-based enrichment of phosphorylated peptides for mass spectrometric analysis. *Analytical Chemistry*, Vol.78, No.6 (Mar 2006), pp.1743–1749, ISSN: 0003-2700

Larsen, M.R.; Thingholm, T.E.; Jensen, O.N.; Roepstorff, P. & Jorgensen, T.J. (2005) Highly selective enrichment of phosphorylated peptides from peptide mixtures using titanium dioxide microcolumns. *Molecular and Cellular Proteomics*, Vol.4, No.7 (Jul 2005), pp.873–886, ISSN: 1535-9476

Leitner, A. (2010) Phosphopeptide enrichment using metal oxide affinity chromatography, *Trends in Analytical Chemistry*, Vol.29, No.2 (Feb 2010), pp.177-185

Loyet, K.M.; Stults, J.T. & Arnott, D. (2005) Mass spectrometric contributions to the practice of phosphorylation site mapping through 2003: a literature review. *Molecular and Cellular Proteomics*, Vol.4, No.3 (Mar 2005), pp.235–245

Lu, P.; Vogel, C.; Wang, R.; Yao, X. & Marcotte, E.M. (2007) Absolute protein expression profiling estimates the relative contributions of transcriptional and translational regulation. *Nature Biotechnology*, Vol.25, No.1, (Jan 2007), pp.117–124

Magi, B.; Bini, L.; Marzocchi, B.; Liberatori, S.; Raggiaschi, R. & Pallini, V. (1999) Immunoaffinity identification of 2-DE separated proteins. *Methods in Molecular Biology*, Vol.112, (Apr 1999), pp.431–443

Mamone, G.; Picariello, G.; Ferranti, P. & Addeo, F. (2010) Hydroxyapatite affinity chromatography for the highly selective enrichment of mono- and multi-phosphorylated peptides in phosphoproteome analysis. *Proteomics*, Vol.10, No.3 (Feb 2010), pp.380–393

Mann, M.; Ong, S.E.; Gronborg, M.; Steen, H.; Jensen, O.N. & Pandey, A. (2002) Analysis of protein phosphorylation using mass spectrometry: deciphering the phosphoproteome. *Trends in Biotechnology*, Vol.20, No.6 (Jun 2002), pp.261–268

Manning, G.; Whyte, D.B.; Martinez, R.; Hunter, T. & Sudarsanam, S. (2002) The Protein Kinase Complement of the Human Genome. *Science*, Vol.298, No.5600 (Dec 2002), pp.1912–1934, ISSN 0036-8075 (print), 1095-9203 (online)

McNulty, D.E. & Annan, R.S. (2008) Hydrophilic interaction chromatography reduces the complexity of the phosphoproteome and improves global phosphopeptide isolation and detection. *Molecular and Cellular Proteomics*, Vol.7, No.5 (May 2008), pp.971-980

Miyagi, M. & Rao, K.C. (2007) Proteolytic O-18-labeling strategies for quantitative proteomics. *Mass Spectrometry Reviews*, Vol.26, No.1 (Jan-Feb 2007), pp.121–136

Mohammed, S. & Heck, A.J.R. (2011) Strong cation exchange (SCX) based analytical methods for the targeted analysis of protein post-translational modifications. *Current Opinion in Biotechnology*, Vol.22, No.1 (Feb 2011), pp.9–16

Molina, H.; Horn, D.M.; Tang, N.; Mathivanan, S. & Pandey, A. (2007) Global proteomic profiling of phosphopeptides using electron transfer dissociation tandem mass spectrometry. *Proceedings National Academy Science USA*, Vol.104, No.7 (Feb 2007), pp.2199–2204, ISSN: 0027-8424

Nelson, L. & Cox, M. (2004) *Lehninger principles of biochemistry*, (4th edition), W.H.Freeman, ISBN 0-7167-4339-6

Nishino, H.; Huang, C.S. & Shea, K.J. (2006) Selective protein capture by epitope imprinting. *Angewandte Chemie International Edition*, Vol.45, No.15 (3 Apr 2006), pp.2392-2396

Nühse, T.S.; Stensballe, A.; Jensen, O.N. & Peck, S.C. (2003) Large-scale analysis of in vivo phosphorylated membrane proteins by immobilized metal ion affinity chromatography and mass spectrometry. Molecular and Cellular Proteomics, Vol.2, No.11 (Nov 2003), pp.1234– 1243, ISSN: 1535-9476

Nühse, T.S.; Stensballe, A.; Jensen, O.N. & Peck, S.C. (2004) Phosphoproteomics of the Arabidopsis plasma membrane and a new phosphorylation site database. The Plant Cell, Vol.16, No.9 (Sep 2004), pp.2394-2405, ISSN: 1040-4651

Oda, Y.; Huang, K.; Cross, F.R.; Cowburn, D. & Chait, B.T. (1999) Accurate quantitation of protein expression and site-specific phosphorylation. Proceedings National Academy Science USA, Vol.96, No.12 (8 Jun 1999), pp.6591–6596.

Oda, Y.; Nagasu, T. & Chait, B.T. (2001) Enrichment analysis of phosphorylated proteins as a tool for probing the phosphoproteome. Nature Biotechnology, Vol.19, No.4 (Apr 2001), pp.379–382

Olsen, J. V., Blagoev, B., Gnad, F., Macek, B.; Kumar, C.; Mortensen, P. & Mann, M. (2006) Global, in vivo, and site-specific phosphorylation dynamics in signaling networks. Cell, Vol.127, No.3 (Nov 2006), pp.635–648, ISSN: 0092-8674

Olsen, J.V.; Macek, B.; Lange, O.; Makarov, A.; Horning, S. & Mann, M. (2007) Higher-energy C-trap dissociation for peptide modification analysis. Nature Methods, Vol.4, No.9 (September 2007), pp.709–712, ISSN: 1548-7091

Olsen, J.V.; Vermeulen, M.; Santamaria, A.; Kumar, C.; Miller, M.L.; Jensen, L.J.; Gnad, F.; Cox, J.; Jensen, T.S.; Nigg, E.A.; Brunak, S. & Mann, M. (2010) Quantitative Phosphoproteomics Reveals Widespread Full Phosphorylation Site Occupancy During Mitosis. Science Signaling, Vol.3, No.104, (Jan 2010), p. ra3

Ong, S.E.; Blagoev, B.; Kratchmarova, I.; Kristensen, D.B.; Steen, H.; Pandey, A. & Mann, M. (2002) Stable isotope labeling by amino acids in cell culture, SILAC, as a simple and accurate approach to expression proteomics, Molecular and Cellular Proteomics, Vol.1, No. 5 (May 2002), pp.376–386, ISSN: 1535-9476

Ong, S.E. & Mann, M. (2005) Mass spectrometry-based proteomics turns quantitative. Nature Chemical Biology, Vol.1, No5. (Oct 2005), pp.252-262

Pan, S.; Zhang, H.; Rush, J.; Eng, J.; Zhang, N.; Patterson, D.; Comb, M.J. & Aebersold, R. (2005) High throughput proteome screening for biomarker detection. Molecular and Cellular Proteomics, Vol.4, No.2 (Feb 2005), pp.182–190

Perkins, D.N.; Pappin, D.J.; Creasy, D.M. & Cottrell, J.S. (1999) Probability-based protein identification by searching sequence databases using mass spectrometry data. Electrophoresis, Vol.20, No.18 (Dec 1999), pp.3551–3567

Pinkse, M.W.; Uitto, P.M.; Hilhorst, M.J.; Ooms, B. & Heck, A.J. (2004) Selective isolation at the femtomole level of phosphopeptides from proteolytic digests using 2D-nanoLC-ESI-MS/MS and titanium oxide precolumns. Analytical Chemistry, Vol.76, No.14 (Jul 2004), pp.3935–3943, ISSN: 0003-2700

Porath, J., Carlsson, J., Olsson, I., Belfrage, G. (1975) Metal chelate affinity chromatography, a new approach to protein fractionation. Nature, Vol.258, No.5536 (Dec 1975), pp.598–599

Posewitz, M.C. & Tempst, P. (1999) Immobilized gallium(III) affinity chromatography of phosphopeptides, Analytical Chemistry, Vol.71, No.14 (Jul 1999), pp.2883–2892, ISSN: 0003-2700

Ptacek, J.; Devgan, G.; Michaud, G.; Zhu, H.; Zhu, X.; Fasolo, J.; Guo, H.; Jona, G.;
    Breitkreutz, A.; Sopko, R.; McCartney, R.R.; Schmidt, M.C.; Rachidi, N.; Lee, S.J.;
    Mah, A.S.; Meng, L.; Stark. M.J.; Stern, D.F.; De Virgilio. C.; Tyers, M.; Andrews, B.;
    Gerstein, M.; Schweitzer, B.; Predki, P.F. & Snyder, M. (2005) Global analysis of
    protein phosphorylation in yeast. *Nature*, Vol.438, No.7068 (1 Dec 2005), pp.679–684
Qi, D.; Lu, J.; Deng, C. & Zhang, X. (2009) Magnetically Responsive Fe(3)O(4)@C@SnO(2)
    Core-Shell Microspheres: Synthesis, Characterization and Application in
    Phosphoproteomics, *Journal of Physical Chemistry C*, Vol.113, No.36 (Sep 2009),
    pp.15854–15861, ISSN: 1932-7447
Rachkov, A. & Minoura, N. (2001) Towards molecularly imprinted polymers selective to
    peptides and proteins. The epitope approach. *Biochimica et Biophysica Acta*,
    Vol.1544, No.1-2 (12 Jan 2001), pp.255-266
Raggiaschi, R.; Gotta, S. & Terstappen G.C. (2005) Phosphoproteome Analysis. *Bioscience
    Reports*, Vol.25, Nos.1-2, (Feb-Apr 2005), pp.33-44
Rappsilber, J.; Ryder, U.; Lamond, A.I. & Mann, M. (2002) Large-scale proteomic analysis of
    the human spliceosome. *Genome Research*, Vol.12, No.8 (Aug 2002), pp.1231–1245.
Reinders, J. & Sickmann, A. (2005) State-of-the-art in phosphoproteomics. *Proteomics*, Vol.5,
    No.16 (Nov 2005), pp.4052–4061
Reynolds, E.C.; Riley, P.F. & Adamson, N.J. (1994) A Selective precipitation purification
    procedure for multiple phosphoseryl-containing peptides and methods for their
    identification. *Analytical Biochemistry*, Vol.217, No.2 (Mar 1994), pp.277–284, ISSN:
    0003-2697
Rigbolt, K.T.; Prokhorova, T.A.; Akimov, V.; Henningsen, J.; Johansen, P.T.; Kratchmarova,
    I.; Kassem, M.; Mann, M.; Olsen, J.V. & Blagoev, B. (2011) System-Wide Temporal
    Characterization of the Proteome and Phosphoproteome of Human Embryonic
    Stem Cell Differentiation. *Science Signaling*, Vol.4, No.164 (Mar 2011), rs3, ISSN:
    1937-9145
Roepstorff, P. & Fohlman, J. (1984) Proposal for a common nomenclature for sequence ions
    in mass spectra of peptides. *Biomedical Mass Spectrometry*, Vol.11, No.11, (Nov 1984)
    pp.601
Ross, P.L.; Huang, Y.N.; Marchese, J.N.; Williamson, B.; Parker, K.; Hattan, S.; Khainovski,
    N.; Pillai, S.; Dey, S.; Daniels, S.; Purkayastha, S.; Juhasz, P.; Martin, S.; Bartlet-
    Jones, M.; He, F.; Jacobson A. & Pappin, D.J. (2004) Multiplexed protein
    quantitation in Saccharomyces cerevisiae using amine-reactive isobaric tagging
    reagents. *Molecular and Cellular Proteomics*, Vol.3, No.12 (Dec 2005), pp.1154–1169
Rush, J.; Moritz, A.; Lee, K.A.; Guo, A.; Goss, V.L.; Spek, E.J.; Zhang, H.; Xiang-Ming Zha;
    Polakiewicz, R.D. & Comb, M.J. (2004) Immunoaffinity profiling of tyrosine
    phosphorylation in cancer cells. *Nature Biotechnology*, Vol.23 (Dec 2004), pp.94-101
Sachs, K.; Perez, O.; Pe'er, D.; Lauffenburger, D.A. & Nolan, G.P. (2005) Causal protein-
    signaling networks derived from multiparameter single-cell data. *Science*, Vol.308,
    No.5721 (22 Apr 2005), pp.523-529
Schmelzle, K. & White, F.M. (2006) Phosphoproteomic approaches to elucidate cellular
    signaling networks. *Current Opinion in Biotechnology*, Vol.17, No. 4 (Aug 2006),
    pp.406-414

Schmidt, A.; Kellermann, J. & Lottspeich, F. (2005) A novel strategy for quantitative proteomics using isotope-coded protein labels. *Proteomics*, Vol.5, No.1 (Jan 2005), pp.4–15

Schmidt, S.R.; Schweikart, F.; Andersson, M.E. (2007) Current methods for phosphoprotein isolation and enrichment. *Journal of Chromatography B*, Vol.849, No.1-2 (Apr 2007), pp.154-162, ISSN: 1570-0232

Schroeder, M. J.; Shabanowitz, J.; Schwartz, J.C.; Hunt, D.F. & Coon, J.J. (2004) A neutral loss activation method for improved phosphopeptide sequence analysis by quadrupole ion trap mass spectrometry. *Analytical Chemistry*, Vol.76, No.13 (1 Jul 2004), pp.3590–3598

Schroeder, M.J.; Webb, D.J.; Shabanowitz, J.; Horwitz, A.F. & Hunt, D.F. (2005) Methods for the detection of paxillin post-translational modifications and interacting proteins by mass spectrometry. *Journal of Proteome Research*, Vol.4, No.5 (Sept-Oct 2005), pp.1832–1841

Schürenberg, M.; Dreisewerd, K.; Hillenkamp, F. (1999) Laser Desorption/Ionization Mass Spectrometry of Peptides and Proteins with Particle Suspension Matrixes. *Analytical Chemistry*, Vol.71, No.1 (Dec 1998), pp.221-229

Schulenberg, B.; Aggeler, R.; Beechem, J.M.; Capaldi, R.A. & Patton, W.F. (2003) Analysis of steady-state protein phosphorylation in mitochondria using a novel fluorescent phosphosensor dye. *Journal of Biological Chemistry*, Vol.278, (Jul 2003), pp.27251–27255, ISSN: 0021-9258

Sheenan, K.M.; Calvert, V.S.; Kay, E.V.; Lu, Y.; Fishman, D.; Espina, V.; Aquino, J.; Speer, R.; Araujo, R.; Mills, G.B.; Liotta, L.A.; Petricoin, E.F.3rd & Wulfkuhle, J.D. (2005) Use of reverse phase protein microarrays and reference standard development for molecular network analysis of metastatic ovarian carcinoma. *Molecular and Cellular Proteomics*, Vol.4, No.4, (Apr 2005), pp.346-355

Sickmann, A. & Meyer, H. E. (2001) Phosphoamino acid analysis. *Proteomics*, Vol.1, No.2, (Feb 2001), pp.200-206

Stannard, C.; Soskic, V. & Godovac-Zimmermann, J. (2003) Rapid changes in the phosphoproteome show diverse cellular responses following stimulation of human lung fibroblasts with endothelin-1. *Biochemistry*, Vol.42, No.47 (Dec 2003), pp.13919–13928, ISSN: 0006-2960

Steen, H.; Kuster, B. & Mann, M. (2001) Quadrupole time-of-flight versus triple-quadrupole mass spectrometry for the determination of phosphopeptides by precursor ion scanning. *Journal Mass Spectrometry*, Vol.36, No.7 (Jul 2001), pp.782–790

Steen, H.; Kuster, B.; Fernandez, M.; Pandey, A.; Mann, M. (2002) Tyrosine phosphorylation mapping of the epidermal growth factor receptor signaling pathway. *Journal of Biological Chemistry*, Vol.277, No.2 (Jan 2002), pp.1031–1039.

Steen, H.; Jebanathirajah, J.A.; Springer, M. & Kirschner, M.W. (2005) Stable isotope-free relative and absolute quantitation of protein phosphorylation stoichiometry by MS. *Proceedings National Academy of Science USA*, Vol.102, No.11 (15 Mar 2005), pp.3948–3953

Steinberg, T.H.; Agnew, B.J.; Gee, K.R.; Leung, W.Y.; Goodman, T.; Schulenberg, B.; Hendrickson, J.; Beechem, J.M.; Haugland, R.P. & Patton, W.F. (2003) Global quantitative phosphoprotein analysis using multiplexed proteomics technology. *Proteomics*, Vol.3, No.7 (Jul 2003), pp.1128–1144, ISSN: 1615-9853

Stensballe, A.; Jensen, O.N.; Olsen, J.V.; Haselmann, K.F. & Zubarev, R.A. (2000) Electron capture dissociation of singly and multiply phosphorylated peptides. *Rapid Communication Mass Spectrometry*, Vol.14, No.19, (Sep 2000), pp.1793–1800

Su, H.C.; Hutchison, C. A. 3rd & Giddings, M.C. (2007) Mapping phosphoproteins in Mycoplasma genitalium and Mycoplasma pneumoniae. *BMC Microbiology*, Vol.7, (Jul 2007), pp.63

Sugiyama, N.; Nakagami, H.; Mochida, K.; Daudi, A.; Tomita, M.; Shirasu, K. & Ishihama, Y. (2008) Large-scale phosphorylation mapping reveals the extent of tyrosine phosphorylation in Arabidopsis. *Molecular Systems Biology*, Vol.4, No.193 (May 2008), ISSN: 1744-4292

Sunner, J.; Dratz, E.; Chen, Y.C. (1995) Graphite surface-assisted laser desorption/ionization time-of-flight mass spectrometry of peptides and proteins from liquid solutions. *Analytical Chemistry*, Vol.67, No.23 (Dec 1995), pp.4335-4342

Swaney, D.L.; Wenger, C.D.; Thomson, J.A.& Coon, J.J. (2009) Human embryonic stem cell phosphoproteome revealed by electron transfer dissociation tandem mass spectrometry. *Proceedings National Academy Science USA*, Vol.106, No.4 (Jan 2009), pp.995-1000

Syka, J.E.; Marto, J.A.; Bai, D.L.; Horning, S.; Senko, M.W.; Schwartz, J.C.; Ueberheide, B.; Garcia, B.; Busby, S.; Muratore, T.; Shabanowitz, J. & Hunt, D.F. (2004) Novel linear quadrupole ion trap/FT mass spectrometer: performance characterization and use in the comparative analysis of histone H3 post-translational modifications. *Journal of Proteome Research*, Vol.3, No.3 (May-Jun 2004), pp.621-626

Syka, J.E.; Coon, J.J.; Schroeder, M.J.; Shabanowitz, J. & Hunt, D.F. (2004) Peptide and protein sequence analysis by electron transfer dissociation mass spectrometry. *Proceedings National Academy Science USA*, Vol.101, No.26 (29 Jun 2004), pp.9528–9533

Tao, W.A.; Wollscheid, B.; O'Brien, R.; Eng, J.K.; Li, X.J.; Bodenmiller, B.; Watts, J.D.; Hood, L. & Aebersold, R. (2005) Quantitative phosphoproteome analysis using a dendrimer conjugation chemistry and tandem mass spectrometry. *Nature Methods*, Vol.2, No.8 (Aug 2005), pp.591–598

Thingholm, T.E.; Jorgensen, T.J.; Jensen, O.N. & Larsen, M.R. (2006) Highly selective enrichment of phosphorylated peptides using titanium dioxide. *Nature Protocols*, Vol.1, No.4, (Nov 2006), pp.1929–1935, ISSN: 1754-2189

Thingholm, T.E.; Jensen, O.N.; Robinson, P.J. & Larsen, M.R. (2008) SIMAC (sequential elution from IMAC), a phosphoproteomics strategy for the rapid separation of monophosphorylated from multiply phosphorylated peptides. *Molecular and Cellular Proteomics*, Vol.7, No.4 (Apr 2008), pp.661-671, ISSN: 1535-9476

Thingholm, T.E.; Larsen, M.R.; Ingrell, C.R.; Kassem, M. & Jensen, O.N. (2008) TiO2-based phosphoproteomic analysis of the plasma membrane and the effects of phosphatase inhibitor treatment. *Journal of Proteome Research*, Vol. 7, No.8 (Aug 2008), pp.3304–3313

Thingholm, T.E.; Jensen, O.N. & Larsen, M.R. (2009) Analytical strategies for phosphoproteomics. *Proteomics*, Vol.9, No.6, (Mar 2009), pp.1451-1468

Thingholm, T.E. & Jensen, O.N. (2009) *Phospho-Proteomics – Methods and Protocols*, Humana Press, Totowa, NJ, Chapter 4, pp.47–56

Thompson, A.; Schafer, J.; Kuhn, K.; Kienle, S.; Schwarz, J.; Schmidt, G.; Neumann, T.; Johnstone, R.; Mohammed, A.K. & Hamon, C. (2003) Tandem mass tags: a novel quantification strategy for comparative analysis of complex protein mixtures by MS/MS. *Analytical Chemistry*, Vol.75, No.8 (15 Apr 2003), pp.1895–1904

Tiselius, A.; Hjerten, S. & Levin, O. (1956) Protein chromatography on calcium phosphate columns. *Archives of biochemistry and biophysics*, Vol.65, No.46 (18 Nov 1956), pp.132–155

Titirici, M.M.; Hall, A.J. & Sellergren, B. (2003) Hierarchical imprinting using crude solid phase peptide synthesis products as templates. *Chemistry of Materials*, Vol.15, No.4 (25 Feb 2003), pp.822–824

Towbin, H.; Staehelin, T. & Gordon, J. (1979) Electrophoretic transfer of proteins from polyacrylamide gels to nitrocellulose sheets: procedure and some applications., *Proceedings National Academy of Science USA*, Vol.76, No.9 (Sep 1979), pp. 4350–4354

Trinidad, J.C.; Specht, C.G.; Thalhammer, A.; Schoepfer, R. & Burlingame, A.L. (2006) Comprehensive identification of phosphorylation sites in postsynaptic density preparations. *Molecular and Cellular Proteomics*, Vol.5, No.5 (May 2006), pp.914–922, ISSN: 1535-9476

Turk, B.E.; Hutti, J.E. & Cantley, L.C. (2006) Determining protein kinase substrate specificity by parallel solution-phase assay of large numbers of peptide substrates. *Nature Protocols*, Vol.1, No.1, (Jun 2006), pp.375–379

Twyman R.M. (2004) *Principles of proteomics*, BIOS Scientific Publishers, Taylor and Francis Group, ISBN 978-1-85996-273-2

Venter, J. C. et al. (2001) The sequence of the human genome. *Science*, Vol.291, No.5507 (Feb 2001), 1304–1351

Villen, J.; Beausoleil, S.A.; Gerber, S.A. & Gygi, S.P. (2007) Large-scale phosphorylation analysis of mouse liver. *Proceedings National Academy of Science USA*, Vol.104, No.5 (Jan 2007), pp.1488–1493, ISSN: 0027-8424

Wang, G., Wu, W.W.; Zeng, W.; Chou, C.L. & Shen, R.F. (2006) Label-free protein quantification using LC-coupled ion trap or FT mass spectrometry: Reproducibility, linearity, and application with complex proteomes. *Journa of Proteome Research*, Vol.5, No.5 (May 2006), pp.1214–1223

Wolf-Yadlin, A.; Hautaniemi, S.; Lauffenburger, D.A. & White, F.M. (2007) Multiple reaction monitoring for robust quantitative proteomic analysis of cellular signaling networks. *Proceedings National Academy of Science USA*, Vol.104, No.14 (3 Apr 2007), pp.5860–5865

Xia, Q.W.; Cheng, D.; Duong, D.M.; Gearing, M.; Lah, J.J.; Levey, A.I. & Peng, J. (2008) Phosphoproteomic analysis of human brain by calcium phosphate precipitation and mass spectrometry. *Journal of Proteome Research*, Vol.7, No.7 (Jul 2008), pp.2845–2851, ISSN: 1535-3893

Yamagata, A.; Kristensen, D.B.; Takeda, Y.; Miyamoto, Y.; Okada, K.; Inamatsu, M. & Yoshizato, K. (2002) Mapping of phosphorylated proteins on two-dimensional polyacrylamide gels using protein phosphatase. *Proteomics*, Vol.2, No.9 (Sep 2002), pp.1267–1276, ISSN: 1615-9853

Yaoi, T.; Chamnongpol, S.; Jiang, X. & Li, X. (2006) Src homology 2 domain-based high throughput assays for profiling downstream molecules in receptor tyrosine kinase pathways. *Molecular Cellular Proteomics*, Vol.5, No.5 (May 2006), pp.959-968

Yeargin, J. & Haas, M. (1995) Elevated levels of wild-type P53 induced by radiolabeling of cells leads to apoptosis or sustained growth arrest. *Current Biology*, Vol.5, No.4 (Apr 1995), pp.423–431, ISSN 0960-9822

Zarei, M.; Sprenger, A.; Metzger, F.; Gretzmeier, C.& Dengjel, J. (2011) Comparison of ERLIC-TiO(2), HILIC-TiO(2), and SCX-TiO(2) for Global Phosphoproteomics Approaches. *Journal of Proteome Research* (in press)

Zhai, B.; Villen, J.; Beausoleil, S.A.; Mintseris, J. & Gygi, S.P. (2008) Phosphoproteome Analysis of Drosophila melanogaster Embryos. *Journal of Proteome Research*, Vol.7, No.4 (March 2008), pp.1675-1682

Zhang, H.; Zha, X.; Tan, Y.; Hornbeck, P.V.; Mastrangelo, A.J.; Alessi, D.R.; Polakiewicz & R.D.; Comb, M.J. (2002) Phosphoprotein analysis using antibodies broadly reactive against phosphorylated motifs. *Journal of Biological Chemistry*, Vol.277, No.42 (Oct 2002), pp.39379-39387

Zhang, K. (2006) From purification of large amounts of phospho-compounds (nucleotides) to enrichment of phospho-peptides using anion-exchanging resin. *Analytical Biochemistry*, Vol.357, No.2 (Oct 2006), pp.225-231

Zhang, X.; Jin, Q.K.; Carr, S.A. & Annan, R.S. (2002) N-Terminal peptide labeling strategy for incorporation of isotopic tags: a method for the determination of site-specific absolute phosphorylation stoichiometry. *Rapid Communication Mass Spectrometry*, Vol.16, No.24, (Nov 2002), pp.2325-2332

Zhang, X.; Ye, J.; Jensen, O.N. & Roepstorff, P. (2007) Highly efficient phosphopeptide enrichment by calcium phosphate precipitation combined with subsequent IMAC enrichment. *Molecular and Cellular Proteomics*, Vol.6, No.11 (Nov 2007), pp.2032–2042, ISSN: 1535-9476

Zhou, H.; Watts, J.D. & Aebersold, R. (2001) A systematic approach to the analysis of protein phosphorylation. *Nature Biotechnology*, Vol.19, No.4 (Apr 2001), pp.375–378

Zhu, X.; Lee, H.; Raina, A.K., Perry, G. & Smith, M. A. (2002) The Role of Mitogen-Activated Protein Kinase Pathways in Alzheimer's Disease. *Neurosignals*, Vol.11, No. 5 (Sep-Oct 2002), pp.270–281

Zubarev, R.A.; Kelleher, N.L. &McLafferty, F.W. (1998) Electron capture dissociation of multiply charged protein cations. A nonergodic process. *Journal of American Chemical Society*, 120, No.13 (8 Apr 1998), pp.3265–3266

9

Lectins: To Combat Infections

Barira Islam and Asad U. Khan

Interdisciplinary Biotechnology Unit, Aligarh Muslim University Aligarh,
India

1. Introduction

The term "lectin" was coined by William Boyd in 1954 from the Greek word "legere" which means "to select" or "to bind". Lectins and hemagglutinins are proteins/glycoproteins of non-immune origin, which have at least one non-catalytic domain that exhibits reversible binding to specific monosaccharides or oligosaccharides (Lis and Sharon, 1986). The lectin-monosaccharide interactions are relatively very weak and the dissociation constants lie in millimolar range. However, in nature for the multimeric sugars the dissociation constants are several folds higher indicating that multiple protein-carbohyrate interactions are involved in the recognition and binding events (Ambrosi et al., 2005). Thus, lectins are multivalent in nature and can bind to the carbohydrate moieties on the surface of erythrocytes and agglutinate the erythrocytes, without altering the properties of the carbohydrates (Lam and Ng, 2011). Lectins are ubiquitous and are extensively distributed in nature. Many hundreds of these lectins have been isolated from varied sources like plants, viruses, bacteria, invertebrates and vertebrates but in all, lectins from different sources show little similarity. Lectins are invaluable tools for the detection, isolation, and characterization of glycoconjugates, primarily of glycoproteins, for histochemistry of cells and tissues and for the examination of changes that occur on cell surfaces during physiological and pathological processes, from cell differentiation to cancer (Sharon and Lis, 2004).

Cell identification and separation
Detection, isolation, and structural studies of glycoproteins
Investigation of carbohydrates on cells and subcellular organelles; histochemistry and cytochemistry
Mapping of neuronal pathways
Mitogenic stimulation of lymphocytes[b]
Purging of bone marrow for transplantation[b]
Selection of lectin-resistant mutants
Studies of glycoprotein biosynthesis

---

[a] Lectins from sources other than plants are rarely in use.
[b] In clinical use.
Source: Sharon and Lis, 2004

Table 1. Major applications of lectins[a]

Lectin Category	Definition
Merolectins	Proteins that consist exclusively of a single carbohydrate-binding domain. They are small, single-polypeptide proteins that are incapable of agglutinating cells because of their monovalent nature.
Hololectins	These are composed exclusively of carbohydrate binding domains, but contain at least two such domains.
Chimerolectins	These are fusion proteins composed of one or more carbohydrate-binding domains and an unrelated domain with a well-defined biological activity. Based on the number of sugar-binding sites, chimerolectins behave as either merolectins or homolectins.

Source: Van Damme and Peumans, 1996.

Table 2. Classification of lectins

Lectin name	Family	Glycan ligands
Plant lectins		
Concanavalin A (Con A; jack bean)	Leguminosae	Man/Glc
Wheat germ agglutinin (WGA; wheat)	Gramineae	(GlcNAc)1–3, Neu5Ac
Ricin (castor bean)	Euphorbiaceae	Gal
Phaseolus vulgaris (PHA; French bean)	Leguminosae	None known
Peanut agglutinin (PNA; peanut)	Leguminosae	Gal, Galb3GalNAca (T-antigen)
Soybean agglutinin (SBA; soybean)	Leguminosae	Gal/GalNAc
Pisum sativum (PSA; pea)	Leguminosae	Man/Glc
Lens culinaris (LCA; lentil)	Leguminosae	Man/Glc
Galanthus nivalus (GNA; snowdrop)	Amaryllidaceae	Man
Dolichos bifloris (DBA; horse gram) Leguminosae		GalNAca3GalNAc, GalNAc
Solanum tuberosum (STA; potato)		(GlcNAc)n
Animal lectins		
Asialoglycoprotein receptor (ASGPR) H1	C-type	Gal
Galectin-3	galectins	Gal
Sialoadhesin	I-type	Neu5Ac
Cation-dependent mannose-6-phosphate receptor (CD-MPR)	P-type	Man6P
C-reactive protein (CRP)	Pentraxins	Gal, Gal6P, galacturonic acid

Source: Ambrosi et al., 2005

Table 3. Examples of lectins, the families to which they belong and their glycan ligand specificities

Lectins are often classified based on saccharide-specificity. Though this conventional method is familiar and practically useful, it is not necessarily relevant for refined specificity. Lectins in the same category (e.g., galactose-specific lectins) show considerably different sugar-binding preferences. Moreover, an increasing number of lectins which never show high affinity to simple saccharides have been found. They can also be categorized according to the overall structures into merolectins, hololectins, chimerolectins and superlectins, or be grouped into different families (legume lectins, type II ribosome-inactivating proteins, monocot mannose-binding lectins, and other lectins).

The first protein showing inhibition of microorganisms was isolated from wheat flour in 1942 (Balls et al., 1942). However, it was found as late as 1980 by Duguid that *E. coli* possesses the ability to agglutinate erythrocytes and this ability is inhibited by mannose and methyl a-mannoside. Erzler in 1986 reported that lectins in higher plants defend against pathogenic bacteria and fungi by recognizing and immobilizing the infecting microorganisms via binding, thereby preventing their subsequent growth and multiplication. The role of lectins as those of the herbaceous Amaranthus in inhibiting bacteria and fungi has long been known (Bolle et al., 1996). Microbes have lectins that help in recognition and the blocking of these can prevent the infection has been established in mouse models. However, success of such treatments in humans has not been achieved yet (Sharon, 2006). Lately, with the emerging problem of multiple drug resistance, research in characterization of newer lectins to combat infections is gaining momentum.

2. Lectins as tools in cell recognition

Lectins agglutinate cells and react with the glycoconjugates present on their surface. Cells at different stages of growth and differentiation express different glycoconjugates on their surface and this made lectins an important tool to investigate the cell pathology and physiology. James Sumner in 1919 isolated concanavalin A in crystalline form and in 1936, together with Howell, reported that it agglutinates cells such as erythrocytes and yeasts and that this agglutination is inhibited by sucrose, thus demonstrating for the first time the sugar specificity of lectins. Walter Morgan and Winifred Watkins in the early 1950s used blood type-specific hemagglutinins to show that the blood type A immunodeterminant is α-linked N-acetylgalactosamine and that the H(O) determinant is α-L-fucose. This was the first demonstration that cell surface carbohydrates can serve as carriers of biological information.

Several basic features of membranes were revealed, or their existence confirmed, with the aid of lectins. Singer and Nicolson using ferritin-conjugated concanavalin A and ricin as an electron microscopic probe found that the lectin derivatives bind specifically to the outer surface of the human and rabbit erythrocyte membrane and concluded that the oligosaccharides of the plasma membrane of eukaryotic cells are asymmetrically distributed (1971). Vincent Marchesi used ferritin-labeled phytohemagglutinin (PHA) and showed that glycophorin is oriented in such a way that its carbohydrate-carrying segment is exposed to the external medium, whereas the other segments of the same molecule are embedded in the lipid bilayer or protrude into the cytoplasm (Sharon, 2007).

It was later found that lectin-induced clustering and patching of the corresponding membrane receptors on lymphocytes and other kinds of cell, as illustrated for example by the treatment with fluorescein-labeled concanavalin A of rat or mouse lymphocytes (Inbar and Sachs, 1973). Reorganization of cell surface carbohydrates was later shown to be required for various activities of lectins on cells such as mitogenic stimulation and induction of apoptosis. There is increasing evidence that changes in cellular glycosylation attend alterations in cell behaviour in both normal and in pathological processes, and that this may be of particular interest in malignancy. The malignant cells differ from the normal cells in the distribution of carbohydrates on the outer surface; many of these have an affinity for lectins (Brooks et al., 2001, Kannan et al., 2003). The reports of Inbar and Sachs also proved that lectins agglutinated malignantly transformed cells but not their normal parental cells (1973). They provided compelling evidence that cancer might be associated with a change in cell surface sugars, an idea that only a few years before had been considered completely unfounded. It was also found that SBA (specific for galactose and N-acetylgalactosamine) also possesses the remarkable ability to distinguish between normal and malignant cells (Sharon, 2007). Numerous subsequent studies have demonstrated that high susceptibility to agglutination by lectins is a property shared by many, albeit not all, malignant cells. The herbal *Viscum album* (Mistletoe) lectin (ML-1), has been shown to have antitumoral activity because of its ability to modulate and activate natural killer cells (Joshi, S, 1993). ML-1 also induces apoptosis in myelomonocytic leukemia (Joshi, S et al., 1994). Another herbal medicine *Agaricus bisporus* lectin (ABL), has also been shown to reverse the proliferation of colorectal and breast cancer cells in humans (Yu L et al., 1983).

Bacterial lectins are typically elongated submicroscopic multi-protein appendages, known as fimbriae (or pili) which mediate their adhesion to glycocalyx. Adhesion appears to prevent bacterium removal by intestinal peristalsis, facilitating colonization of the small intestine. Lectins from human pathogens like *E.coli, Actinomyces naeslundii, C. jejuni, E. cloacae, H. influenzae, H. pylori, K. pneumoniae, N. gonorrhoea, N. meningitides, S. mutans* and *P. aeruginosa* with diverse specificities have been isolated and characterized. An individual bacterium may co-express more than one lectin, e.g., certain strains of *E. coli* are both mannose and galabiose specific and those of *H. pylori* recognize simultaneously the tri- and tetrasaccharide. The major function of the enterobacterial surface lectins, as that of similar lectins of other microorganisms, is to mediate the adhesion of the organisms to host cells, an initial stage of infection. This has been extensively demonstrated both in vitro, in studies with isolated cells and cell cultures, and in vivo in experimental animals, and is supported in some cases also by clinical data. It is best documented for *E. coli* type 1 and P fimbriae (Bergsten et al., 2005). P-fimbriae have been shown to enhance the early establishment of *E. coli* in the human urinary tract, and a strong association has been found between the presence of P-fimbriae with disease severity, suggesting that adherence mediated by these organelles has a direct effect on mucosal inflammation in vivo (Sauer et al., 2000). Concerning the type 1 fimbriae, it has been reported that mutants of *E. coli* deficient in Fim H, the carbohydrate-binding subunit of the fimbriae, are unable to cause cystitis in monkeys (Sauer et al., 2000). Attachment of a pathogen to a tissue does not of itself initiate disease. It must be coupled to specific responses that lead to infection. Adherence of P-fimbriated *E. coli* or of the isolated P fimbriae to galabiose of uroepithelial cells induces a two-way flow of biological crosstalk via the lectin bridge, affecting both partners. Following adherence, the target cells are activated, with resultant production of cytokines that engender acute inflammation and other symptoms of disease, while in the bacteria the interaction leads to up-regulation of signal transduction systems that allow responses to the changing environment (Sharon, 2006).

The FimH subunits of both *E. coli* and *K. pneumoniae* are 88% homologous, yet they have different specificities (Gupta et al., 2009). They mediate not only bacterial adhesion, but also invasion of human bladder and intestinal, respectively. In contrast, adhesion mediated by PapG, the lectin subunit of P fimbriae, did not initiate bacterial internalization. *E. coli* strains that cause urinary tract infections are not strictly extracellular pathogens and FimH can directly trigger host cell signalling cascades that lead to bacterial internalization. Type 1 fimbriae are instrumental also in the attachment of *E. coli* to human polymorphonuclear cells and human and mouse macrophages, in the absence of opsonins. This is often followed by the ingestion and killing of the bacteria, a phenomenon named "lectinophagocytosis" (Ofek and Sharon, 2000), an early example of innate immunity; it may function in vivo, for example in sites poor in opsonins, and in the peritoneal cavity. Indeed, injection of type 1 fimbriated *E. coli* into the peritoneal cavity of mice led to the activation of the peritoneal macrophages; no activation was observed in the presence of methyl α-mannoside or when non-fimbriated bacteria were used (Bernhard et al., 1992). Enterobacteria can attach by their surface lectins to mast cells as well, with resultant activation of the target cells and production of high levels of certain cytokines, in particular TNF-α (Malviya and Abraham, 2001). Activation of mast cells can also be induced by purified type 1 fimbriae, and by FimH. The cytokines released by the activated mast cells cause rapid recruitment of neutrophils into the site of infection, resulting in early clearance of the bacteria. As expected, mice lacking mast cells were significantly less efficient in clearing intranasal or intraperitoneal infection caused by *K. pneumoniae*.

Specific binding of lectin to Chlamydial cell wall structures is demonstrated by the binding of Galanthus nivalis lectin (GNA). Binding of sialic acid residues to peanut agglutinin (PNA), and jackfruit lectin (JFL), were also found in two Chlamydial glycopeptides (Siridewa, et al., 1993). The study suggests that lectins may be of use as therapeutic agents to keep Chlamydial organisms from entering human cells, thus rendering them more susceptible to immune system elimination.

3. Antibacterial effect of lectins

3.1 Direct inhibition of lectin

Quite recently novel lectins are usually tested for any potential antimicrobial activity. A novel galactoside binding lectin from *Bothrops leucurus* snake venom was purified and it exhibited antibacterial effect against the human pathogenic Gram positive bacteria *Staphylococcus aureus*, *Enterococcus faecalis* and *Bacillus subtilis* (Nunes et al., 2011). *Archidendron jiringa* seed lectin showed inhibitory activity against *B. subtilis* and *S. aureus* but did not show any activity against *E.coli* and *P. aeruginosa* (Charungchitrak et al., 2010). Lectins have been islated from serum, plasma, skin mucus and egg of fishes (Jensen et al., 1997;, Ottinger et al., 1999; Dong et al., 2004; Tasumi et al., 2004). A galactose binding lectin has been isolated from Indian catfish *Clarias batrachus*. The lectin agglutinated *E.coli*, *P aeruginosa* and *Klebsiella* strains (Dutta et al., 2005).

Plant (tissue)	Lectin specificity	Antibacterial activity
Eugenia uniflora (seeds)	Carbohydrate complex	*Bacillus subtilis, Corynebacterium bovis, Escherichia coli, Klebsiella sp., Pseudomonas aeruginosa, Streptococcus sp., Staphylococcus aureus*
Myracrodruon urundeuva (heartwood)	GlcNAc	*B. subtilis, Corynebacterium callunae, E. coli, Klebsiella pneumoniae, P. aeruginosa, S. aureus, Streptococcus faecalis.*
Phthirusa pyrifolia (leaf)	Fru-1,6-P2	*B. subtilis, K. pneumoniae, Staphylococcus epidermidis, S. faecalis*

Table 4. Lectins with antibacterial activity, Source: Paiva et al., 2010.

3.2 Lectins against the virulence properties of pathogenic bacteria

The use of lectins in antiadhesion therapy has already been proposed in the literature (Ofek et al., 2003; Mody et al., 2005).This may be of particular importance for controlling diseases where opportunistic pathogens are involved and bacterial adhesion is critical, followed by attainment of sedentary mode of bacterial lifestyle (biofilms) like in oral infections. The acquired enamel pellicle is an organic and acellular film formed by selective adsorption of salivary molecules to the teeth (Yin et al., 2005). Oral bacteria adhere to this pellicle during the initial events of dental plaque formation (Saxton 1973; Yao et al., 2001), a crucial event to dental caries, pathology that represents a health expenditure of several billion dollars per year in the United States alone (Global Oral Health 2006). As bacterial adhesion to the acquired pellicle is one of the primary stages of plaque formation which may lead to caries (Scheie, 1994), it is reasonable to suppose that avoiding adhesion could be a good method to prevent this disease at early stages. Lectins may be good candidates to carry out this approach, as the adherence of bacteria to host cells is, in many cases, mediated by lectin-like adhesins on the bacterial surface that recognize carbohydrate receptors (Ofek and Sharon 1990; Hytonen et al., 2000). Marine algal lectins are especially interesting for biological applications because they have generally lower molecular masses as compared with most land plant lectins. An additional benefit might be that small algal lectin molecules may be expected to be less antigenic than the larger land plant lectins (Rogers and Hori, 1993). Further, they possess great stability on account of their several disulfide bridges and present high specificity for complex carbohydrates and glycoconjugates, especially for mucins (Ainouz et al., 1995; Sampaio et al., 1998; Nagano et al., 2005). The ability of two algal lectins BSL and BTL to bind to the SHA beads and their effectiveness in decreasing the adhesion of streptococci to the pellicle: BSL showed statistically significant results (<0.01), especially for *S. mutans*, whose adhesion was decreased almost totally; while BTL achieved this type of result for only two strains (*S. sobrinus* and *S. mitis*) (Teixeira et al., 2007)

The streptococcal cell wall contains four major antigenic polymers: peptidoglycan, group and type-specific polysaccharides, proteins and the glycerol form of teichoic and lipoteichoic acids. We studied lectins from edible sources and different specificities to the different components of the cell wall of oral pathogen, *Streptococcus mutans*. Lectins from *Canavalia ensiformis* (ConA), *Trigonella foenumgraecum* (TFA), *Triticum aestivum* (WGA), *Arachis hypogaea* (PNA), *Cajanus cajan* (CCL), *Phaseolus vulgaris* (PHA) and *Pisum sativum* (PSA) were tested against the growth and biofilm formation of *S. mutans* on saliva coated surface. None of these lectins inhibit the bacterial growth even up to a concentration of 1000mg/ml. However, all the lectins inhibited the biofilm formation by *S.mutans in-vitro*. Amongst these, lectins with Mannose/Glucose (ConA, TFA, CCL and PSA) specificity showed the highest inhibitory effect on the biofilm formation while lectins with N-acetylglucosamine specificity (WGA and PHA) and N-acetylgalactosamine specificity (PNA) also showed inhibition, albeit to a lesser degree (Islam et al., 2009).

Organism	Target tissue	Carbohydrate	Form[b]
C. jejuni[c]	Intestinal	Fucα2GalβGlcNAc	GP
E. coli Type 1	Urinary	Manα3Manα6Man	GP
P	Urinary	Galα4Gal	GSL
S	Neural	NeuAc (α2–3)Galβ3GalNAc	GSL
CFA/1	Intestinal	NeuAc (α2–8)–	GP
F1C[d]	Urinary	GalNAcβ4Galβ	GSL
F17[e]	Urinary	GlcNAc	GP
K1	Endothelial	GlcNAcβ4GlcNAc	GP
K99	Intestinal	NeuAc(α2–3)Galβ4Glc	GSL
H. influenzae	Respiratory	[NeuAc(α2–3)]$_{0,1}$Galβ4GlcNAcβ3Galβ4GlcNAc	GSL
H. pylori	Stomach	NeuAc(α2–3)Galβ4GlcNAc	GP
		Fucα2Galβ3(Fucα4)Gal	GP
K. pneumoniae	Respiratory	Man	GP
N. gonorrhoea	Genital	Galβ4Glc(NAc)	GSL
N. meningitidis	Respiratory	[NeuAc(α2–3)]$_{0,1}$Galβ4GlcNAcβ3Galβ4GlcNAc	GSL
P. aeruginosa[f]	Respiratory	L-Fuc	GP
	Respiratory	Galβ3Glc(NAc)β3Galβ4Glc	GSL
S. typhimurium	Intestinal	Man	GP
S. pneumoniae	Respiratory	[NeuAc(α2–3)]$_{0,1}$Galβ4GlcNAcβ3Galβ4GlcNAc	GSL
S. suis	Respiratory	Galα4Galβ4Glc	GSL

Source: Gupta et al., 2009.

Table 5. Carbohydrates as attachment sites for bacterial pathogens on animal tissues[a]

A surface glycoprotein of *S.mutans* of 60 kDa (with mannose and N-acetylgalactosamine) has been known to involve in saliva and bacterial interaction. The lesser adherence in the presence of glucose/mannose and galactosamine specific lectins could be because of the interaction with this protein. The PHA and WGA lectin binds to a constituent of the peptidoglycan of the cell wall (Sharon and Lis 2003). The attachment of bacteria is mediated by glucan binding lectin (GBL) and the presence of lectin in the growth media perhaps leads to competition between GBL of bacteria and plant lectins for the attachment sites on salive-coated plates resulting in less binding of the cells. With regard to bacterial surface lectins that often play a role in the initial step of adherence, plant lectins by interfering in this process show a promising future as anti-adherence agents (Islam et al., 2009). A schematic description of how lectins might inhibit attachment of bacteria to the host tissue is shown in Figure 1 (Ghazarian et al., 2011).

Use of bacterial lectin inhibitors such as mannose to prevent the adhesion of *Eschericia coli* to bladder epithelial cells has been employed in clinical practice for some time. Other bioglycans, such as that from *Crenomytalus grayanus* (mussels), has been found to considerably decrease the adhesion of the bacteria *Eschericia coli*, *Staphylococcus aureus* and *Pseudomonas aeruginosa* (Zaporozhets et al., 1994). Plant lectins such as those from *Datura stramonium*, *Robinia pseudoacacia* and *Dolichos biflorus* agglutinated Streptococcal Group C bacterial cells (Kellens et al., 1994) which prevents them from adhering to human cell surfaces.

4. Antiviral effect of lectin

The surfaces of retroviruses such as human immunodeficiency virus (HIV) and many other enveloped viruses are covered by virally-encoded glycoproteins. Glycoproteins gp120 and gp41 present on the HIV envelope are heavily glycosylated, with glycans estimated to contribute almost 50% of the molecular weight of gp120 (Mizuochi et al., 1988; Ji et al., 2006). The antiviral activity of lectins appears to depend on their ability to bind mannose-containing oligosaccharides present on the surface of viral envelope glycoproteins. Agents that specifically and strongly interact with the glycans may disturb interactions between the proteins of the viral envelope and the cells of the host (Botos & Wlodawer, 2005; Balzarini, 2006). Sugar-binding proteins can crosslink glycans on the viral surface (Sacchettini et al., 2001; Shenoy et al., 2002) and prevent further interactions with the co-receptors. Unlike the majority of current antiviral therapeutics that act through inhibition of the viral life cycle, lectins can prevent penetration of the host cells by the viruses. Antiviral lectins are best suited to topical applications and can exhibit lower toxicity than many currently used antiviral therapeutics. Additionally, these proteins are often resistant to high temperatures and low pH, as well as being odorless, which are favorable properties for potential microbicide drugs. Antiviral activity of a number of lectins that bind high-mannose carbohydrates has been described in the past. Examples of such lectins include jacalin (O'Keefe et al., 1997), concanavalin A (Hansen et al., 1989), Urtica diocia agglutinin (Balzarini et al., 1992), Myrianthus holstii lectin (Charan et al., 2000), and Narcissus pseudonarcissus lectin (Balzarini et al., 1991). However, lectins derived from marine organisms, a rich source of natural antiviral products (Tziveleka et al., 2003), such as CV-N (Boyd et al., 1997), SVN (Bokesch et al., 2003), MVL (Bewley et al., 2004) and GRFT (Mori et al., 2005), exhibit the highest activity among the lectins that have been investigated so far (Ziółkowska NE and Wlodawer A 2006). Some lectins found in algae, such as cyanovirin-N (CV-N) (Boyd et al., 1997; Esser et al., 1999; Barrientos et al., 2003; O'Keefe et al., 2003; Helleet al., 2006); scytovirin (SVN) (Bokesch et al., 2003), Microcystis viridis lectin (MVL) (Bewley et al., 2004), and griffithsin (GRFT) (Mori et al., 2005; Ziółkowska et al., 2006) exhibit significant activity against human immunodeficiency virus (HIV) and other enveloped viruses, which makes them particularly promising targets for the development as novel antiviral drugs (De Clercq, 2005; Reeves & Piefer, 2005)

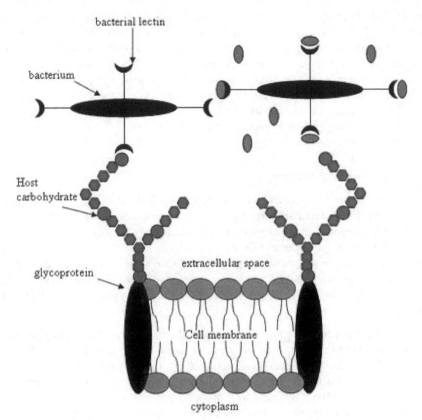

Fig. 1. Representation of bacterial lectins binding to the host cell (left) and specific lectins, used as drug interfering with this bacteria-host interaction (right)

Keyaerts et al., (2007) described the antiviral activity of plant lectins with specificity for different glycan structures against the severe acute respiratory syndrome coronavirus (SARS-CoV) and the feline infectious peritonitis virus (FIPV) in vitro. The SARS-CoV emerged in 2002 as an important cause of severe lower respiratory tract infection in humans, and FIPV infection causes a chronic and often fatal peritonitis in cats. A unique collection of 33 plant lectins with different specificities were evaluated. The plant lectins possessed marked antiviral properties against both coronaviruses with EC50 values in the lower microgram/ml range (middle nanomolar range), being non-toxic (CC50) at 50–100 µg/ml. The strongest anti-coronavirus activity was found predominantly among the mannose-binding lectins. In addition, a number of galactose-, N-acetylgalactosamine-, glucose-, and N-acetylglucosamine-specific plant agglutinines exhibited anti-coronaviral activity. A significant correlation (with an r-value of 0.70) between the EC50 values of the 10 mannose-specific plant lectins effective against the two coronaviruses was found. In contrast, little correlation was seen between the activities of other types of lectins. Two targets of possible antiviral intervention were identified in the replication cycle of SARS-CoV. The first target is located early in the replication cycle, most probably viral attachment, and the second target is located at the end of the infectious virus cycle (Keyaerts et al., 2007).

The carbohydrate binding profile of the red algal lectin KAA-2 from *Kappaphycus alvarezii* was studied by Sato et al (2011). They tested the anti-influenza virus activity of KAA-2 against various strains including the recent pandemic H1N1-2009 influenza virus. KAA-2 inhibited infection of various influenza strains with EC50s of low nanomolar levels. Immunofluorescence microscopy using an anti-influenza antibody demonstrated that the antiviral activity of KAA-2 was exerted by interference with virus entry into host cells. This mechanism was further confirmed by evidence of direct binding of KAA-2 to a viral envelope protein, hemagglutinin (HA), using an ELISA assay. These results indicate that this lectin could be a useful antiviral agent (Sato Y et al., 2011).

5. Antifungal effects of lectins

Despite the large numbers of lectins and hemagglutinins that have been purified, only a few of them manifested antifungal activity (Table 5). The expression of *Gastrodia elata* lectins in the vascular cells of roots and stems was strongly induced by the fungus *Trichoderma viride*, indicating that lectin is an important defense protein in plants (Sá et al., 2009). Following insertion of the precursor gene of stinging nettle isolectin I into tobacco, the germination of spores of *Botrytis cinerea*, *Colletotrichum lindemuthianum*, and *T. viride* was significantly reduced (Does et al., 1999). Thus, lectins may be introduced into plants to protect them from fungal attack.

Plant lectins can neither bind to glycoconjugates on the fungal membranes nor penetrate the cytoplasm owing to the cell wall barrier. It is not likely that lectins directly inhibit fungal growth by modifying fungal membrane structure and/or permeability. However, there may be indirect effects produced by the binding of lectins to carbohydrates on the fungal cell wall surface. Chitinase-free chitin-binding stinging nettle (*Urtica dioica* lectin) impeded fungal growth. Cell wall synthesis was interrupted because of attenuated chitin synthesis and/or deposition (Van Parijs et al., 1991). The effects of nettle lectin on fungal cell wall and hyphal morphology suggest that the nettle lectin regulates endomycorrhizal colonization of the rhizomes. Several other plant lectins inhibit fungal growth. The first group includes small chitin-binding merolectins with one chitin-binding domain, e.g., hevein from rubber tree latex (Van Parijs et al., 1991) and chitin-binding polypeptide from *Amaranthus caudatus* seeds (Broekaert et al., 1992). The only plant lectins that can be considered as fungicidal proteins are the chimerolectins belonging to the class I chitinases. However, the antifungal activity of these proteins is ascribed to their catalytic domain.

6. Lectins and the immune system

To initiate immune responses against infection, the surface receptors on antigen presenting cells must recognise the corresponding molecules on infectious agents. Pattern recognition receptors (PRR) which include C-type lectin like receptor (CLR) recognise and interact with carbohydrate moieties of many pathogens. Despite the presence of a highly conserved domain, C-type lectins are functionally diverse and have been implicated in various processes including cell adhesion, tissue integration and remodelling, platelet activation, complement activation, pathogen recognition, endocytosis, and phagocytosis.

Natural source of lectin	Fungal species inhibited	Sugar specificity	Reference
Amaranthus viridis (Green Amaranth) seeds	*Botrytis cinerea, Fusarium oxysporum*	Asialofetuin, fetuin, T-antigen, N-acetyl-d-lactosamine, N-acetyl-d-galactosamine	Kaur et al.2006
Astragalus mongholicus (huangqi) roots	*Borrytis cinerea, Colletrichum sp., Droschslara turia, Fusarium oxysporum*	d-galactose, lactose	Yan et al.2005
Capparis spinosa (caper) seeds	*Valsa mali*	D(+)galactose, α-lactose, raffinose, rhamnose, L(+)-arabinose, D(+)glucosamine	Lam et al.2009
Capsicum frutescens (red cluster pepper) seeds	*Aspergillus flavus, Fusarium moniliforme*	d-mannose, glucose	Ngai and Ng 2007
Curcuma amarissima Roscoe (wei ji ku jiang-huang) Rhizomes	*Colectrotrichum cassiicola, Exserohilum turicicum, Fusarium oxysporum*	Not found	Kheeree et al. 2010
Dendrobium findlayanum (orchid) pseudobulbs	*Alternaria alternata, Colletrichum sp.*	Not found	Sattayasai et al. 2009
Phaselous vulgaris cv " flageolet bean" seeds	*Mycosphaerella arachidicola*	Not found	Xia and Ng 2005
Phaselous vulgaris cv "French bean 35" seeds	*Valsa mali*	Not found	Lam and Ng 2010
Phaseolus coccineus seeds	*Gibberalla sanbinetti, Helminthosporium maydis, Rhizoctonia solani, Sclerotinia sclerotiorum*	Sialic acid	Chen et al.2009
Phaseolus vulgaris cv "red kidney bean" seeds	*Coprinus comatus, Fusarium oxysporum, Rhizoctonia solani*	Lactoferrin, ovalbumin, thyroglobulin	Ye et al. 2001
Pouteria torta (pouteria trees/eggfruits) seeds	*Saccharomyces carevisiae, C. musae, Fusarium oxysporum*	Fetuin, asialofetuin, heparin, orosomucoid, ovoalbumin	Boleti et al.2007
Talisia esculenta (pitomba) seeds	*Microsporum canis*	d-mannose	Pinheiro et al. 2009

Natural source of lectin	Fungal species inhibited	Sugar specificity	Reference
Withania somnifera (Ashwagandha/Indian ginseng/Winter cherry/Ajagandha/Kanaje Hindi/Amukkuram) leaves	*Fusarium moniliforme, Macrophomina phaseolina*	Not found	Ghosh 2009
Zea mays (maize) endosperm	*Aspergillus flavus*	D(+)galactose	Baker et al.,2009

Table 6. Examples of lectins with antifungal activity, Source: Lam and Ng, 2011

6.1 Mannose Receptor

The MR binds a broad array of microorganisms, including *Candida albicans, Pneumocystis carinii, Leishmania donovani, Mycobacterium tuberculosis,* and capsular polysaccharides of *Klebisella pneumoniae* and *Streptococcus pneumonia* (Chakraborty et al., 2001; Ezekowitz et al., 1991; Marodi et al., 1991; O'Riordan et al., 1995; Schlesinger, 1993; Zamze et al., 2002). The receptor recognises mannose, fucose or N-acetylglucosamine sugar residues on the surfaces of these microorganisms (Largent et al., 1984) and carbohydrate recognition is mediated by CTLDs 4–8 (Taylor et al., 1992). The MR has been implicated in the phagocytic uptake of pathogens, but there are limited examples actually demonstrating MR-dependent phagocytosis.

6.2 Dectin-1

Dectin-1 is a type II transmembrane protein that is classified as a Group V non-classical C-type lectin and lacks the conserved residues involved in the ligation of calcium that are usually required to co-ordinate carbohydrate binding. Dectin-1 was initially identified as a dendritic cell specific receptor that modulates T cell function through recognition of an unidentified ligand (Ariizumi et al., 2000; Grunebach et al., 2002). It was subsequently reidentified as a receptor for β-glucans, which are carbohydrate polymers found primarily in the cell walls of fungi, but also in plants and some bacteria (Brown and Gordon, 2001, 2003). Dectin-1 can recognise a number of fungal species, including *C. albicans, P. carinii, Saccharomyces cerevisiae, Coccidioides posadasii* and *Aspergillus fumigatus* (Brown et al., 2003; Gersuk et al., 2006; Saijo et al., 2007; Steele et al., 2003, 2005; Taylor et al., 2007; Viriyakosol et al., 2005).The ligation of Dectin-1 also triggers intracellular signalling resulting in a variety of cellular responses, including phagocytosis.

6.3 DC-SIGN (CD209)

DC-SIGN is a type II transmembrane protein that is classified as a Group II C-type lectin. DC-SIGN was originally identified as a receptor for intercellular adhesion molecule-3 (ICAM-3) that facilitates DC-mediated T-cell proliferation and binds HIV-1 (Geijtenbeek et al., 2000a, b). It has since been reported that the receptor interacts with a range of pathogens, including *M. tuberculosis, C. albicans, Helicobacter pylori, Schistosoma mansoni* and *A. fumigatus* (Appelmelk et al., 2003; Cambi et al., 2008; Geijtenbeek et al., 2000b, 2003; Serrano-Gomez et al., 2004; Tailleux et al., 2003; van Die et al., 2003). There have been no reports of a mechanism for DC-SIGN mediated phagocytosis. However, activation of DC-SIGN triggers Rho-GTPase (Hodges et al., 2007) making it conceivable that Rho could be involved in phagocytosis mediated by this receptor.

6.4 Mannose-binding lectin (MBL)

Mannose-binding lectin (MBL) is a Group III C-type lectin belonging to the collectins (Holmskov et al., 2003), which are a group of soluble oligomeric proteins containing collagenous regions and CTLDs. MBL is secreted into the blood stream as a large multimeric complex and is primarily produced by the liver, although other sites of production, such as the intestine, have been proposed (Uemura et al., 2002). It recognises carbohydrates such as mannose, glucose, l-fucose, N-acetyl-mannosamine (ManNAc), and N-acetyl-glucosamine (GlcNAc). Oligomerisation of MBL enables high avidity binding to repetitive carbohydrate ligands, such as those present on a variety of microbial surfaces, including *E. coli, Klebisella aerogenes, Neisseria meningitides, Staphylococcus aureus, S. pneumoniae, A. fumigatus* and *C. albicans* (Davies et al., 2000; Neth et al., 2000; Schelenz et al., 1995; Tabona et al., 1995; van Emmerik et al., 1994).MBL has also been proposed to function directly as an opsonin by binding to carbohydrates on pathogens and then interacting with MBL receptors on phagocytic cells, promoting microbial uptake and stimulating immune responses (Kuhlman et al., 1989). It was shown in a recent study that MBL modifies cytokine responses through a novel cooperation with TLR2/6 in the phagosome (Ip et al., 2008).

7. Lectins and drug delivery

The concept of lectin-mediated specific drug delivery was proposed by Woodley and Naisbett in 1988 (Bies et al., 2004). Delivery of targeted therapeutics via direct and reverse drug delivery systems (DDS) to specific sites provides numerous advantages over traditional non-targeted therapeutics (Rek et al., 2009). Targeted drug delivery increases the efficacy of treatment by enhancing drug exposure to targeted sites while limiting side effects of drugs on normal and healthy tissues (Rek et al., 2009). Furthermore, specific drug delivery increases the uptake and internalization of therapeutics that have reduced cellular permeability (Rek et al., 2009). Lectin based drug-targeting can be done in two ways. In the first approach, carbohydrate moieties form a part of DDS. The carbohydrate tag drives the drug to the endogenous lectins present on the cell surface. In the second approach, lectins are present on the drug surface and it interacts with the glycosylated surfaces of the cells (Gabor et al., 2004). Considering the fact that epithelial cells contain a thin layer of mucus which has mucins that are highly glycosylated proteins, the lectin-encapsulated drug strategy offers great potential. As non-specific interactions are susceptible to changes in pH and to interactions with food digesta, which probably reduce the mucoadhesive effect, specific mucoadhesiva of the second generation seem to be preferable. The second target is the glycocalyx of the absorptive epithelium. In case of identical oligosaccharide structures of the mucin and the glycocalyx, partitioning of the formulation to the cell surface is facilitated due to full reversibility of the mucin–lectin interaction. In case of lectin-matching carbohydrates only at the glycocalyx, the formulation has to penetrate the mucuos layer. Both pathways result in fixation of the drug delivery system closer to the site of absorption. That way cytoadhesion will increase the concentration gradient between the extracellular and intracellular compartment, which facilitates at least passive diffusion of the drug into the cell. The third target is represented by glycosylated receptors at the cell membrane. The binding of some lectins, such as WGA to the EGF-receptor, induces active receptor mediated endocytosis, which can improve cytoinvasion of prodrugs as well as nanoscaled carrier systems (Gabor, 2004).

In an approach towards pulmonary delivery, lectinised liposomes (130–170 nm in diameter) were screened for binding to alveolar type II epithelial cells (Bruck et al., 2001). As compared to plain liposomes, the binding to A549 cells increased 6–11-fold upon surface modification with wheat germ agglutinin (WGA), Concanavalin A (ConA) or soybean agglutinin. The binding was not affected by a synthetic lung surfactant and no cytotoxic effect of the free lectins or the lectinised liposomes was observed. Upon incubation with primary cultured human alveolar epithelial cells, which exhibit barrier functions, the WGA-liposomes were not only bound but also taken up into the cells. In search for non-viral vectors for gene therapy of cystic fibrosis and as a basis for lectin-mediated gene transfer, 32 lectins were screened for binding and uptake into living human airway epithelium (Yi et al., 2001). Whereas ConA was internalised within 1 h, the lectins from *Erythrina cristagalli* and *Glycine max*, peanut lectin, and Jacalin were taken up into the epithelium within 4 h. The endocytosis of WGA was minimal even after 4 h. Irrespective of the specificity of the lectin–carbohydrate interaction; the internalised lectins exhibited a non-selective binding pattern on the epithelium. Only peanut lectin bound to subpopulations of ciliated and non-ciliated cells.

Owing to their remarkable specificities, plant lectins with affinities for the carbohydrates on microbial cell surface are already well characterised. Given the potential of porphyrins to act as antimicrobials it is pertinent to ask whether lectins could be used *in vivo* to specifically deliver porphyrins into pathogenic microbial cells, thereby improving the efficacy of the treatment, reducing the concentration of the drug required to be introduced into the system and thereby reducing the possible side-effects. In particular, lectins could be successful oral and mucosal drug delivery agents. Not only are a large number of lectins part of our everyday diet, but also several of them are known to survive the harsh conditions of human gastro-intestinal tract. Similarly, attempts have been made to use lectins in ocular drug delivery. Specific hydrophobic binding sites on lectins provide the ideal opportunity to expand the use of these molecules in targeted therapy (Komath et al., 2006).

8. Conclusions

Lectins are ubiquitous in nature and have garnered much attention due to specificity of its interaction with the carbohydrates. Glycosylation is a key step in many cellular processes and with more reports about the change in cell-surface carbohydrates in different pathological conditions, research about exploiting lectins as a therapeutic tool is now at the forefront. Lectins are now routinely used in the identification and purification of glycoproteins. Their use in blood typing as well as in clinical diagnostics is well established. Many lectins show antibacterial, antiviral or antifungal activities in-vitro. However, clinical trials need to be done for establishing their therapeutic effect and optimising their dosage delivery. As microbes use their surface lectins for attachment to the host tissue, dietary/therapeutic lectins may interfere in this interaction. Thus lectins can be used anti-adhesion agents and prevent the colonization of the microbe and hence the establishment of the infection. In the immune system, endogenous lectins play a role in ligand recognition and hence are an important component of the host's defense against microbes. Given their ability to specifically target different cell types, they have always been looked upon as useful candidates for targeted drug delivery. Research utilizing lectins as carriers of monoclonal antibodies or specific chemotherapeutic agents has been conducted. Alongwith the beneficial effect, lectins have been reported to have caused severe allergic reactions. Most of the information on the acute toxicity of lectins in humans has been derived from observations of incidences of accidental poisoning. Since no experimental data is available to show the possible adverse effects of lectins on humans but can be inferred from experiments with laboratory animals. Although results obtained with mice, rats or pigs cannot simply be extrapolated to humans, the observed effects on the gut and other organs of these animals demonstrate the possible toxicity of the lectins. Thus lectin-based therapeutics for combating infections is very promising owing to its highly selective nature, provided the dosage is well below the toxic limits.

9. References

[1] Ambrosi M, Cameron NR & Davis BG (2005). Lectins : tools for molecular understanding of the glycocode. Org Biol Chem. (3), 1593-1608.

[2] Appelmelk BJ, van Die I, van Vliet SJ, Vandenbroucke-Grauls CM, Geijtenbeek TB & van Kooyk Y (2003). Carbohydrate profiling identifies new pathogens that interact with dendritic cell-specific ICAM-3 grabbing nonintegrin on dendritic cells. J Immunol. 170, 1635–1639.

[3] Ariizumi K, Shen GL, Shikano S, Xu S, Ritter R, Kumamoto T, Edelbaum D, Morita A, Bergstresser PR & Takashima A (2000). Identification of a novel, dendritic cell-associated molecule, dectin-1, by subtractive cDNA cloning. J. Biol. Chem. 275, 20157–20167.

[4] Baker RL, Brown RL, Chen ZY, Cleveland TE, Fakhoury AM (2009). A maize lectin-like protein with antifungal activity against Aspergillus flavus. J Food Prot. 72, 120–127.

[5] Balls AK, Halle WS and Harris TH (1942). A crystalline protein obtained from a lipoprotein of wheat flour. Cereal Chem.19, 279-288.

[6] Balzarini J (2006).Large-molecular-weight carbohydrate-binding agents as HIV entry inhibitors targeting glycoprotein gp120. Curr Opin HIV AIDS, 1(5), 355-360.

[7] Balzarini J, Neyts J, Schols D, Hosoya M, Van Damme E, Peumans W & De Clercq E (1992). The mannose-specific plant lectins from cymbidium hybrid and epipactis helleborine and the (N-acetylglucosamine)n-specific plant lectin from Urtica dioica are potent and selective inhibitors of human immunodeficiency virus and cytomegalovirus replication in vitro. Antiviral Res. 18, 191–207.

[8] Balzarini J, Schols D, Neyts J, Van Damme E, Peumans W& De Clercq E (1991). Alpha-(1-3)- and alpha-(1-6)-d-mannose-specific plant lectins are markedly inhibitory to human immunodeficiency virus and cytomegalovirus infections in vitro. Antimicrob. Agents Chemother. 35, 410–416.

[9] Bergsten G, Wullt B, Svanborg C (2005). Escherichia coli, fimbriae, bacterial persistence and host response induction in the human urinary tract. Int J Med Microbiol., 295(6-7), 487-502.

[10] Bernhard W, Gbarah A and Sharon N (1992). Lectinophagocytosis of type 1fimbriated (mannose-specific) Escherichia coli in the mouse peritoneum. J. Leukocyte Biol. 52, 343–348., Cai M, Ray S, Ghirlando R, Yamaguchi M & Muramoto K (2004).New carbohydrate specificity and HIV-1 fusion blocking activity of the cyanobacterial

protein MVL: NMR, ITC and sedimentation equilibrium studies. J Mol Biol. 339(4), 901-914.

[11] Bokesch HR, O'Keefe BR, McKee TC, Pannell LK, Patterson GM, Gardella RS, Sowder RC 2nd, Turpin J, Watson K, Buckheit RW Jr & Boyd MR (2003). A potent novel anti-HIV protein from the cultured cyanobacterium Scytonema varium. Biochemistry. 42(9), 2578-2584.

[12] Boleti AP, Freire MG, Coelho MB, Silva W, Baldasso PA, Gomes VM, Marangoni S, Novello JC & Macedo ML (2007). Insecticidal and antifungal activity of a protein from Pouteria torta seeds with lectin-like properties. J Agric Food Chem. 55, 2653-2658.

[13] Boyd MR, Gustafson KR, McMahon JB, Shoemaker RH, O'Keefe BR, Mori T, Gulakowski RJ, Wu L, Rivera MI, Laurencot CM, Currens MJ, Cardellina JH 2nd, Buckheit RW Jr, Nara PL, Pannell LK, Sowder RC 2nd & Henderson LE (1997). Discovery of cyanovirin-N, a novel human immunodeficiency virus-inactivating protein that binds viral surface envelope glycoprotein gp120: potential applications to microbicide development.Antimicrob Agents Chemother. 41(7),1521-1530.

[14] Broekaert WF, Mariën W, Terras FR, De Bolle MF, Proost P, Van Damme J, Dillen L, Claeys M, Rees SB, Vanderleyden J, et al (1992). Antimicrobial peptides from Amaranthus caudatus seeds with sequence homology to the cysteine/glycine-rich domain of chitin-binding proteins. Biochemistry 31(17), 4308-4314.

[15] Brooks SA, Hall DM & Buley I (2001). GalNAc glycoprotein expression by breast cell lines, primary breast cancer and normal breast epithelial membrane. Br J Cancer 85(7), 1014-1022.

[16] Brown GD & Gordon S (2001). Immune recognition. A new receptor for beta-glucans. Nature 413, 36-37.

[17] Brown GD & Gordon S (2003). Fungal beta-glucans and mammalian immunity. Immunity 19, 311-315.

[18] Brown GD, Herre J, Williams DL, Willment JA, Marshall AS & Gordon S (2003). Dectin-1 mediates the biological effects of beta-glucans. J. Exp. Med. 197, 1119-1124.

[19] Bruck A, Abu-Dahab R, Borchard G, Schafer UF & Lehr CM (2001). Lectin-functionalized liposomes for pulmonary drug delivery: interaction with human alveolar epithelial cells, J. Drug Targeting 9 (2001) 241- 251.

[20] Cambi A, Netea MG, Mora-Montes HM, Gow NA, Hato SV, Lowman DW, Kullberg BJ, Torensma R, Williams DL & Figdor CG (2008). Dendritic cell interaction with Candida albicans critically depends on N-linked mannan. J. Biol. Chem. 283, 20590-20599.

[21] Chakraborty P, Ghosh D & Basu MK (2001).Modulation of macrophage mannose receptor affects the uptake of virulent and avirulent Leishmania donovani promastigotes. J Parasitol 87(5), 1023-1027.

[22] Charan RD, Munro MH, O'Keefe BR, Sowder RCII, McKee TC, Currens MJ, Pannell LK, Boyd MR (2000). Isolation and characterization of Myrianthus holstii lectin, a potent HIV-1 inhibitory protein from the plant Myrianthus holstii(1). J Nat Prod., 63(8),1170-1174.

[23] Charungchitrak S, Petsom A, Sangvanich P & Karnchanatat A (2011). Antifungal and antibacterial activities of lectin from the seeds of Archidendron jiringa Nielsen. Food Chemistry 126, 1025-1032.

[24] Chen J, Liu B, Ji N, Zhou J, Bian HJ, Li CY, Chen F& Bao JK (2009). A novel sialic acid-specific lectin from Phaseolus coccineus seeds with potent antineoplastic and antifungal activities. Phytomedicine 16, 352–360.

[25] Davies J, Neth O, Alton E, Klein N & Turner M (2000). Differential binding of mannose-binding lectin to respiratory pathogens in cystic fibrosis. Lancet. 355, 1885–1886.

[26] De Bolle MF, Osborn RW, Goderis IJ, Noe L, Acland D, Hart CA, Torrekens S, van Leuven F, Broekart NF(1996). Antimicrobial properties from Mirablis jalapa and *Amaranthus caudalus*: Expression, processing, localization and biological activity in transgenic tobacco. Plant Mol. Biol. 31, 993-1008.

[27] Does MP, Houterman PM, Dekker HL & Cornelissen BJ (1999) Processing, targeting, and antifungal activity of stinging nettle agglutinin in transgenic tobacco. Plant Physiol 120, 421–432.

[28] Dong CH, Yang ST, Yang ZA, Zhang L & Gui JF (2004). A C-type lectin associated and translocated with cortical granules during oocyte maturation and egg fertilization in fish. Dev. Biol. 265, 341–354.

[29] Duguid, JP and Old DC (1980) in *Bacterial Adherence*, Beachey EH, ed, pp 185-217, Chapman and Hall, London.

[30] Dutta S, Sinha B, Bhattacharya B, Chatterjee B, Mazumder S (2005). Characterization of a galactose binding serum lectin from the Indian catfish, Clarias batrachus: Possible involvement of fish lectins in differential recognition of pathogens. Comparative Biochemistry and Physiology Part C 141, 76 – 84.

[31] Etzler, ME (1986). Distribution and function of plant lectins. In *The Lectins. Properties, Function, and Applications in Biology and Medicine*, Liener IE, Sharon N, Goldstein IJ, eds; pp. 371-435, Academic Press: Orlando, FL, USA.

[32] Ezekowitz RA, Williams DJ, Koziel H, Armstrong MY, Warner A, Richards FF, Rose RM (1991). Uptake of Pneumocystis carinii mediated by the macrophage mannose receptor. Nature 351(6322), 155-158.

[33] Gabius HJ, Walzel H, Joshi SS, Kruip J, Kojima S, Gerke V, Kratzin H & Gabius S (1992). The immunomodulatory beta-galactoside-specific lectin from mistletoe: partial sequence analysis, cell and tissue binding, and impact on intracellular biosignalling of monocytic leukemia cells. Anticancer Res. 12(3), 669-75.

[34] Gabius S, Joshi S, Kayser K & Gabius H (1992). The galactoside-specific lectin from mistletoe as biological response modifier. Int J Oncol. 1(6), 705-708.

[35] Gaynor CD, McCormack FX, Voelker DR, McGowan SE & Schlesinger LS. Pulmonary surfactant protein A mediates enhanced phagocytosis of Mycobacterium tuberculosis by a direct interaction with human macrophages.J Immunol. 155(11), 5343-5351.

[36] Geijtenbeek TB, Torensma R, van Vliet SJ, van Duijnhoven GC, Adema GJ, van Kooyk Y & Figdor CG (2000). Identification of DC-SIGN, a novel dendritic cell-specific ICAM-3 receptor that supports primary immune responses. Cell. 100, 575–585.

[37] Geijtenbeek TB, Van Vliet SJ, Koppel EA, Sanchez-Hernandez M, Van denbroucke-Grauls CM, Appelmelk B & Van Kooyk Y (2003). Mycobacteria target DC-SIGN to suppress dendritic cell function. J. Exp. Med. 197, 7–17.

[38] Gersuk GM, Underhill DM, Zhu L & Marr KA (2006). Dectin-1 and TLRs permit macrophages to distinguish between different Aspergillus fumigatus cellular states. J. Immunol. 176, 3717–3724.

[39] Ghosh M (2009). Purification of a lectin-like antifungal protein from the medicinal herb, *Withania somnifera*. Fitoterapia 80, 91–95.

[40] Global Oral Health (2006) http://www.whocollab.od.mah.se.

[41] Grunebach F, Weck MM, Reichert J & Brossart P (2002). Molecular and functional characterization of human Dectin-1. Exp. Hematol. 30, 1309–1315.

[42] Gupta A, Gupta RK & Gupta GS (2009). Targeting cells for drug and gene delivery: Emerging applications of mannans and mannan binding lectins. J of Sci & Indus Res. 68, 465-483.

[43] Hansen JE, Nielsen CM, Nielsen C, Heegaard P, Mathiesen LR, Nielsen JO. Correlation between carbohydrate structures on the envelope glycoprotein gp120 of HIV-1 and HIV-2 and syncytium inhibition with lectins. AIDS. 1989, 3(10), 635-641.

[44] Hodges A, Sharrocks K, Edelmann M, Baban D, Moris A, Schwartz O, Drakesmith H, Davies K, Kessler B, McMichael A & Simmons A (2007). Activation of the lectin DC-SIGN induces an immature dendritic cell phenotype triggering Rho-GTPase activity required for HIV-1 replication. Nat. Immunol. 8, 569–577.

[45] Holmskov U, Thiel S & Jensenius JC (2003). Collections and ficolins: humoral lectins of the innate immune defense. Annu. Rev. Immunol. 21, 547–578.

[46] Hytonen J, Haataja S, Isomaki P and Finne J (2000) Identification of a novel glycoprotein-binding activity in Streptococcus pyogenes regulated by the mga gene. Microbiology 146, 31–39.

[47] Inbar M & Sachs L (1973). Mobility of carbohydrate-containing sites on the surface membranes in relation to control of cell growth. FEBS Lett. 32, 124–128.

[48] Ip WK, Takahashi K, Moore KJ, Stuart LM. & Ezekowitz RA (2008). Mannose-binding lectin enhances Toll-like receptors 2 and 6 signaling from the phagosome. J. Exp. Med. 205,169–181.

[49] Islam B, Khan SN, Naeem A, Sharma V, Khan AU (2009). Novel effect of plant lectins on the inhibition of Streptococcus mutans biofilm formation on saliva-coated surface. J Appl Microbiol. 106(5), 1682-1689.

[50] Jensen LE, Hiney MP, Shields DC, Uhlar, CM, Lindsay, AJ, Whitehead AS (1997). Acute phase protein in salmonids—evolutionary analyses and acute phase response. J. Immunol. 158, 384–392.

[51] Ji X, Chen Y, Faro J, Gewurz H, Bremer J, Spear GT (2006). Interaction of human immunodeficiency virus (HIV) glycans with lectins of the human immune system. Curr Protein Pept Sci., 7(4), 317-324.

[52] Kannan S, Lakku RA, Niranjali D, Jayakumar K, Steven AH, Taralakshmi VV, Chandramohan S, Balakrishnan R, Schmidt C & Halagowder D (2003). Expression of peanut agglutinin-binding mucin-type glycoprotein in human esophageal squamous cell carcinoma as a marker. Mol Cancer. 2, 38.

[53] Kaur N, Dhuna V, Kamboj SS, Agrewala JN & Singh J (2006). A novel antiproliferative and antifungal lectin from Amaranthus viridis Linn seeds. Protein Pept Lett. 13, 897–905.

[54] Kellens J, Jacobs J, Peumans W, Stobberingh E (1994). Agglutination of "Streptococcus milleri" by lectins. J Med Microbiol 41: 1, 14-19 .

[55] Keyaerts E, Vijgen L, Pannecouque C, Van Damme E, Peumans W, Egberink H & Balzarini J, Van Ranst M (2007). Plant lectins are potent inhibitors of coronaviruses by interfering with two targets in the viral replication cycle. Antiviral Res 75(3), 179-87.

[56] Kheeree N, Sangvanich P, Puthong S & Karnchanatat A (2010). Antifungal and antiproliferative activities of lectin from the rhizomes of *Curcuma amarissima* Roscoe. Appl Biochem Biotechnol. 162, 912–925.

[57] Komath SS, Kavitha M & Swamy MJ (2006). Beyond carbohydrate binding: new directions in plant lectin research. Org. Biomol. Chem., 4, 973–988.

[58] Kuhlman M, Joiner K & Ezekowitz RA (1989). The human mannose-binding protein functions as an opsonin. J. Exp. Med. 169,1733–1745.

[59] Lam SK & Ng TB (2011). Lectins production and practical applications. Appl Microbiol Biotechnol. 89; 45-55.

[60] Lam SK & Ng TB (2010). Isolation and characterization of a French bean hemagglutinin with antitumor, antifungal, and anti-HIV-1 reverse transcriptase activities and an exceptionally high yield. Phytomedicine 17, 457–462.

[61] Lam SK, Han QF &Ng TB (2009). Isolation and characterization of a lectin with potentially exploitable activities from caper (Capparis spinosa) seeds. Biosci Rep. 29, 293–299.

[62] Largent BL, Walton KM, Hoppe CA, Lee YC & Schnaar RL (1984). Carbohydrate-specific adhesion of alveolar macrophages to mannose-derivatized surfaces. J Biol Chem. 10, 259(3), 1764-1769.

[63] Lis H & Sharon N (1986). Lectins as molecules and as tools. Ann Rev Biochem. 55; 35-67.

[64] Malaviya R and Abraham SN (2001). Mast cell modulation of immune responses to bacteria. Immunol. Rev. 179, 16–24.

[65] Maródi L, Korchak HM & Johnston RB Jr (1991). Mechanisms of host defense against Candida species. I. Phagocytosis by monocytes and monocyte-derived macrophages.J Immunol.146(8), 2783-2789.

[66] Morgan WT and Watkins WM (2000). Unraveling the biochemical basis of blood group ABO and Lewis antigenic specificity. Glycoconj. J. 17, 501–530.

[67] Mori T, O'Keefe BR, Sowder RC 2nd, Bringans S, Gardella R, Berg S, Cochran P, Turpin JA, Buckheit RW Jr, McMahon JB & Boyd MR (2005). Isolation and characterization of griffithsin, a novel HIV-inactivating protein, from the red alga Griffithsia sp. J Biol Chem. 280(10), 9345-9353.

[68] Nagano CS, Debray H, Nascimento KS, Pinto VPT, Cavada BS, Saker-Sampaio S, Farias WRL, Sampaio AH, et al. (2005). HCA and HML isolated from Hypnea cervicornis and Hypnea musciformis define a novel lectin family. Protein Sci 14, 2167–2176.

[69] Neth O, Jack DL, Dodds AW, Holzel H, Klein NJ & Turner MW (2000). Mannose-binding lectin binds to a range of clinically relevant microorganisms and promotes complement deposition. Infect. Immun. 68, 688–693.

[70] Ngai PH & Ng TB (2007). A lectin with antifungal and mitogenic activities from red cluster pepper (Capsicum frutescens) seeds. Appl Microbiol Biotechnol. 74, 366–371.

[71] Nicholson GL and Singer SJ (1971). Ferritin-conjugated plant agglutinins as specific saccharide stains for electron microscopy: application to saccharides bound to cell membranes Proc. Natl. Acad. Sci. U. S. A. 68, 942-945.

[72] Nunes EdS, de Souza MAA, Vaz AFdM, Santana GMdS, Gomes FS, Coelho LCBB, Paiva PMG, da Silva RML, Silva-Lucca RA, Oliva MLV, Guarnieri MC & Correia MTdS (2011).Purification of a lectin with antibacterial activity from Bothrops leucurus snake venom. Comparative Biochemistry and Physiology Part B: Biochemistry and Molecular Biology 159 (1), 57-63.

[73] Ofek I & Sharon N (1990). Adhesins as lectins: specificity and role in infection. Curr Top Microbiol Immunol 151, 91–113.

[74] Ofek I and Sharon N(1988). Lectinophagocytosis: a molecular mechanism of recognition between cell surface sugars and lectins in the phagocytosis of bacteria. Infect. Immun. 56, 539–547.

[75] Ofek I, Hasty DL and Sharon N (2003) Anti-adhesion therapy of bacterial diseases: prospects and problems. FEMS Immunol Med Microbiol. 38, 181–191.

[76] O'Keefe BR, Beutler JA, Cardellina JH 2nd, Gulakowski RJ, Krepps BL, McMahon JB, Sowder RC 2nd, Henderson LE, Pannell LK, Pomponi SA, Boyd MR. Isolation and characterization of niphatevirin, a human-immunodeficiency-virus-inhibitory glycoprotein from the marine sponge Niphates erecta.Eur J Biochem. 1997, 245(1), 47-53.

[77] O'Riordan DM, Standing JE &Limper AH (1995). Pneumocystis carinii glycoprotein A binds macrophage mannose receptors. Infect Immun. 63(3):779-784.

[78] Ottinger CA, Johnson SC, Ewart KV, Brown LL, Ross NW (1999).Enhancement of anti-Aeromonas salmonicida activity in Atlantic salmon (Salmo salar) macrophages by a mannose binding lectin. Comp. Biochem. Physiol. C 123, 53–59.

[79] Pinheiro AQ, Melo DF, Macedo LM, Freire MG, Rocha MF, Sidrim JJ, Brilhante RS, Teixeira EH, Campello CC, Pinheiro DC & Lima MG (2009). Antifungal and marker effects of Talisia esculenta lectin on Microsporum canis in vitro. J Appl Microbiol. 107, 2063–2069.

[80] Reeves JD & Piefer AJ (2005). Emerging drug targets for antiretroviral therapy. Drugs. 65(13),1747-1766.

[81] Rek A, Krenn E & Kungl AJ (2009). Therapeutically targeting protein-glycan interactions. Br J Pharmacol. 157 (5), 686-694.

[82] Rogers DJ and Hori K (1993). Marine algal lectins: new developments. Hydrobiology 260, 589–593.

[83] Sá RA, Santos ND, da Silva CS, Napoleão TH, Gomes FS, Cavada BS, Coelho LC, Navarro DM, Bieber LW& Paiva PM (2009). Larvicidal activity of lectins from Myracrodruon urundeuva on Aedes aegypti. Comp Biochem Physiol C Toxicol Pharmacol.. 149(3), 300-306.

[84] Sacchettini JC, Baum LG, Brewer CF (2001). Multivalent protein-carbohydrate interactions. A new paradigm for supermolecular assembly and signal transduction. Biochemistry 40(10), 3009-3015.

[85] Saijo S, Fujikado N, Furuta T, Chung SH, Kotaki H, Seki K, Sudo K, Akira S, Adachi Y, Ohno N, Kinjo T, Nakamura K, Kawakami K & Iwakura Y (2007). Dectin-1 is required for host defense against Pneumocystis carinii but not against Candida albicans. Nat. Immunol. 8, 39–46.

[86] Sato Y, Morimoto K, Hirayama M & Hori K (2011). High mannose-specific lectin (KAA-2) from the red alga Kappaphycus alvarezii potently inhibits influenza virus infection in a strain-independent manner.Biochem Biophys Res Commun. 405(2), 291-296.

[87] Sattayasai N, Sudmoon R, Nuchadomrong S, Chaveerach A, Kuehnle AR, Mudalige-Jayawickrama RG & Bunyatratchata W (2009). Dendrobium findleyanum agglutinin: production, localization, anti-fungal activity and gene characterization. Plant Cell Rep. 28, 1243–1252.

[88] Sauer FG, Mulvey MA, Schilling JD, Martinez JJ, Hultgren SJ (2000). Bacterial pili: molecular mechanisms of pathogenesis. Curr Opin Microbiol., 3(1), 65-72.

[89] Scheie AA (1994). Mechanisms of dental plaque formation. Adv Dent Res 8, 246–253.

[90] Schelenz S, Malhotra R, Sim RB, Holmskov U & Bancroft GJ (1995). Binding of host collectins to the pathogenic yeast Cryptococcus neoformans: human surfactant protein D acts as an agglutinin for acapsular yeast cells. Infect. Immun. 63, 3360–3366.

[91] Sela BA, Lis H, Sharon N and Sachs L. (1970) Different locations of carbohydrate-containing sites at the surface membrane of normal and transformed mammalian cells. J. Membr. Biol. 3, 267–279.

[92] Serrano-Gomez D, Dominguez-Soto A, Ancochea J, Jimenez-Heffernan JA, Leal JA & Corbi AL (2004). Dendritic cell-specific intercellular adhesion molecule 3-grabbing nonintegrin mediates binding and internalization of Aspergillus fumigatus conidia by dendritic cells and macrophages. J. Immunol. 173, 5635–5643.

[93] Sharon N & Lis H (2004). History of lectins: from hemagglutinins to biological recognition molecules. Glycobiology 14(11):53R-62R.

[94] Sharon N (2006). Carbohydrates as future anti-adhesion drugs for infectious diseases. Biochem. Biophys. Acta 1760, 527-537.

[95] Sharon N (2007). Lectins: Carbohydrate-specific reagents and biological recgnition molecules. J of Biol Chem 282, 2753-2764.

[96] Shenoy SR, Barrientos LG, Ratner DM, O'Keefe BR, Seeberger PH, Gronenborn AM, Boyd MR (2002). Multisite and multivalent binding between cyanovirin-N and branched oligomannosides: calorimetric and NMR characterization. Chem Biol. 9(10), 1109-1118.

[97] Siridewa K, Fröman G, Hammar L & Mårdh PA (1993). Characterization of glycoproteins from Chlamydia trachomatis using lectins. APMIS 101(11), 851-7.

[98] Steele C, Marrero L, Swain S, Harmsen AG, Zheng M, Brown GD, Gordon S, Shellito JE & Kolls JK (2003). Alveolar macrophage-mediated killing of Pneumocystis carinii f. sp. muris involves molecular recognition by the dectin-1 beta-glucan receptor. J. Exp. Med. 198, 1677–1688.

[99] Steele C, Rapaka RR, Metz A, Pop SM, Williams DL, Gordon S, Kolls JK & Brown GD (2005). The beta-glucan receptor dectin-1 recognizes specific morphologies of Aspergillus fumigatus. PLoS Pathogens. 1, e42.

[100] Sumner JB and Howell SF (1936). The identification of the hemagglutinin of the jack bean with concanavalin. A. J. Bacteriol. 32, 227–237.

[101] Tabona P, Mellor A & Summerfield JA (1995). Mannose binding protein is involved in first-line host defence: evidence from transgenic mice. Immunology. 85, 153–159.

[102] Tailleux L, Schwartz O, Herrmann JL, Pivert E, Jackson M, Amara A, Legres L, Dreher D, Nicod LP, Gluckman JC, Lagrange PH, Gicquel B & Neyrolles O (2003). DC-SIGN is the major Mycobacterium tuberculosis receptor on human dendritic cells. J. Exp. Med. 197, 121–127.

[103] Tasumi S, Yang WJ, Usami T, Tsutsui S, Ohira T, Kawazoe I, Wilder MN, Aida K & Suzuki Y (2004). Characteristics and primary structure of a galectin in the skin mucus of the Japanese eel (Anguilla japonica). Dev. Comp. Immunol. 28, 325– 335.

[104] Taylor ME, Bezouska K, Drickamer K (1992). Contribution to ligand binding by multiple carbohydrate-recognition domains in the macrophage mannose receptor. J Biol Chem. 267(3),1719-1726.

[105] Taylor PR, Tsoni SV, Willment JA, Dennehy KM, Rosas M, Findon H, Haynes K, Steele C, Botto M, Gordon S & Brown GD (2007). Dectin-1 is required for beta-glucan recognition and control of fungal infection. Nat. Immunol. 8, 31–38.

[106] Teixeira E, Napimoga M, Carneiro V, De Oliveira T, Nascimento, K, Nagano C, Souza J, Havt A, Pinto V, Gonçalves R, Farias W, Saker-Sampaio S, Sampaio A and Cavada B (2007). In vitro inhibition of oral streptococci binding to the acquired pellicle by algal lectins. Journal of Applied Microbiology 103, 1001-1006.

[107] Tziveleka LA, Vagias C & Roussis V (2003). Natural products with anti-HIV activity from marine organisms. Curr Top Med Chem., 3(13), 1512-1535.

[108] Uemura K, Saka M, Nakagawa T, Kawasaki N, Thiel S, Jensenius JC & Kawasaki T (2002). L-MBP is expressed in epithelial cells of mouse small intestine. J. Immunol. 169, 6945-6950.

[109] van Die I, van Vliet SJ, Nyame AK, Cummings RD, Bank CM, Appelmelk B, Geitjenbeek TB, van Kooyk Y (2003). The dendritic cell-specific C-type lectin DC-SIGN is a receptor for Schistosoma mansoni egg antigens and recognizes the glycan antigen Lewis x. Glycobiology. 13, 471-478.

[110] van Emmerik LC, Kuijper EJ, Fijen CA, Dankert J & Thiel S (1994). Binding of mannan-binding protein to various bacterial pathogens of meningitis. Clin. Exp. Immunol. 97, 411-416.

[111] Van Parijs J, Broekaert WF, & Peumans WJ (1991). Hevein: an antifungal protein from rubber three (Hevea brasiliensis) latex. Planta 183, 258-264.

[112] Van Samme EJM & Peumans WJ (1996). Prevalence, biological activity and genetic manipulations of lectins in foods. Trends in Food Science and Technology 7: 132-138.

[113] Viriyakosol S, Fierer J, Brown GD & Kirkland TN (2005). Innate immunity to the pathogenic fungus Coccidioides posadasii is dependent on toll-like receptor 2 and dectin-1. Infect. Immun. 73, 1553-1560.

[114] Xia L & Ng TB (2005). An antifungal protein from flageolet beans. Peptides. 26, 2397-2403.

[115] Yan Q, Jiang Z, Yang S, Deng W & Han L (2005). A novel homodimeric lectin from A stragalus mongholicus with antifungal activity. Arch Biochem Biophys. 442, 72-81.

[116] Yao Y, Grogan J, Zehnder M, Lendenmann U, Nam B, Wu Z, Costello CE & Oppenheim FG (2001) Compositional analysis of human acquired enamel pellicle by mass spectrometry. Arch Oral Biol 46, 293-303.

[117] Ye XY, Ng TB, Tsang PW & Wang J (2001). Isolation of a homodimeric lectin with antifungal and antiviral activities from red kidney bean (Phaseolus vulgaris) seeds. J Protein Chem. 20, 367-375.

[118] Yi SM, Harson RE, Zabner J and Welsh MJ (2001). Lectin binding and endocytosisat the apical membrane of human airway epithelia. Gene Ther. 8, 1826-1832.

[119] Yin A, Margolis HC, Yao Y, Grogan J & Oppenheim FG (2005) Multi-component adsorption model for pellicle formation: the influence of salivary proteins and non-salivary phosphoproteins on the binding of histatin 5 onto hydroxyapatite. Arch Oral Biol 51, 102-110.

[120] Yu LG, Fernig DG, Smith JA, Milton JD, Rhodes JM (1993) Cancer Res. 53:4627-4632.

[121] Zamze S, Martinez-Pomares L, Jones H, Taylor PR, Stillion RJ, Gordon S & Wong SY (2002). Recognition of bacterial capsular polysaccharides and lipopolysaccharides by the macrophage mannose receptor. J Biol Chem. 1, 277(44), 41613-41623.

[122] Zaporozhets T, Besednova N, Ovodova R, Glazkova V (1994). The lectin activity of mytilan, a bioglycan from mussels, and its effect on microbial adhesion to macroorganism cells. Zh Mikrobiol Epidemiol Immunobiol.3, 86-88.

Permissions

The contributors of this book come from diverse backgrounds, making this book a truly international effort. This book will bring forth new frontiers with its revolutionizing research information and detailed analysis of the nascent developments around the world.

We would like to thank Rizwan Ahmad, PhD, for lending his expertise to make the book truly unique. He has played a crucial role in the development of this book. Without his invaluable contribution this book wouldn't have been possible. He has made vital efforts to compile up to date information on the varied aspects of this subject to make this book a valuable addition to the collection of many professionals and students.

This book was conceptualized with the vision of imparting up-to-date information and advanced data in this field. To ensure the same, a matchless editorial board was set up. Every individual on the board went through rigorous rounds of assessment to prove their worth. After which they invested a large part of their time researching and compiling the most relevant data for our readers. Conferences and sessions were held from time to time between the editorial board and the contributing authors to present the data in the most comprehensible form. The editorial team has worked tirelessly to provide valuable and valid information to help people across the globe.

Every chapter published in this book has been scrutinized by our experts. Their significance has been extensively debated. The topics covered herein carry significant findings which will fuel the growth of the discipline. They may even be implemented as practical applications or may be referred to as a beginning point for another development. Chapters in this book were first published by InTech; hereby published with permission under the Creative Commons Attribution License or equivalent.

The editorial board has been involved in producing this book since its inception. They have spent rigorous hours researching and exploring the diverse topics which have resulted in the successful publishing of this book. They have passed on their knowledge of decades through this book. To expedite this challenging task, the publisher supported the team at every step. A small team of assistant editors was also appointed to further simplify the editing procedure and attain best results for the readers.

Our editorial team has been hand-picked from every corner of the world. Their multi-ethnicity adds dynamic inputs to the discussions which result in innovative outcomes. These outcomes are then further discussed with the researchers and contributors who give their valuable feedback and opinion regarding the same. The feedback is then collaborated with the researches and they are edited in a comprehensive manner to aid the understanding of the subject.

Apart from the editorial board, the designing team has also invested a significant amount of their time in understanding the subject and creating the most relevant covers. They scrutinized every image to scout for the most suitable representation of the subject and create an appropriate cover for the book.

The publishing team has been involved in this book since its early stages. They were actively engaged in every process, be it collecting the data, connecting with the contributors or procuring relevant information. The team has been an ardent support to the editorial, designing and production team. Their endless efforts to recruit the best for this project, has resulted in the accomplishment of this book. They are a veteran in the field of academics and their pool of knowledge is as vast as their experience in printing. Their expertise and guidance has proved useful at every step. Their uncompromising quality standards have made this book an exceptional effort. Their encouragement from time to time has been an inspiration for everyone.

The publisher and the editorial board hope that this book will prove to be a valuable piece of knowledge for researchers, students, practitioners and scholars across the globe.

List of Contributors

Anthony P. Timerman
University of Wisconsin-Stevens Point, USA

William Ward
Rutgers University, New Brunswick NJ, USA

Luana C. B. B. Coelho, Thiago H. Napoleão, Maria T. S. Correia and Patrícia M. G. Paiva
Universidade Federal de Pernambuco, Centro de Ciências Biológicas, Departamento de Bioquímica, Av, Prof. Moraes Rego, Recife-PE, Brazil

Andréa F. S. Santos
University of Minho, IBB-Institute for Biotechnology and Bioengineering, Centre of Biological Engineering, Campus de Gualtar, Braga, Portugal

Tove Boström, Johan Nilvebrant and Sophia Hober
Royal Institute of Technology, Stockholm, Sweden

Giovanni Magistrelli, Pauline Malinge, Greg Elson and Nicolas Fischer
NovImmune SA, 14 Chemin des Aulx, Plan-les-Ouates, Switzerland

Di Xiang and Wei Han
Laboratory of Regeneromics, School of Pharmacy, Shanghai Jiaotong University, Shanghai, China

Yan Yu
Shanghai Municipality Key Laboratory of Veterinary Biotechnology, School of Agriculture and Biology, Shanghai Jiaotong University, Shanghai, China

Kin-Kwan Lai, Clara Vu, Ricardo B. Valladares, Anastasia H. Potts and Claudio F. Gonzalez
University of Florida, USA

Francesco Lonardoni and Alessandra Maria Bossi
Verona University, Italy

Barira Islam and Asad U. Khan
Interdisciplinary Biotechnology Unit, Aligarh Muslim University Aligarh, India

Printed in the USA
CPSIA information can be obtained
at www.ICGtesting.com
JSHW011420221024
72173JS00004B/603

9 781632 390547